THE STRUCTURE AND FUNCTION OF RYANODINE RECEPTORS

THE STRUCTURE AND FUNCTION OF RYANODINE RECEPTORS

Editors

R. Sitsapesan

A. J. Williams

Cardiac Medicine
National Heart & Lung Institute
Imperial College School of Medicine

ICP

Imperial College Press

Published by

Imperial College Press
203 Electrical Engineering Building
Imperial College
London SW7 2BT

Distributed by

World Scientific Publishing Co. Pte. Ltd.
P O Box 128, Farrer Road, Singapore 912805
USA office: Suite 1B, 1060 Main Street, River Edge, NJ 07661
UK office: 57 Shelton Street, Covent Garden, London WC2H 9HE

British Library Cataloguing-in-Publication Data
A catalogue record for this book is available from the British Library.

THE STRUCTURE AND FUNCTION OF RYANODINE RECEPTORS

ISBN 1-86094-145-1

Printed in Singapore.

Preface

The ryanodine receptor (RyR) plays a pivotal role in excitation-contraction coupling in striated muscle and in the signalling processes of many other types of cells. For this reason the Ca^{2+}-release channel properties of RyR have been studied with increasing interest in the last 15 years. This period has witnessed enormous advances in our understanding of the molecular mechanisms and structures that underlie membrane ion channel function. Molecular cloning of surface membrane K^+, Na^+ and Ca^{2+} channels, and the investigation of the properties of channel proteins with modified structures, have revealed details of the processes of ion handling and gating in these channels. By comparison, progress with RyR has been slow. This relative paucity of information is not due to laziness on the part of workers in the field; rather, it reflects the comparative novelty of the subject (the first electrophysiological recordings of RyR being published by Gerhard Meissner's laboratory in 1985) and the technical difficulties associated with working with an intracellular membrance channel with a molecular mass in the region of 2.5 M Da. Nevertheless, much progress has been made and the aim of this book is to bring together current opinion on the structure of the RyR molecule and its function as a Ca^{2+}-release channel, both in isolation and *in situ*.

In attempting to assemble opinion in such an active area of research, it is inevitable that some investigations and views will have been missed and we apologise in advance for these omissions. Overall, we believe that the book provides a very useful commentary on the current state of our knowledge of RyR.

We are greatly indebted to all the contributing authors for their patience and support during the preparation the book.

Rebecca Sitsapesan and Alan Williams
London, July 1998

Contents

List of Contributors

Gerhard Meissner
Departments of Biochemistry and Biophysics and Physiology
University of North Carolina, Chapel Hill
NC 27599-7260 U.S.A.
919-966—5021
Fax 919-966-2852
Email:gmeissne@uvnhs.unc.edu

Donald M. Bers
Department of Physiology, Loyola University Chicago
Stritch School of Medicine, Maywood
IL. 60514, U.S.A.
708-216-1018
Fax 708-216-6308
Email:dbers@wpo.it.luc.edu

Irina I. Serysheva and Susan L. Hamilton
Department of Molecular Physiology and Biophysics
Baylor College of Medicine, Houston
Texas 77030, U.S.A.
713-798-5704
Fax 713-798-3475
Email:susanh@bcm.tmc.edu

Varda Shoshan-Barmatz and Wei Feng
Department of Life Sciences, Ben-Gurion University of the Negev
Beer-Sheva, Israel
972-7-6461336
Fax 972-7-6472992

William Welch
Department of Biochemistry, University of Nevada, Reno, NV 89557, U.S.A.
Fax 702-784-1419
Email:welch@scs.unr.edu

Yasuo Ogawa and Takashi Murayama
Department of Pharmacology, Juntendo University School of Medicine
Tokyo 113, Japan
81-3-5802-1034
Fax 81-3-5802-0419
Email:ysogawa@med.juntendo.ac.jp

Judith A. Airey and Rong He
Dept. of Pharmacology, University of Nevada, Reno, NV 89557, U.S.A.
702-784-4651
Fax 702-784-1620
Email:jaairey@med.unr.edu

Georgina B. Gurrola and Hector H. Valdivia
Dept. of Physiology, University of Wisconsin, Madison, WI 53706, U.S.A.
608-265-5960
Fax 608-265-5512
Email:hhvaldiv@facstaff.wisc.edu

Charles F. Louis[1,2] and James R. Mickelson[1]
Depts. of Veterinary PathoBiology[1] and Biochemistry[2]
University of Minnesota, St Paul
Minnesota 55108, U.S.A.
612-624-4202
Fax 612-625-0204
Email:louis@maroon.tc.umn.edu

Patricio Vélez, Rafael Mejia-Alvarez and Michael Fill
Dept. of Physiology, Loyola University, 2160 South First Ave.
Maywood, IL 60153-5500. U.S.A.
708-216-6305
Fax 708-216-6308
Email:mfill@luc.edu

Elena V. Orlova and Marin van Heel
Dept. of Biochemistry, Imperial College of Science
Technology and Medicine, London SW7 2AZ, U.K.
0171 594 45317
fax 0171 225-0960
Email:m.vanheel@ic.ac.uk

Rebecca Sitsapesan, National Heart and Lung Institute
Imperial College of Science, Technology and Medicine
Dovehouse St., London SW3 6LY, U.K.
0171-352-8121x3309
Fax 0171-823-3392
Email:r.sitsapesan@ic.ac.uk

Alan J. Williams, National Heart & Lung Institute
Imperial College of Science, Technology and Medicine
Dovehouse St., London SW3 6LY, U.K.
0171-352-8121x3308
Fax 0171-823-3392
Email:a.j.williams@ic.ac.uk

Graham D. Lamb and Derek R. Laver
School of Zoology, La Trobe University, Bundoora, Victoria 3083
Australia
61-3-94792249
Fax 61-3-94791551
Email:zoogl@zoo.latrobe.edu.au

Wah Chiu, Verna & Marrs McLean
Dept. of Biochemistry, Baylor College of Medicine
Houston, Texas 77030, U.S.A.

Chapter 1

An Overview of Ryanodine Receptor Structure and Function

Rebecca Sitsapesan and Alan J. Williams

Ryanodine receptors (RyR) are a class of Ca^{2+}-selective ion channels located in the intracellular membrane systems (sarcoplasmic reticulum and endoplasmic reticulum) of animal cells. These channels provide regulated pathways for the release of sequestered Ca^{2+} to raise cytosolic levels of this cation in response to stimuli. The aim of this book is to review our current understanding of the structure and function of RyR in isolation and to discuss how its function is integrated into its best recognised physiological role in muscle excitation-contraction coupling.

Three mammalian isoforms of RyR exist and homologues of these receptors have been identified in non-mammalian and invertebrate cells. In Chapter 2, Ogawa and Murayama address the relationship of the mammalian type 3 receptor to its invertebrate homologue and discuss the sites of expression, the structure, function and biological significance of these proteins.

The functional RyR channel is a homotetramer with each monomer consisting of approximately 5000 amino acids. Details of the structure of the channel have been derived from 3 dimensional reconstructions of images obtained from frozen purified receptors. In Chapter 3, Orlova and colleagues describe the methods used to obtain this information and to provide images of different gating states of the rabbit skeletal muscle RyR.

The properties of individual RyR channels have been studied following the incorporation of either native or purified channel proteins into planar phospholipid bilayers. In this system the mechanisms underlying channel gating and ion handling can be studied under conditions in which the ionic environment on both sides of the channel and the trans-membrane voltage can be controlled. The influence of a range of physiological and pharmacological regulators of channel gating are discussed in Chapter 4 and the characteristics of ion conduction and selectivity of RyR are described in Chapter 5.

As is implied from its name, the RyR channel binds the plant alkaloid ryanodine. Each channel protein has both high and low affinity binding sites for this ligand and in Chapter 6, Serysheva and Hamilton discuss the use of [^{3}H] ryanodine binding in the assessment of the functional state of RyR1. In addition they provide information on the location of ryanodine binding sites within the channel protein.

Quantitative molecular modeling techniques have been used to assess three-dimensional structure-activity relationships for the interaction of a number of natural and synthetic ryanoids with the channel. The structural elements of the ligand involved in the high affinity binding of ryanodine have been investigated by monitoring the interaction of these ryanoids with the channel. Similarly, the structural elements of the ryanoids responsible for altering single channel ion handling have been investigated. The findings of these studies are described by Welch in Chapter 7.

In Chapter 8, Airey and He discuss protein kinase-mediated phosphorylation of various isoforms of RyR and the influence of phosphorylation on channel function. In addition they discuss the phosphorylation of non RyR SR proteins that affect RyR function and consider if phosphorylation of RyR or accessory proteins is of physiological importance.

The contraction of striated muscle cells is initiated by the release of SR Ca^{2+} via RyR. In cardiac muscle, RyR2 is believed to open in response to an influx of Ca^{2+} to the cell via dihydropyridine-sensitive Ca^{2+} channels in the cell surface membrane; a process referred to as Ca^{2+}-induced Ca^{2+} release. In skeletal muscle cells the opening of RyR1 appears to be mediated by a direct coupling of RyR to the dihydropyridine-sensitive Ca^{2+} channel, with the latter channel acting as a voltage-sensor rather than a Ca^{2+} channel. In Chapter 9 Meissner describes the consequences of the interaction of dihydropyridine-sensitive Ca^{2+} channel-derived peptides with RyR channels.

In Chapter 10 Gurrola and Valdivia discuss the identification, purification and structure of novel peptide toxins with high affinity for RyR. They go on to discuss the usefulness of these peptides as probes for RyR in disrupted and intact cells.

Covalent modification by chemical probes has been used extensively to identify amino acid side chains involved in various aspects of protein function. In Chapter 11 Shoshan-Barmatz and Feng review studies in which a wide range of chemical modifiers (group-directed, binding site-directed and cross-linking reagents) have been used to probe RyR.

Mutations of RyR1 are associated with the inherited muscle diseases malignant hyperthermia and central core disease. The clinical features of these diseases and the identification of the mutations involved are discussed in Chapter 12 by Louis and Mickelson. The authors also give descriptions of the functional alterations of SR Ca^{2+}-release and RyR channel activity associated with malignant hyperthermia.

The control of RyR channel gating and particularly the mechanisms leading to decreases in channel activity have generated much interest over the past five years. Velez, Mejia-Alvarez and Fill, in Chapter 13, discuss the evidence for adaptation of reconstituted RyR channel activity to rapid changes in cytosolic Ca^{2+} concentration and whether adaptation is a physiological process. In Chapter 14, Lamb and Laver describe alternative explanations for adaptation of RyR gating and outline how RyR activity can be activated and inhibited by various other mechanisms.

Finally, in Chapter 15 Bers focuses on the function of RyR channels under physiological conditions. He describes how SR Ca^{2+} release is regulated in intact cardiac cells and details the major factors which influence spontaneous localized Ca^{2+} release events (Ca^{2+} sparks).

Chapter 2

Characteristics of Ryanodine Receptor Type 3 Isoform (RyR3) and Its Homologues

Yasuo Ogawa and Takashi Murayama

Ryanodine receptor (RyR) is an intracellular Ca^{2+} release channel protein, primarily on the sarcoplasmic reticulum of striated muscles, and was given its name because of its high affinity for a litmus-neutral alkaloid, ryanodine.[1–4] In mammals there are three genomically distinct isoforms, RyR1-3, the chromosomal loci of which are identified in human and murine.[3,5] RyR1 and RyR2 are predominant isoforms in skeletal and cardiac muscles, respectively, and their molecular characterization has been performed with purified proteins.[1–3] RyR3, in contrast, is identified by RNA or its cDNA in various tissues including neurons and smooth muscles, but there are few reports on properties of the protein.[3,5] In many non-mammalian vertebrate skeletal muscles, two distinct protein isoforms of RyR, referred to as α- and β-RyR, are usually detected in almost equal amounts.[3,4] Sequencing of their cDNAs reveals that α- and β-RyR are homologues of RyR1 and RyR3, respectively.[3,4] In this review article, RyR1 and RyR3 refer to the isoforms in mammals, whereas α- and β-RyR refer to those in non-mammalian vertebrates where the chromosomal loci are not yet identified (Table 2.1). We will primarily discuss here the properties of the protein isoform

Table 2.1 Nomenclature of RyR Isoforms with Their Primarily Expressed Sites

Mammals	Vertebrates other than mammals
RyR1 (Skeletal muscle)	α-RyR (skeletal muscle)
RyR2 (cardiac muscle, brain)	cardiac RyR (cardiac muscle)
RyR3	β-RyR (skeletal muscle)

RyR3 or β-RyR. Many review articles including those in other sections of this book can be consulted for general information on RyR and related topics.[1–13]

HISTORICAL OVERVIEW

Giannini *et al.* (1992)[14] first identified a novel RyR isoform in mink lung epithelial cells Mv1Lu which had been treated with TGF-β. The isoform was distinct from RyR1 or RyR2 in the partial sequence of the C-terminal region and characteristically lacked a stretch containing the putative trans-membrane segment of RyR that approximately corresponded to the "M1 (4564–4580) after Numa's Model"[15] or "M5 (4554–4575) after MacLennan's Model"[16] in rabbit RyR1. This isoform was characterized as a ryanodine-sensitive, caffeine-insensitive Ca^{2+} release channel on the Ca-stores. At almost the same time, Hakamata *et al.* (1992)[17] sequenced a novel RyR cDNA isolated from rabbit brain cDNA library and deduced the amino acid sequence to be composed of 4872 residues. They found that 100 or more amino acid residues after 1300 in RyR1 were deleted in RyR3 and that the region immediately preceding the M1 after Numa's Model was markedly divergent (see Table 2.3). They confirmed that the partial sequence which Giannini *et al.* determined in mink lung epithelial cell corresponded to the C-terminal one-eighth in the rabbit brain RyR isoform with 94% identity. While RyR2 was the main isoform expressed ubiquitously in the brain,[7,9] the novel isoform RyR3 is localized in some specific regions such as corpus striatum, thalamus and hippocampus. The RNA for RyR3 was also detected in smooth muscle tissues: the aorta, esophagus, taenia coli, urinary bladder, ureter and uterus.

RyR3 was assumed to be a ryanodine sensitive, caffeine insensitive Ca^{2+} release channel which was characteristically localized in various smooth muscle cells and restricted regions of the brain. Hakamata *et al.* (1994)[18] also reported that human Jurkat T-lymphocyte expressed only mRNA for RyR3 which may function as a ryanodine sensitive, caffeine insensitive Ca^{2+} release channel. They further observed that 100 mM ryanodine stimulated proliferation and altered the growth pattern of cultured human T-cells when added together with FK 506.

Oyamada *et al.* (1994)[19] sequenced cloned cDNAs for α- and β-RyR in bullfrog skeletal muscles and found them most homologous to RyR1 and RyR3, respectively, with 80% or higher identity in the amino acid sequences. Ottini *et al.* (1996)[20] recently determined the full sequence of the cDNA of chicken β-RyR and reached the same conclusion. Therefore, β-RyR in non-mammalian vertebrate skeletal muscles may be concluded to be the homologue of RyR3. β-RyRs are found to be caffeine sensitive.[21,22] Takeshima *et al.* (1994)[23] reported caffeine-induced Ca^{2+} release from the sarcoplasmic reticulum of *ryr1*-knocked out mouse. Since mRNA for RyR3 was detected in the knock-out mouse, the caffeine-induced Ca^{2+} release was attributed to the RyR3.[24] Giannini *et al.* (1995)[25] examined distribution of ryanodine receptor gene products in various tissues of mouse and showed that the expression pattern of each of three isoforms was tissue-specific as described below. They observed the RNA for RyR3 in diaphragm and thigh muscle by the RNase protection method, but failed to detect the protein by Western blot analysis, indicating a low level of RyR3 protein in skeletal muscle. They acknowledged that the lack of the response to caffeine might be due to the scarcity of RyR3 rather than to its intrinsic property. Nowadays we understand that RyR3 and β-RyR are similar to each other in their primary sequences, while they are distinct from the other isoforms. Although β-RyR occurs in an amount similar to α-RyR in skeletal muscle, RyR3 does little not only in skeletal muscle, but also in other various tissues. Insensitivity to caffeine is not necessarily the characteristic unique to the isoform.

EXPRESSION SITES OF RyR3 AND β-RyR

Giannini *et al.* (1995)[25] determined the expression of RyR3 in the following tissues besides diaphragm and thigh muscle of mouse by the RNase protection method using an anti-sense probe: cardiac Purkinje fiber, cerebrum, cerebellum, lung, esophagus, spleen, gut, kidney, stomach, submaxillary gland, thymus, testis, ovary, and adrenal gland. Although RyR3 was characteristically detected in neurons and smooth muscles, they denied the possibility that its expression in non-excitable tissues such as glands, testis and ovary might be due to the contaminating neurons and vascular smooth muscles. They showed that mRNAs not only for RyR3 but also for the other isoforms were detected in non-excitable cells including several cell lines of different lineage (fibroblast, epithelial cells and hematopoietic cells) in addition to excitable cells of neurons and muscle cells. However, they concluded that the actual amounts of RyR proteins present in many cells were low.

Wide distribution of β-RyR was also found by RT-PCR analysis in skeletal muscles, atrium (but not ventricle), brain, lung and stomach of bullfrog, while α-RyR was detectable exclusively in skeletal muscles and weakly in brain.[19]

Since the distributions of RyR3 and β-RyR in the brain and skeletal muscles have been examined in detail, some interesting findings will be discussed below.

Brain

Because of the history of discovery of RyR3, it is natural to closely examine its distribution in the brain. While RyR2 is the main isoform in the brain and is ubiquitously distributed throughout this organ, mRNA for RyR3 was selectively detected in the restricted regions of corpus striatum, hippocampus and thalamus by Northern blot[17] and by in situ hybridization.[25,26] The preferential distribution of RyR3 protein was confirmed by Western blot[25] and by immunoprecipitation[27]: a greater amount in hippocampus, diencephalon, and corpus striatum, and a lesser amount in cerebral cortex and mesencephalon, but not in cerebellum of rabbit.[27] The in situ hybridization also revealed differential distributions of RyR3 and the other isoforms among neurons in a specified area. For example, RyR3 was abundant in a CA1 pyramidal cell of hippocampus, while RyR2 was abundant in the CA1-4 pyramidal cells. In cerebellum, RyR1 was characteristically detected in Purkinje cells, while RyR2 was more broadly distributed in the granular cell layer.[25,26] RyR3 in the cerebellum, however, differed between rabbit and mouse: in rabbit it was minimal,[26,27] while it was moderate in murine.[25]

Skeletal Muscle

We stated above that non-mammalian vertebrate skeletal muscles have an almost equal amount of α- and β-RyR, while RyR1 is the predominant isoform in mammalian skeletal muscles. This may be true of most skeletal muscles in avian, frog and fish.[21,28,29] Even in fish whose body skeletal muscles have the two isoforms, however, rapid contracting muscles such as extra-ocular muscles and swim-bladder muscle that produces a sound by vibrating have a single isoform of α-RyR.[29] β-RyR isoform cannot be detected there. The rattlesnake tail-shaker muscle which contracts rapidly also has a single isoform of α-RyR.[29] Among reptiles, lizards and snakes showed a single band of a-RyR, while turtles and crocodiles showed two bands composed of α- and β-RyR.[29]

Protein staining of the SDS-polyacrylamide gel electrophoresis (PAGE) pattern of SR from mammalian skeletal muscles shows a single band of RyR1. The band for RyR3, however, is invisible, suggesting its minuscule amount. The RyR3 may be detected by Western blot or more clearly by immunoprecipitation. The content of RyR3 was variable among different muscles in rat:[30] higher levels in diaphragm and soleus, lower levels in abdominal muscles and tibialis anterior, and no detectable amounts in the extensor digitorum longus. Preferential high content of RyR3 in the diaphragm was observed in several mammals examined (rat, mouse, rabbit, and cow).[30] Murayama and Ogawa (1997)[31] recently confirmed that RyR3 protein was found in diaphragm of rabbit but that it was undetected in back muscles, material commonly used for biochemical preparations.

In view of the fact that the RyR3 protein is difficult to detect because of its minute amount, it is desirable to reconfirm by immunoprecipitation if the nonmammalian vertebrate skeletal muscles which showed α-RyR alone do not express β-RyR.

CONTENTS OF RyR3

Because the quantity of RyR3 is so little that its detection is difficult without immunoprecipitation, it is helpful for further discussion to get a general idea of its contents.

Table 2.2 Relative Contents of [^{3}H]Ryanodine Binding Sites in Rabbit

Skeletal Muscle	RyR1	100%
	RyR3	≤ 0.6%
Cardiac Muscle	RyR2	30%
Brain	RyR2	3%
	RyR3	~0.06%

Under the assumption that any RyR isoform can bind one mol [^{3}H]ryanodine per one mol of tetramer, the relative contents of [^{3}H]ryanodine binding sites in rabbit organs were obtained. RyR3 was estimated from immunoprecipitated products. RyR1 content in skeletal muscle is assumed to be 100%.

Immunoprecipitated RyR3 showed Ca^{2+} dependent [^{3}H]ryanodine binding activity, and its affinity for ryanodine (K_d=1.6 nM) was similar to those for RyR1, RyR2, α-RyR and β-RyR (K_d=2-5 nM).[3,31] Therefore, we may assume that RyR3 binds 1 mol of [^{3}H]ryanodine per 1 mol of homotetramer as is true of the other RyR isoforms. The Bmax value thus reflects the number of ryanodine binding sites in any preparation. Table 2.2 shows the relative contents of RyR isoforms thus obtained in some tissues of rabbit (see also Ref. 27). RyR3 in skeletal muscle is at most about 0.6% of RyR1. Since the average content of RyR2 in the brain was about 3% of RyR1 in the skeletal muscle, RyR3 in the brain was estimated to be as little as 0.06% of RyR1 in the skeletal muscle. Although this estimate was an average throughout the brain, the content of the area preferentially expressed would be greater. However it may not be so much larger because of longer periods of exposure for RyR3 (14 days) than for RyR2 (5 days) to detect isoforms in the brain by in situ hybridization.[26]

STRUCTURE OF RyR3

Primary Structure — Potential Variants of RyR3

Table 2.3 summarizes the major differences in the amino acid sequences among the three isoforms in mammals.[3] The difference between α- and β-RyR in frog skeletal muscles is similar to that between RyR1 and RyR3.[19] We may tacitly assume that this is also the case with chicken α- and β-RyR, although the definite answer must await determination of the full sequence of α-RyR. Consistent with different molecular masses between RyR1 homologues (RyR1 and α-RyR) and RyR3 homologues (RyR3 and β-RyR), the latter moved faster than the former on SDS-PAGE.[21,25,31,32] RyR2[25] and chicken cardiac isoform[32] were intermediate in mobility between RyR1 and RyR3 homologues.

Table 2.3 Comparisons among RyR Isoforms in Rabbit

	RyR1	RyR2	RyR3
Amino acid residues	5037	4976	4872
Calculated MW	565 kDa	565 kDa	552 kDa
Marked differences[a]	1302–1408	1315–1408	deleted
	4249–4626	4210–4562	4100–4400

[a]These stretches correspond approximately to D2 and D1 of Sorrentino and Volpe.[57]

Marziali *et al.*[33] recently determined the complete sequence of the RyR3 cDNA cloned from the mink lung epithelial cell line CCL64. Although the deduced amino acid sequence shows close similarity to that of rabbit RyR3, numerous putative alternative splicing variants were detected, some of which were tissue specific (Fig. 2.1). Two cases are those already shown in RyR1 of human[34] and RyR2 of rabbit.[35] They are insertion of 5 amino acids (AMQVK) (ASI) after residue K3335 and of 6 amino acids (LIVRER) (ASII) after residue T3710 of mink epithelial RyR3, respectively. A third splicing site (ASIII), unique to RyR3, is defined by a mutually exclusive splicing mechanism, in which exon A or exon B can be alternatively included in the mature RyR3 mRNA. This occurred around 3496–3526 of the epithelial RyR3 which might contain potential regulatory and modulatory sites. The distribution of ASIII is tissue specific: the mRNA containing exon A was detected in all tissues analyzed (mink brain, cerebellum, skeletal muscle, diaphragm, ovary, ileum and CCL64), while the mRNA containing exon B was recognized in brain, cerebellum, skeletal muscle, diaphragm, and ileum, but not in ovary or CCL64. Furthermore, some clones of mink RyR3 lacked the

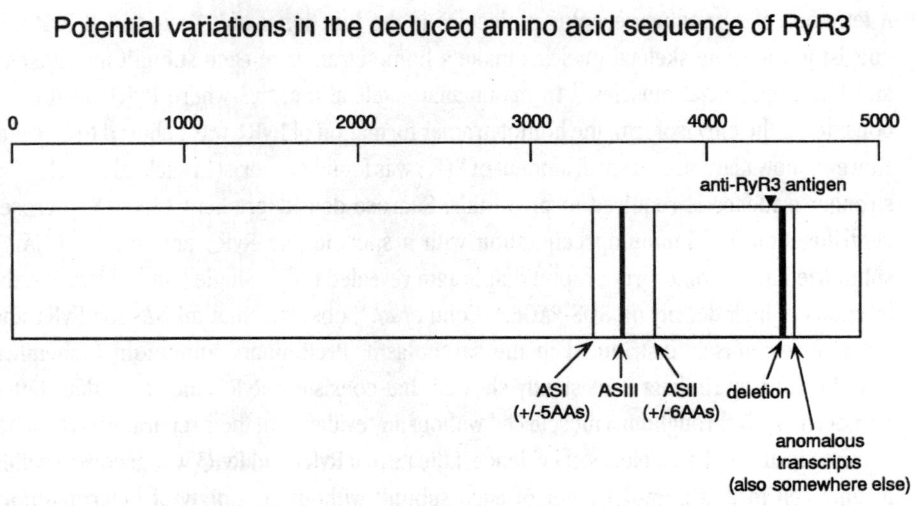

Fig 2.1 Potential variations in the deduced amino acid sequence of RyR3. The sites of potential variations refer to the position or stretch in the deduced amino acid sequence of the rabbit brain RyR3, although most of the variations metioned here were found in mink lung epithelial cells.[33] While Marziali *et al.*[33] did not describe the position of anomalous transcripts, Miyatake *et al.*[37] clearly pointed it out in the mouse brain RyR3, and Chen *et al.*[42] suggested it to be somewhere around ASI-ASII in rabbit uterus. ASI-III stands for alternative splicing site I-III.

region encoding the trans-membrane sequence, M5 after MacLennan's Model (see also Ref. 18). Marziali and colleagues also observed the existence of RyR3 anomalous transcripts with some exons skipped. As a consequence of this process, termination codons were introduced, so that these mRNAs would, if translated, encode shorter proteins unless decoded by a mechanism similar to programmed ribosomal frameshifting. Similar findings have been reported by Chen (1996)[36] for rabbit smooth muscle RyR3 (uterus, aorta and vas deferens) and by Miyatake *et al.* (1996)[37] after 4476 of mouse brain RyR3 (Fig. 2.1).

These findings suggest that RyR3 in the brain may not be identical with RyR3 in the lung epithelial cell and indicate that properties of RyR3 may be tissue-specific. Therefore, the possibility cannot be excluded that RyR3 in lung epithelial cells may be insensitive to caffeine, while that in the brain and skeletal muscles is sensitive. Since the effect of ryanodine was detected in epithelium RyR3, the scarcity of RyR3 cannot be the reason for the failure to observe the caffeine response. More detailed experiments are required for a final conclusion.

Homotetramer as a Functional Unit

A tetramer of ryanodine receptor molecules is the functional unit.[38] α- and β-RyR that coexist in the same skeletal muscles make a homotetramer of each subunit in chicken[39] and bullfrog skeletal muscles.[21] In mammalian skeletal muscles where RyR1 used to be considered the only isoform, the homotetramer formation of RyR1 was believed to be valid. However, nowadays since a small amount of RyR3 was found to coexist in skeletal muscles,[24,25] stronger evidence is required to prove this. Sucrose density gradient (5–20% sucrose) centrifugation and immunoprecipitation with a specific anti-RyR3 antibody of CHAPS-solubilized microsomes from rabbit diaphragm revealed only a single band of RyR3 in the fractions of high density on SDS-PAGE.[31] Conti *et al.*[30] observed that mRNAs for RyR1 and RyR3 were diffusely distributed in the sarcoplasm. Preliminary immunohistochemical examinations by Hijikata consistently showed that coexisting RyR3 and RyR1 distributed homogeneously throughout a muscle cell without any evidence of their separation (personal communication). These pieces of evidence indicate that RyR1 and RyR3 which coexist within a same cell form a homotetramer of each subunit without possibility of heterotetramer formation.[31] This is also the case with RyR2 and RyR3 in the brain.[27] RyR2 is the major isoform in the brain where it is ubiquitously expressed. RyR3, a minor component, is localized in specific regions of the brain. It can be tacitly assumed that RyR2 and RyR3 coexist in the same neuron, although there is no strict evidence available for this. RyR2 and RyR3 in the brain also form homotetramers of each subunit.[27] At present we cannot say, however, whether differently spliced variants of RyR3, if they coexist, can form a tetramer.

The impossibility of heterotetramer formation among different isoforms of RyR is in marked contrast to the case with IP3 receptor. There are at least three IP3 receptor isoforms. Monkawa *et al.*[40] and Nucifora *et al.*[41] presented evidence for heterotetramer formation among these isoforms which might underlie the diversified properties of IP3-gated Ca^{2+} release channel on the Ca-store (endoplasmic reticulum).

FUNCTIONAL CHARACTERISTICS OF RyR3

As stated, there are many potential splicing variants of RyR3 in various tissues which may show diverse functional properties. Here, the description will be limited to RyR3 in the brain and skeletal muscles and β-RyR in skeletal muscles, because they have been directly examined. Diverse properties have very recently been claimed with RyR3 of various origins.[42, 43].

General properties common to those of Ca^{2+}-induced Ca release (CICR) were observed in [^{3}H]ryanodine binding to RyR3 in the brain[27] and skeletal muscles[31] and β-RyRs from bullfrog skeletal muscles,[21, 44] and in channel activities in lipid bilayer membranes of β-RyR from chicken[22] and bullfrog[44] skeletal muscles. These activities are concentration-dependently stimulated by Ca^{2+} lower than 0.1 mM, while they are inhibited by Ca^{2+} at mM concentrations or higher; they are stimulated by adenine nucleotide and caffeine, and inhibited by Mg^{2+}, procaine (or other local anesthetics), and ruthenium red. Lipid bilayer experiments showed Ca^{2+}-activated ion channel activities which had unit conductances for RyR3 and β-RyR similar in magnitude to those of RyR1 and α-RyR.[21, 22, 42]

Although RyR3 and β-RyR were similar in that they showed the general characteristics of CICR which were also shared by RyR1 and α-RyR, RyR3 and/or β-RyR showed particular specific properties, as discussed below.

Vulnerability to Some Perturbation

When we first isolated α-RyR and β-RyR from the sarcoplasmic reticulum of bullfrog skeletal muscle, we determined their [^{3}H]ryanodine binding activities in a medium containing 1 M NaCl, because a high concentration of monovalent salt caused an enhanced activity.[21] Both isoforms showed monophasically Ca^{2+} dependent [^{3}H]ryanodine binding. β-RyR was about 20 times more sensitive to Ca^{2+} than α-RyR which showed a Ca^{2+} sensitivity similar to that of SR vesicles. Later we found that β-RyR was made more sensitive to Ca^{2+} by combination of

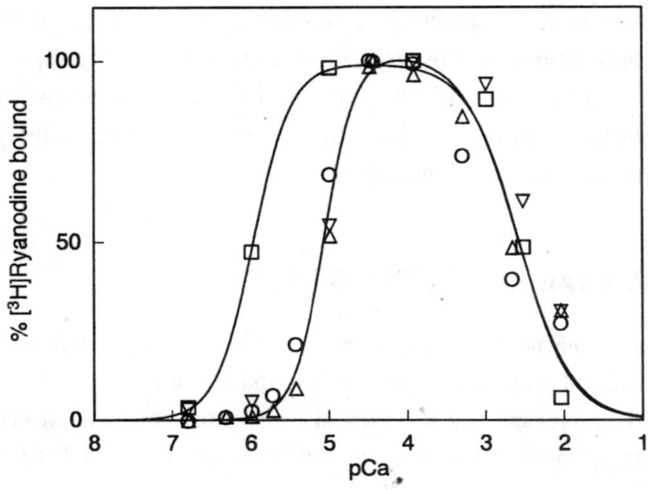

Fig 2.2 Ca^{2+}dependent [^{3}H]ryanodine binding activities for various RyR isoforms. The results for RyR1 (squares) and RyR3 (inverted triangles) are simultaneously determined with solubilized SR vesicles from rabbit diaphragm and immunoprecipitated products, respectively. The results for bullfrog α-RyR (circles) and β-RyR (triangles) were determined with the purified proteins (see also Ref. 44).

1 M NaCl and CHAPS/phospholipids, of which the underlying mechanism is distinct from that of caffeine.[45] Experiments with various combinations of monovalent anions and cations indicated that the change in the Ca^{2+} sensitivity cannot be ascribed simply to the effect of ionic strength or Cl^- anion.[44]

In an isotonic medium containing 0.17 M NaCl medium, however, both isoforms showed almost the same biphasic Ca^{2+} dependence: Ca^{2+} lower than 0.1 mM was stimulatory, while higher Ca^{2+} was inhibitory (see Fig. 2.2).[44] Their Ca^{2+} sensitivities were consistent with those for SR vesicles and for Ca^{2+} induced Ca^{2+} release (CICR) from SR in skinned frog skeletal muscles.

This change in the Ca^{2+} sensitivity and Ca^{2+} dependence by a high concentration of NaCl was also observed in RyR3 from rabbit brain[27] and diaphragm.[31] This may be common among RyR3 and its homologues.

There were claims that RyR3 and its homologues might be characteristically insensitive to the inhibitory action of high concentrations of Ca^{2+}, which is similar to RyR2.[46] However, we would like to insist that RyR3 in the isotonic medium was as sensitive to the inhibitory effect of high concentrations of Ca^{2+} as RyR1 (see Fig. 2.2). The lack of inhibitory effect of high Ca^{2+} was seen in the medium containing 1 M NaCl; however, in this medium RyR1 showed biphasic Ca^{2+} dependence, while that of α-RyR was monophasic (Murayama, T. and Ogawa, Y., unpublished observation). RyR1 from rabbit skeletal muscles may be more sensitive to the inhibitory effect of Ca^{2+} than α-RyR from bullfrog.

Ca^{2+} Sensitivity

Takeshima *et al.* (1995)[24] showed that CICR in wild-type mouse skeletal muscle (largely RyR1) showed EC_{50} of about 0.7 μM Ca^{2+}, while CICR in dyspedic mouse skeletal muscle (probably RyR3) showed EC_{50} of about 5 μM; they concluded that RyR3 was ~10 times less sensitive to Ca^{2+} than RyR1. They claimed that the lower sensitivity to Ca^{2+} was characteristic of RyR3. Our recent simultaneous determination of Ca^{2+} dependent [^{3}H]ryanodine binding to RyR1 and RyR3 of rabbit diaphragm[31] showed EC_{50}s of 1.2 and 7.9 μM Ca^{2+}, respectively, which was not incompatible with their results (Fig. 2.2). The EC_{50}s for activation of α- and β-RyR were 9.1 and 8.7 μM, respectively, under similar conditions (Fig. 2.2).[44] Published values of EC_{50} for CICR were about 1 μM or less with skinned mammalian skeletal muscles, and about 10 μM with skinned frog or Xenopus skeletal muscles.[47] These results show that RyR1 is more sensitive to Ca^{2+} than α-RyR, while RyR3 is similar to β-RyR. α-RyR and β-RyR are very similar to each other in their Ca^{2+} sensitivity (Fig. 2.2). Therefore, RyR3 and its homologues show comparable Ca^{2+} sensitivity, while RyR1 and its homologues show species specificity. In this respect, it is interesting to mention that the anti β-RyR antibody against frog β-RyR showed broader cross-reactivity with β-RyRs from chicken, frog, toad and carp, while the anti α-RyR antibody against frog α-RyR was so narrow that the reaction was specific only to *Rana*.[48] Furthermore, mobilities of α-RyRs were variable on SDS-PAGE among animal species, while those of β-RyRs were similar.[48]

The following reservation, however, may be necessary: it is quite possible that RyR3 from some tissues may be similar to RyR1 in Ca^{2+} sensitivity and have different properties from those in the brain or skeletal muscles, because there may be tissue specific splicing variants (see Ref. 42).

Effect of Mg^{2+}

RyR3 in rabbit brain and skeletal muscle is unique in the respect that it is somewhat resistant to the inhibitory effect of Mg^{2+}. With increase in Mg^{2+} concentration up to 5 mM,

[^{3}H]ryanodine binding activity of RyR3 was reduced to about 75% of the control and no further diminution was observed at higher Mg^{2+} concentrations. The activity of RyR1, on the other hand, decreased monotonically to close to nil as Mg^{2+} concentration was increased up to 20 mM (Murayama and Ogawa, unpublished results). This immunity to Mg^{2+} of RyR3 has not been seen with α-RyR from non-mammalian vertebrates except for the case in 1 M NaCl medium where β-RyRs from fish and frog were somewhat resistant to Mg^{2+} and local anesthetics.[44,49] Since RyR3 was inhibited by millimolar Ca^{2+} to as great an extent as RyR1 (see Fig. 2.2), the inhibitory sites for Ca^{2+} and Mg^{2+} in RyR3 might be distinct, whereas they were considered to be the same in RyR1. Alternatively, RyR3 might be composed of at least two different populations, i.e., Mg^{2+} sensitive and Mg^{2+} insensitive ones.

BIOLOGICAL SIGNIFICANCE OF RyR3 AND β-RyR

Extra-ocular muscles and swim-bladder muscle of fish, and rattlesnake tail-shaker muscle, all of which contract very rapidly, show a single band of α-RyR on SDS-PAGE. On the basis of this finding, O'Brien *et al.*[29] claimed that the absence of β-RyR or the presence of α-RyR alone might be important for rapidity of contraction. Ogawa,[3] however, suggested that a device for rapid relaxation, for example greater abundance of Ca-ATPase protein would be more critical for rapid contraction. Later, Rome *et al.*[50] showed that swim-bladder muscle has faster cross-bridge kinetics probably due to a myosin isoform and greater dissociation rate constant of troponin C for Ca^{2+} in addition to a large and very fast Ca^{2+} transient. In mammalian skeletal muscles, no correlation has been found among RyR3 content, fiber types and speed of muscle contraction.[30]

Skeletal muscles from the crooked neck dwarf mutant of chicken which lack α-RyR fail to contract with electrical or neuronal stimulation, but there is an increase in myoplasmic Ca^{2+} concentration with the application of caffeine.[51–53] This strongly indicates that α-RyR may function as the depolarization-induced Ca^{2+} release channels in non-mammalian vertebrate skeletal muscles as is true of RyR1 in mammals. Although β-RyR can serve as CICR channel, little is known, about its physiological role.

Several groups have reported the properties of two RyR isoforms in non-mammalian vertebrate skeletal muscles. Bull and Marengo[54] reported two types of Ca^{2+} release channels in lipid bilayer experiments with distinct Ca^{2+} dependencies in SR vesicles isolated from frog skeletal muscle: one of biphasic dependence with marked inhibition at high Ca^{2+} concentrations that is similar to RyR1, and the other of monophasic dependence with little inhibition that was more sensitive to Ca^{2+}, i.e. RyR2-like. Similar results were obtained from

SR vesicles of toad fish white swimming muscles by O'Brien *et al.*[49] Our results on [^{3}H]ryanodine binding to α- and β-RyRs in 1 M NaCl medium would be on this line except that both isoforms were monophasically Ca^{2+}-dependent.[21] Percival *et al.*[22] investigated the ion channel properties of α- and β-RyRs partially purified from SR of chicken skeletal muscle and reported that the two isoforms embodied Ca^{2+} channels with similar conductances but different gating properties on activation by Ca^{2+} and ATP, and on inactivation by high concentrations of Ca^{2+}. Thus these investigators have concluded that the two RyR isoforms in non-mammalian vertebrates have distinct CICR properties. In this respect, recent results on modulation of RyR1 activity during NEM titration may be helpful in considering the functional states of RyR isoforms. With titration of -SH residues, RyR1 underwent a complex transition from temporarily enhanced activity to final reduced activity.[55] We must be careful of the apparently functionally changed states of the same isoform. In contrast, our recent results on [^{3}H]ryanodine binding to the two isoforms in 0.17 M NaCl medium showed that α- and β-RyR are almost equally potent and that they are very similar to each other as CICR channels with similar Ca^{2+} sensitivities.[44] The CICR activity in skinned muscle fibers or [^{3}H]ryanodine binding to isolated SR vesicles which contain the two isoforms is considered to be the sum of the contribution from each isoform. In frog, [^{3}H]ryanodine binding to isolated SR vesicles and CICR activity of SR in skinned skeletal muscle fibers are consistent with [^{3}H]ryanodine binding to α- and β-RyR in 0.17 M NaCl medium.[44] [^{3}H]ryanodine binding to solubilized SR vesicles in 1 M NaCl medium was also explained by those to α- and β-RyR in the corresponding medium.[45] Under the physiological conditions, the possibility that β-RyR might preferentially serve as an amplifier of Ca^{2+} release through CICR is very low. At the present time, the biological significance of β-RyR is not known. Disposition of α- and β-RyR on the junctional face of the terminal cisternae would shed light on this.

While skeletal muscle in *ryr-1*-targeted dyspedic mouse failed to contract on electrical stimulation as expected,[23, 24] skeletal muscle in RyR3-knocked out mouse showed almost normal contraction.[56] As stated, the content of RyR3 in mammalian skeletal muscle is 0.6% or less. Although the channel property of RyR3 is not known in detail, it can be tacitly assumed that RyR3 and RyR1 may be comparable in their unit conductances. Even if RyR3 is more resistant to Mg^{2+} than RyR1, the difference would not be significant because the sarcoplasmic concentration of Mg^{2+} is around 1 mM. Therefore, the contribution of RyR3 would be insignificant even if it is active. Takeshima *et al.*[56] also reported that knock-out of RyR3 does not affect any function of T-lymphocyte or smooth muscles. This is unexpected because we thought that RyR3 might be expressed in a larger amount in smooth muscles than the other isoforms. Even in smooth muscles, the isoforms other than RyR3 might be more abundant. RyR itself might not be involved in the regulation of cytoplasmic Ca^{2+} in smooth muscles although potentially truncated, non-functional RyR3 might exist. Ca^{2+} entry, Ca^{2+} pump out and Na, Ca-exchange through the cytoplasmic membrane may be more significant in addition to Ca^{2+} release from the IP3 sensitive stores. Identification and quantitative estimate of RyR isoforms in smooth muscles are required.

Knockout of RyR3 caused enhanced locomotor activity in mutated animals. RyR3 in the brain, therefore, may take part in the integrity of the function of the central nervous system. Although there is also only a minute amount of RyR3 in the brain, it localizes in such critical areas as hippocampus, corpus striatum, and thalamus. The specific localization of RyR3 in the neurons in those areas may be critical in its effect on signal transmission by specific transmitter(s). Determination of the distribution of RyR3 in neuronal cells is critically important to understand the functional role of this isoform in the brain.

CONCLUSIONS

The biological significance of β-RyR still remains to be determined in nonmammalian vertebrates. To date we have assumed that RyR3 might function in mammalian tissues as RyR1 and RyR2 do in striated muscles when mRNA or cDNA of RyR3 was detected. The actually expressed amount of RyR3 protein, however, is found to be minute, resulting in a negligible functional contribution in most cases. Because the primary amino acid sequences of RyR3 may be variable, some might be functional with distinct tissue-specific properties, and some might be non-functional with truncated residues. It is very interesting that RyR3 in brain and skeletal muscles, however, is very similar among various animal species in the face of potential tissue-specificity. To understand the role of RyR3, the properties, amount and expression site of the expressed RyR3 protein inside the cell must be determined with its complete sequence.

Note added in proof

We have recently purified RyR3 from rabbit diaphragm. The experiments using the isolated RyR3 revealed that there was no difference in the inhibitory effect of Mg^{2+} between RyR1 and RyR3, being at variance with the results of Fig. 9 in our paper.[31]

ACKNOWLEDGEMENTS

We thank Dr. S. R. Wayne Chen for his helpful comments on variations of RyR3 sequence, and we acknowledge the skilful secretarial assistance of Ms. Naomi Ariji.

REFENCES

1. Coronado R, Morrissette J, Sukhareva M *et al.* Structure and function of ryanodine receptors. *Am J Physiol* 1994; **266**:C1485–C1504.
2. Meissner G. Ryanodine receptor Ca^{2+} release channels and their regulation by endogenous effectors. *Ann Rev Physiol* 1994; **56**:485–508.
3. Ogawa Y. Role of ryanodine receptor. *CRC Critical Rev Biochem Mol Biol* 1994; **29(4)**: 229–274.
4. Sutko JL, Airey JA. Ryanodine receptor Ca^{2+} release channels: does diversity in form equal diversity in function? *Physiol Rev* 1996; **76(4)**:1027–1071.
5. Sorrentino V, ed. *Ryanodine Receptors.* N. Y., London, Tokyo: CRC Press, Boca Raton, 1996.
6. Fleischer S, Inui M. Biochemistry and biophysics of excitation-contraction coupling. *Ann Rev Biophys Biophys Chem* 1989; **18**:333–364.
7. MacLennan DH. International lecture: Molecular tools to elucidate problems in excitation-contraction coupling. *Biophys J* 1990; **58**:1355–1365.
8. Rios E, Pizarro G. Voltage sensor of excitation-contraction coupling in skeletal muscle. *Physiol Rev* 1991; **71**:849–908.
9. McPherson PS, Campbell KP. The ryanodine receptor/Ca^{2+} release channel. *J Biol Chem* 1993; **268**:13766–13768.
10. Schneider MF. Control of calcium release in functioning skeletal muscle fibers. *Ann Rev Physiol* 1994; **56**:463-484.
11. Sitsapesan R, McGarry SJ, Williams AJ. Cyclic ADP-ribose, the ryanodine receptor and Ca^{2+} release. *Trends Pharmacol Sci* 1995; **16(11)**:386–391.
12. Wagenknecht T, Radermacher M. Three-dimensional architecture of the skeletal muscle ryanodine receptor. *FEBS Lett* 1995; **369**:43–46.
13. Zucchi R, Ronca-Testoni S. The sarcoplasmic reticulum Ca^{2+} channel/ryanodine receptor: Modulation by endogenous effectors, drugs and disease states. *Pharmcol Rev* 1997; **49(1)**: 1–51.
14. Giannini G, Clementi E, Ceci R *et al.* Expression of a ryanodine receptor-Ca^{2+} channel that is regulated by TGF-β. *Science* 1992; **257**:91–94.
15. Takeshima H, Nishimura S, Matsumoto T *et al.* Primary structure and expression from complementary DNA of skeletal muscle ryanodine receptor. *Nature* 1989; **339**:439–445.
16. Zorzato F, Fujii J, Otsu K *et al.* Molecular cloning of cDNA encoding human and rabbit forms of the Ca^{2+} release channel (ryanodine receptor) of skeletal muscle sarcoplasmic reticulum. *J Biol Chem* 1990; **265**:2244–2256.
17. Hakamata Y, Nakai J, Takeshima H *et al.* Primary structure and distribution of a novel ryanodine receptor/ calcium release channel from rabbit brain. *FEBS Lett* 1992; **312(2-3)**:229–235.
18. Hakamata Y, Nishimura S, Nakai J *et al.* Involvement of the brain type of ryanodine receptor in T-cell proliferation. *FEBS Lett* 1994; **352**:206–210.
19. Oyamada H, Murayama T, Takagi T *et al.* Primary structure and distribution of ryanodine-binding protein isoforms of the bullfrog skeletal muscle. *J Biol Chem* 1994; **269**:17206–17214.

20. Ottini L, Marziali G, Conti A *et al.* α and β isoforms of ryanodine receptor from chicken skeletal muscle are the homologues of mammalian RyR1 and RyR3. *Biochem J* 1996; **315**:207–216.
21. Murayama T, Ogawa Y. Purification and characterization of two ryanodine-binding protein isoforms from sarcoplasmic reticulum of bullfrog skeletal muscle. *J Biochem* 1992; **112**: 514–522.
22. Percival AL, Williams AJ, Kenyon JL *et al.* Chicken skeletal muscle ryanodine receptor isoforms: ion channel properties. *Biophys J* 1994; **67**:1834–1850.
23. Takeshima H, Iino M, Takekura H *et al.* Excitation-contraction uncoupling and muscular degeneration in mice lacking functional skeletal muscle ryanodine-receptor gene. *Nature* 1994; **369**:556–559.
24. Takeshima H, Yamazawa T, Ikemoto T *et al.* Ca^{2+}-induced Ca^{2+} release in myocytes from dyspedic mice lacking the type-1 ryanodine receptor. *EMBO J* 1995; **14**:2999–3006.
25. Giannini G, Conti A, Mammarella S *et al.* The ryanodine receptor/calcium channel genes are widely and differentially expressed in murine brain and peripheral tissues. *J Cell Biol* 1995; **128**:893–904.
26. Furuichi T, Furutama D, Hakamata Y *et al.* Multiple types of ryanodine receptor/Ca^{2+} release channels are differentially expressed in rabbit brain. *J Neurosci* 1994; **14(8)**:4794–4805.
27. Murayama T, Ogawa Y. Properties of RyR3 ryanodine receptor isoform in mammalian brain. *J Biol Chem* 1996; **271(9)**:5079–5084.
28. Olivares EB, Tanksley SJ, Airey JA *et al.* Nonmammalian vertebrate skeletal muscles express two triad junctional foot protein isoforms. *Biophys J* 1991; **59**:1153–1163.
29. O'Brien J, Meissner G, Block BA. The fastest contracting muscles of nonmammalian vertebrates express only one isoform of the ryanodine receptor. *Biophys J* 1993; **65**:2418–2427.
30. Conti A, L.Gorza, Sorrentino V. Differential distribution of ryanodine receptor type 3 (RyR3) gene product in mammalian skeletal muscles. *Biochem J* 1996; **316**:19–23.
31. Murayama T, Ogawa Y. Characterization of type 3 ryanodine receptor (RyR3) of sarcoplasmic reticulum from rabbit skeletal muscle. *J. Biol. Chem.* 1997; **272(38)**: 24030–24037.
32. Airey JA, Grinsell MM, Jones LR *et al.* Three ryanodine receptor isoforms exist in avian striated muscle. *Biochemistry* 1993; **32**:5739–5745.
33. Marziali G, Rossi D, Giannini G *et al.* cDNA cloning reveals a tissue specific expression of alternatively spliced transcripts of the ryanodine receptor type 3 (RyR3) calcium release channel. *FEBS Lett* 1996; **394**:76–82.
34. Zhang Y, Chen HS, Khanna VK *et al.* A mutation in human ryanodine receptor gene associated with central core disease. *Nature Genetics* 1993; **5**:46–50.
35. Nakai J, Imagawa T, Hakamata Y *et al.* Primary structure and functional expression from cDNA of the cardiac ryanodine receptor calcium release channel. *FEBS Lett* 1990; **271**:169–177.
36. Chen SRW. Alternatively spliced variants of the brain Ca^{2+}-release channel (ryanodine receptor) expressed in smooth muscle tissues. *Biophys J* 1996; **70**:A390.
37. Miyatake R, Furukawa A, Matsushita M *et al.* Tissue-specific alternative splicing of mouse brain type ryanodine receptor/calcium release channel mRNA. *FEBS Lett* 1996; **395**:123–126.

38. Lai FA, Misra M, Xu L *et al.* The ryanodine receptor-Ca^{2+} release channel complex of skeletal muscle sarcoplasmic reticulum: Evidence for a cooperatively coupled, negatively charged homotetramer. *J Biol Chem* 1989; **264**:16776–16785.
39. Airey JA, Beck CF, Murakami K *et al.* Identification and localization of two triad junctional foot protein isoforms in mature avian fast twitch skeletal muscle. *J Biol Chem* 1990; **265**: 14187–14194.
40. Monkawa T, Miyawaki A, Sugiyama T *et al.* Heterotetrameric complex formation of inositol 1,4,5-trisphosphate receptor subunits. *J Biol Chem* 1995; **270**:14700–14704.
41. Nucifora FC, Jr., Sharp AH, Milgram SL *et al.* Inositol 1,4,5-trisphosphate receptors in endocrine cells: localization and association in hetero- and homotetramers. *Mol Biol Cell* 1996; 7: 949–960.
42. Chen SRW, Li X, Zhang L. Functional expression of cDNA encoding the type 3 ryanodine receptor of rabbit uterus in HEK293 cells. *Biophys J* 1997; **72**:A13.
43. DiJulio DH, Buck JD, Watson EL *et al.* Ryanodine receptor type III (Ry3R) identification in mouse parotid acini. *Biophys J* 1997; **72**:A377.
44. Murayama T, Ogawa Y. Similar Ca^{2+} dependences of $[^3H]$ryanodine binding to α- and β-ryanodine receptors purified from bullfrog skeletal muscle in an isotonic medium. *FEBS Lett* 1996; **380(3)**:267–271.
45. Murayama T, Ogawa Y. Potential factors which affect the Ca^{2+}-sensitivity of ryanodine receptors. *Biomed Res* 1993; **14 (Supplement 2)**:71–72.
46. Lai FA, Liu QY, Xu L *et al.* Amphibian ryanodine receptor isoforms are related to those of mammalian skeletal or cardiac muscle. *Amer J Physiol* 1992; **263**:C365–C372.
47. Endo M. Calcium release from sarcoplasmic reticulum. *Cur Top Memb Trans* 1985; **25**: 181–230.
48. Murayama T, Ogawa Y. Relationships among ryanodine receptor isoforms expressed in vertebrate skeletal muscles based on immunologic cross-reactivities. *J Biochem* 1994; **116(5)**: 1117–1122.
49. O'Brien J, Valdivia HH, Block BA. Physiological differences between the α and β ryanodine receptors of fish skeletal muscle. *Biophys J* 1995; **68(2)**:471–482.
50. Rome LC, Syme DA, Hollingworth S *et al.* The whistle and the rattle: The design of sound producing muscles. *Proc Natl Acad Sci USA* 1996; **93(15)**:8095–8100.
51. Airey JA, Baring MD, Beck CF *et al.* Failure to make normal α-ryanodine receptor is an early event associated with the crooked neck dwarf (cn) mutation in chicken. *Develop Dyn* 1993; **197(3)**:169–188.
52. Airey JA, Deerinck TJ, Ellisman MH *et al.* Crooked neck dwarf (cn) mutant chicken skeletal muscle cells in low density primary cultures fail to express normal alpha-ryanodine receptor and exhibit a partial mutant phenotype. *Develop Dyn* 1993; **197(3)**:189–202.
53. Ivanenko A, McKemy DD, Kenyon JL *et al.* Embryonic chicken skeletal muscle cells fail to develop normal excitation-contraction coupling in the absence of the α ryanodine receptor: Implications for a two-ryanodine receptor system. *J Biol Chem* 1995; **270**:4220–4223.
54. Bull R, Marengo JJ. Sarcoplasmic reticulum release channels from frog skeletal muscle display two types of calcium dependence. *FEBS Lett* 1993; **331**:223–227.

55. Aghdasi B, Zhang J-Z, Wu Y *et al.* Multiple classes of sulfhydryls modulate the skeletal muscle Ca^{2+}release channel. *J Biol Chem* 1997; **272(6)**:3739–3748.
56. Takeshima H, Ikemoto T, Nishi M *et al.* Generation and characterization of mutant mice lacking ryanodine receptor type 3. *J Biol Chem 1996;* **271(33)**: 19649–19652.
57. Sorrentino V, Volpe P. Ryanodine receptors: how many, where and why? *Trends Pharmacol Sci* 1993; **14**:98–103.

Chapter 3

The Skeletal Muscle Calcium-Release Channel Visualised By Electron Cryomicroscopy and Angular Reconstitution

Elena V. Orlova, Irina I. Serysheva, Susan L. Hamilton, Wah Chiu and Marin van Heel

In recent years we have witnessed a rapid development in three-dimensional (3D) electron microscopy of frozen-hydrated specimens. By rapidly freezing a solution of macromolecules to liquid nitrogen temperatures, the solution becomes vitreous thus creating an almost natural, hydrated embedding medium for the biological macromolecules.[1] The macromolecules, such as the calcium-release channel, exhibit random orientations in solution and the different orientations in the sample may be exploited to find the 3D structure that best fits to all molecular images in a set of micrographs. The angular reconstitution approach[2] allows one to find the different orientations of the molecules in the data set and to thus calculate the 3D structure of the molecule without growing crystals or without tilting the specimen in the microscope. Although first proposed in 1987, this approach only matured recently and the calcium-release channel was one of the first structures to be visualised using this technique.[3] Given the experimental simplicity of the angular reconstitution approach, we are now in a position to image the calcium-release channel in different functional and structural states[4] as well as to study the different isoforms of the "ryanodine receptor".

INTRODUCTION

It has been known since the 1940's, that muscle contraction is regulated by calcium ions.[5] The contractile unit of muscle cells is the myofibril which is composed of actin and myosin filaments. Contraction is caused by sliding of these two classes of filaments past each other, as a result of the temporary connection between myosin molecules in the thick filaments and actin subunits in the thin filaments. Upon activation, Ca^{2+} ions bind to troponin and a chain of subsequent reactions initiates the actin-myosin interaction (see Ref. 6). This interaction is suppressed in the absence of Ca^{2+}. Nature has created a subtle mechanism for simultaneous delivery of calcium ions inside the muscle fiber. The mechanism is based on the interplay of two networks of membranous structures: the sarcoplasmic reticulum (SR) and transverse tubules. The rapid release of Ca^{2+} ions into the cytoplasm is regulated by the transmission of the activity signal from voltage-sensitive protein (the dihydropyridine receptor) in the transverse tubule membrane to the Ca^{2+}-release channel protein, which is located in the junctional face membrane of the terminal cisternae of the SR.[7–10] This transmission causes the Ca^{2+}-release channel to open, allowing the flow of Ca^{2+} ions stored in the SR into the cytoplasm. The Ca^{2+}-release channel complex contains four identical large polypeptides each with a molecular mass of 565 kDa.[11,12] The entire structure of the Ca^{2+}-release channel, including associated small proteins, is about 2.3 million Dalton[13] and the full tetrameric organisation is necessary for channel activity.[14]

The Ca^{2+} channel changes its state according to the physiological conditions. A number of 'blockers' and 'openers' of the channel have been characterised to date.[15,17–20] In its inactive state, which we term "closed", Ca^{2+} ions are not transmitted across the membrane. Depletion of Ca^{2+} from the environment of the channel was found to shift it into the closed state.[15,16] In contrast, in its active "open" state, Ca^{2+} ions can pass freely through the channel. The plant alkaloid ryanodine was found to bind the open Ca^{2+}-release channel and to produce dose-dependent effects on Ca^{2+}-release and thus on contraction (see Ref. 17). We designate the ryanodine modified open state, which we have studied , as the 'open' state.

The molecular mechanism of action of the Ca^{2+}-release channel can only be fully understood once atomic-resolution 3D structures of the channel are available in all relevant functional states. In reality, solving all these structures by X-ray crystallography alone, is not easily accomplished since this would require the availability of good crystals - diffracting to high resolution (~3 Å) - for each of the channel's functional states and it is already extremely difficult to grow good crystals of any membrane protein. Obtaining good crystals of this very large membrane protein in different functional states may thus remain a structural biologist's dream. Nuclear Magnetic Resonance (NMR) spectroscopy[21] can be used for 3D structure determination of macromolecules in solution without growing crystals. At the present state of the technology, however, NMR is only suitable for studying small proteins of less than ~300 residues (~30,000 Dalton) and may be used to study the atomic structure of the small associated proteins. However, NMR is not suited for studying the entire Ca^{2+}-release channel tetramer with its mass of 2.3 MDa.

In recent years electron (cryo-) microscopy has become an important tool for structural analysis of macromolecules organised as thin crystals or as single particles (see Ref. 21). Traditionally, electron microscopy was used for the visualisation of biological macromolecules by embedding the macromolecules in heavy-metal salts ("negative stains" such as uranyl-acetate, ammonium molybdate, potassium phosphotungstate *etc.* (for a review see 22). This layer was first air-dried and then further dehydrated in the vacuum of the electron microscope. The conventional negative-stain preparation is a relatively easy and fast technique which allows the direct visualisation of macromolecules. However, the dry negative-stain technique has serious disadvantages such as: the molecules are often flattened and distorted by dehydration; the molecules frequently exhibit strong preferential attachment to the supporting carbon; the staining is often inhomogeneous thus complicating the interpretation of the images. The properties of negative staining techniques have been studied and reviewed by several authors.[23–26]

The modern alternative to the use of dry negative stain as an embedding medium is to use hydrated, rapidly-frozen biological specimens. When a thin layer of solution (thinner than ~100nm) is rapidly cooled to liquid Nitrogen temperature, the water transfers to a vitreous phase directly, bypassing the crystalline ice phases. These transitions would otherwise damage the molecules in solution.[1,21,27] For all practical purposes this vitreous phase of water behaves like a liquid and it represents an almost natural hydrated environment for biological macromolecules. Now that the methodology of freezing, and cryo-transferring the thin films to the electron microscope, and of imaging these at liquid Nitrogen temperature is in place, biological macromolecules may be routinely imaged in this structure-preserving frozen-hydrated state.

The electron microscope provides (very noisy) projection images of the molecules and historically, 3D information was obtained by tilting the specimen holder in the microscope in order to take multiple pictures of the same object in different orientations.[28–31] The different projection images of the object can then be combined in a 3D reconstruction using tomographic reconstruction algorithms. Tilting the specimen holder, in order to obtain high-resolution 3D reconstructions of biological macromolecules,[29,32] however, has a number of disadvantages which reduce the achievable level of resolution.[31] Of more practical value are methods in which the relative orientation of the molecular images are determined *a posteriori*, i.e., methods in which one exploits the random orientations of the macromolecules in the EM specimen itself, to extract the 3D information without an explicit tilting of the EM specimen in the microscope.

The first such technique was the "common lines" technique applicable to objects with icosahedral symmetry such as spherical viruses.[33] The angular reconstitution approach[2] combines pattern recognition methodologies developed earlier[34,35] to find the characteristic projection images present in the micrographs, with a real-space method to determine the relative orientation of the individual molecular views. The pattern recognition aspects of the approach allow the extraction of systematic information ("characteristic views"[36]) from a very large number of noisy images of the randomly oriented macromolecules. The determination of the relative orientation of the "characteristic views" then permit us to calculate the 3D densities distribution of the molecules. The structures of many different macromolecules and aggregates have now been determined ranging from the entirely asymmetric bacterial ribosome,[37,38] to spherical viruses.[39] Here we describe 3D reconstructions of the Ca^{2+}-release channel in different functional and structural states using the angular reconstitution approach. Knowledge of the 3D structure of the Ca^{2+}-release channel in different states at the intermediate resolution levels of 10–30 Å may form the basis for assessing models of the mechanism of conduction and gating of the channel.

CRYOMICROSCOPY OF THE CA^{2+}-RELEASE CHANNEL

The Ca^{2+}-release channel was purified from rabbit fast-twitch skeletal muscle as described previously.[40] For driving the great majority of the Ca^{2+}-release channels to the steady open state,[16] the purified sample was incubated in 100 μM Ca^{2+} and 100 nM ryanodine for 3 hours at 20°C. The closed state was obtained by depleting the channels of Ca^{2+} with 1 mM EGTA in 300 mM KCl, 1 mM MDT, 10 mM MOPS (pH 7.4).[16] Single drops of the channel solutions in these different buffers were applied to holey grids covered with thin carbon film (made hydrophilic by air glow discharge in a Denton vacuum evaporator). After blotting, they were frozen rapidly by freeze-plunging.[1] The frozen, hydrated specimens were kept at −160° C, and transferred to a Gatan cryo-holder for imaging at 30,000 magnification in a JEOL1200 electron microscope, operated at 100 kV under minimal dose conditions (~7 e/Å^2).[41]

Any transmission electron microscope (TEM) "distorts" the images in the sense that the fine detail ("high-frequency" detail) may be strongly attenuated, or may appear with reversed contrast with respect to the large-scale ("low-frequency") details ("black is white" and vice versa).[42] Theoretically speaking, the effect can be described by a multiplication of the Fourier Transform (FT) of the ideal image of the object, by the contrast transfer function ("CTF") of the electron microscope at the defocus value chosen for the imaging. For the imaging conditions used here, the first zero of the CTF was at 26–30 Å and no explicit CTF correction was necessary (see Ref. 38). The defocus values of the images were determined computationally from the position of the minima and maxima of the contrast transfer function rings seen in power spectra of the images from each micrograph.

DENSITOMETRY AND IMAGE PROCESSING

Electron micrographs of ice-embedded Ca^{2+}-release channel - or of any other biological macromolecule sample - have low signal contrast levels and, at the same time, have very high noise levels (Fig. 3.1). This is true even when the focus is chosen optimally (i.e. the CTF) for imaging the details one is interested in. The high noise levels are essentially dictated by the radiation sensitivity of the biological material: the sensitive molecules are completely destroyed by the high electron exposure levels needed to shed enough light on the matter in order to elucidate the relevant details. The Signal-to-Noise-Ratio (SNR) of such images is thus necessarily very low. Averaging of the raw molecular images allows the *a posteriori* (that is, after the micrographs have been collected) enhancement of the SNRs of the molecular images. Many thousands of molecular images must be available in order to provide the necessary information on the molecule seen from all possible orientations.

For the sake of an efficient data collection, each micrograph should optimally contain many hundreds of macromolecular images. In our case, seven micrographs of the channel specimen in the open state and four micrographs of the channel in the closed state (each data set chosen to have similar defocus settings) were digitised in a Perkin Elmer microdensitometer (step size of 20 mm per pixel corresponding to 6.7 Å in the object scale). From the digitised micrographs, 5700 channel particles were selected interactively

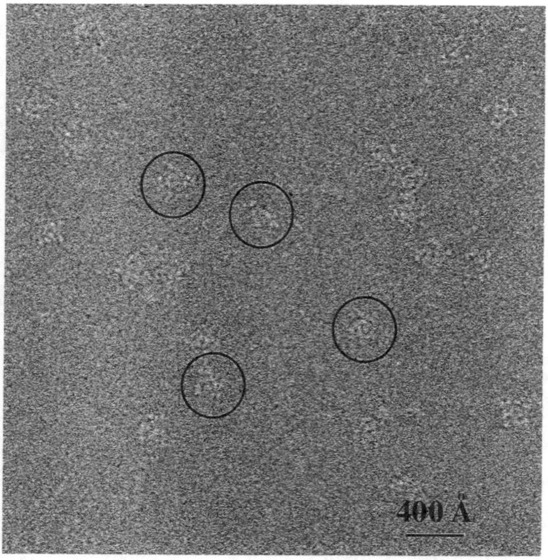

Fig. 3.1 Images of a frozen-hydrated Ca^{2+}-release channel protein sample Biological macromolecules are extremely sensitive to radiation damage exerted by the electrons used for the imaging in the electron cryo-microscope. Compared to room temperature, cooling the specmens by liquid nitrogen temperature allows one to use an approximately ten-fold higher electron dose for a given level of radiation damage. Several molecular images of the individual channel proteins seen in different orientations are circled. In this print the protein is seen as a light structure against a dark background.

from the seven"open-state" micrographs, and 3000 particles from the four "closed-state" micrographs. In both cases, individual molecular images were extracted as 80×80 pixel fields, and processed as single data sets. All data processing was performed using the IMAGIC-5 software system on eighter Silicon Graphics Indigo or DEC Alpha workstations.[43,44]

IMAGE ALIGNMENT AND CHARACTERISTIC VIEWS

The individual macromolecules have three translational degrees of freedom in the specimen (x,y,z), although only two of these, the translations in the plane of the grid (x,y), are of relevance because the microscope projects our molecules along the beam direction (z). The molecules also have three rotational degrees of freedom described by the Euler angles (α,β,γ). Whereas the angle (α) represents the rotation of the image in the plane of the grid and thus leads to equivalent but rotated projection images, the remaining two Euler angles

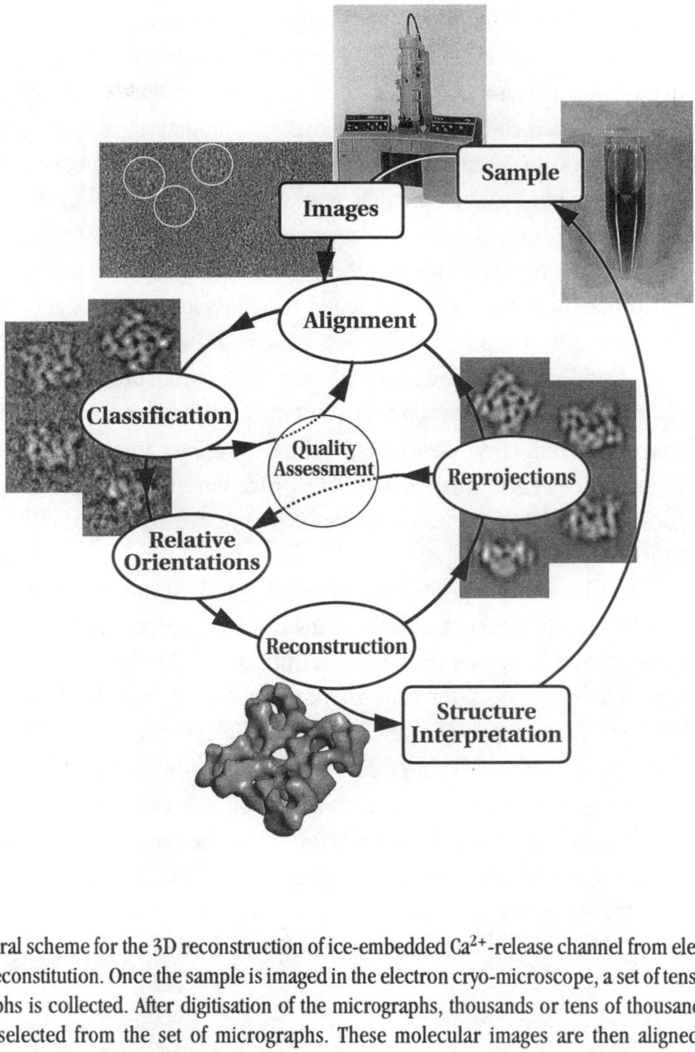

Fig. 3.2 Overal scheme for the 3D reconstruction of ice-embedded Ca^{2+}-release channel from electron images by angular reconstitution. Once the sample is imaged in the electron cryo-microscope, a set of tens or hundreds of micrographs is collected. After digitisation of the micrographs, thousands or tens of thousands of particle images are selected from the set of micrographs. These molecular images are then aligned by iterative multireference alignment techniques and sorted into homogeneous classes of similar images by automatic multivariate statistical classification. All molecular images belonging to the same class are averaged to produce noise-reduced class averages or charactinstic views. Four class-averages are shown on the left-hand-side of this illustration, representative of ~200 class averages used in this study. An Euler angle orientation is assigned to each class average by angular reconstitution. Subsequently, a first 3D reconstruction is computed, which is used to produce a set of reprojected images such as those shown in the right-hand-side of this picture. Reprojections are used as reference images for realignment of the original raw image data set and to improve the accuracy of the Eular angle determinations. The four reprojections shown here correspond to the same angular orientations as were assigned to the original four classes shown on the left. The final 3D map is shown viewed at an oblique angle from the cytoplasmic side of the channel.

(β,γ) represent rotations of the molecule out of the plane of the grid and these lead to fundamentally different projection images. It is only advantageous to average images of molecules in the same orientation in terms of (x,y,α,β,γ). Averaging the noisy input images into class average or "characteristic views"[36] boosts the common signal while simultaneously reducing the random noise. Much higher signal-to-noise ratios are achieved in the averages than in raw molecular images and this reduction of noise facilitates all further processing. The best results were obtained with class averages consisting of ~20 molecular images although this number is typically varied in the course of the procedures (Fig. 3.2).

Alignment procedures have been designed to eliminate the translational (x,y) and rotational in-plane differences (α) between images of similar appearance based on measurements of similarity of images using the correlation function.[45–47] Because we are dealing with mixed data sets containing fundamentally different projection images corresponding to different (β,γ) directions, multiple reference images are needed for the alignment in Multi-Reference Alignment (MRA) procedures.[36] The reference images for the MRA procedures in later phases of the analysis were reprojections of the current 3D reconstructions (see below).

In the starting phase of the procedure, however, no 3D reconstructions are available yet, and the first set of reference images stem from the reference-free alignment-by-classification approach.[48] In that procedure all images of the data set are first centred translationally with respect to the rotationally averaged total sum of all the images: this (iterative) procedure corresponds to the placing of the centre of mass of the projection image in the centre of the frame without any rotation at all. Multivariate-Statistical Analysis (MSA) data compression[35] and automatic classification procedures[34,49] are then used to find similar images in similar angular orientations. The class-averages resulting from this procedure are used as first reference images for the MRA of the full data set. At this stage of the procedure, the symmetry properties of the molecule are also clearly revealed.[48] The 4-fold symmetry of the Ca^{2+} channel was seen distinctly as the first pair of eigenimages of the data set (Fig. 3.3).

EULER ANGLE ASSIGNMENT

From a sufficiently large set of 2-dimensional (2D) projection images (characteristic views) of a 3D structure in different projection directions it is possible to calculate a 3D reconstruction using an appropriate "computerised tomography" reconstruction scheme.[29,50,51] However, since we are exploiting the random orientations of particles in the embedding medium without any instrumental tilt, we do not know the relative orientations of the characteristic views, and this orientational information is essential for the 3D

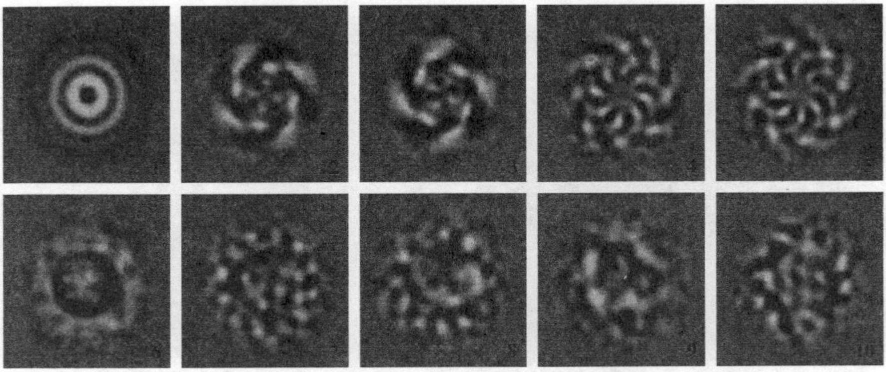

Fig. 3.3 Eigenimages of the set centred raw images. Eigenimage #1 corresponds to the total sum of the images in the set #2 and #3 reveal the 4-fold symmetry of the Ca^{2+}-release channel. Note that eigenimage #3 is identical to eigenimage #2 rotated over 45°. This behaviour is indicative of 4-fold symmetry as explained in Ref. 48. Images #4 and #5 reveal a residual 8-fold symmetry component.

reconstruction algorithms. We use the angular reconstitution technique[2] to *a posteriori* determine these Euler angular orientations (Fig. 3.2). The angular reconstitution method is based on the common line projection (CLP) theorem, stating that any two 2D projections of a 3D object share at least one one-dimensional (1D) line projection. The 1D line projection is calculated as a total sum of all densities of a 2D projection (here, a characteristic view) in a given direction.[2,52]

The search for the best relative Euler angle assignments for a set of 2D projection images is performed by first calculating all possible 1D projections (the collection of all possible line projections constitutes a "sinogram"; see Fig. 3.4) from each of the 2D projection images.[2,52] To find the CLPs between two 2D projection images, their sinograms are compared line-by-line through the calculation of their cross correlation coefficients (CCC). The collection of all possible such CCCs between two sinograms is the "sinogram correlation function" (Fig. 3.4). The number of peaks in the sinogram correlation function depends on the pointgroup symmetry of the structure. For the 4-fold symmetric Ca^{2+}-release channel, there are four pairs of CLPs. This leads to a total of eight peaks in the sinogram correlation functions (because of the 2-fold redundancy of the sinogram correlation function) (Fig. 3.4).[3,52,53]

For a 4-fold symmetric structure, we start the procedures with one projection image with low error residuals among the symmetry-related peaks (see below) in the self sinogram correlation function ("SSCF", Fig. 3.4g). The Euler angle determination for the further projection images, to be added to the set, is performed with respect to all earlier projection

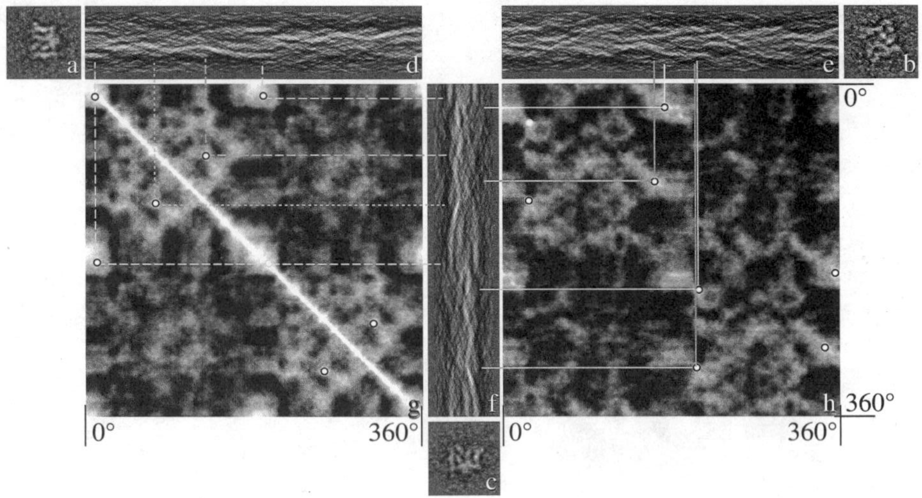

Fig. 3.4 Sinograms and Sinogram correlation functions. Figure 3.4a and 3.4b: Two dfferent class averages (projection images through the 3D structure of the tetrameric channel protein) chosen from the full set of ~ 200 class averages used in this study. The projection image Fig. 3.4c, is the same as image 3.4a, but has been rotated by 90° clockwise. For each of these images the corresponding sinograms have been calculated (d, e and f). The sinogram of a 2D image is the collection of all possible 1D projections through the 2D density distribution. The first 1D projection of the 2D image 3.4a, i.e. the first vertical line of the sinogram 3.4d on the left side is generated by summing all horizontal lines of the image in 3.4a. The second line of the sinogram is the 1D projection in a direction of 1° away from the first projection direction, the third line is the 1D projection in a direction of 2° away from the first, etc. A sinogram correlation function is the line by line comparison of two sinograms. Each point of the sinogram correlation function is calculated as a line by line correlation of the sinograms as illustrated by the lay-out of the montage. There are two kinds of sinogram correlation functions: the first kind is that between a projection image (3.4a) and itself (3.4c), the Self Sinogram Correlation Function (SSCF), such as shown in Fig. 3.4g; the second kind (3.4h) is that between two different projection images (like between projection images 3.4e and 3.4c), the Cross Sinogram Correlation Function (CSCF). Sinogram correlation functions calculated over 0°–360° have a two-fold redundancy because any line projection of a 2D image is the mirror of the line projection in the opposite direction. The eight open circles mark symmetry-related peaks in the SSCF. They exist as pairs separated by 180°. Lines identify the corresponding 1D projections in the sinograms. The SSCF of any 2D projection has a diagonal line of maxima (see Fig. 3.4g), since every 1D line projection matches itself perfectly. SSCFs are mainly important for the first Eular angle assignments of (highly) symmetric particles. The eight open circles in the CSCF (Fig. 3.3h) indicate the optimal relative orientations of these two projections assuming 4-fold symmetry. In the final phases of the "angular reconstitution" analysis all Euler angle assignments are performed using CSCFs with respect to reprojections of previously calculated 3D reconstruction.

images with an already determined Euler angle orientation, and with respect to the new projection image itself. We perform a simultaneous search for symmetry-related peaks in all Cross Sinogram Correlation Functions ("CSCF") of the new projection with respect to the "old" projection images which already have been assigned Euler angle orientations and in the SSCF of the new projection itself.

Because of the 4-fold redundancy inherent in the 4-fold symmetry of the Ca^{2+}-release channel (pointgroup "C4"), the Euler angles search may be limited to a one fourth segment of the unit sphere ranging from the north pole to the south pole and spanning 90° along the equator. The angle γ, in our definition of Euler angles, thus ranges from 0° to 90°. The

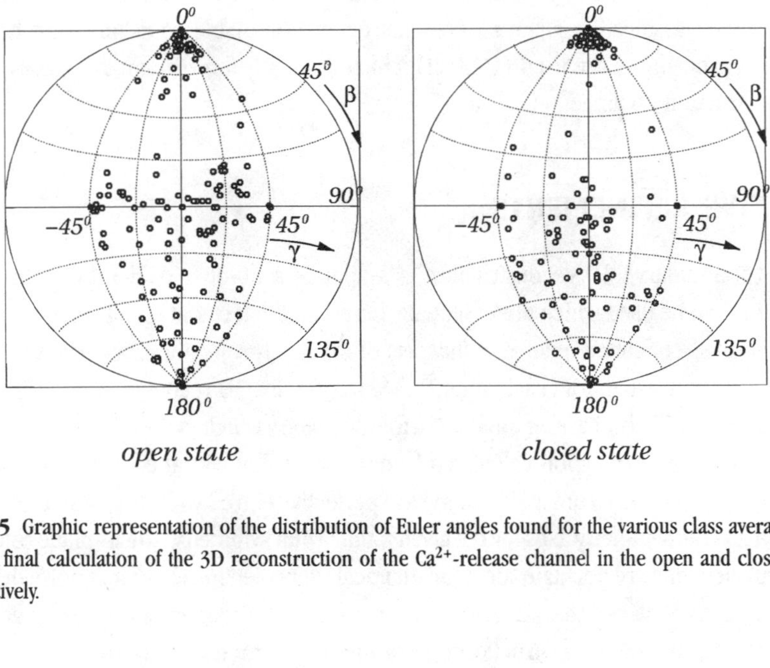

Fig. 3.5 Graphic representation of the distribution of Euler angles found for the various class averages used for the final calculation of the 3D reconstruction of the Ca^{2+}-release channel in the open and closed states respectively.

distribution of angles of the molecular projections over this segment (the "asymmetric triangle") is illustrated in Fig. 3.5. Such a distribution is indicative of the interactions between the solubilised channels and the surfaces of the thin water film and/or the carbon support film. Ideally, one hopes to have a uniform distribution of projection direction covering the whole asymmetric triangle. However, such a uniform distribution is not an absolute necessity for achieving a good uniform coverage of the 3D information.[53,54]

3D RECONSTRUCTION

Having assigned Euler orientations to a set of projections, a 3D density distribution of the object is calculated using the exact-filter back projection algorithm.[29,51] This algorithm correctly weights the contributions of all available projection images in the sense that if certain areas of the unit sphere are heavily sampled with projections (i.e. certain Euler angle orientations are over-represented in the data set), these areas are automatically down-weighted. For our first 3D reconstruction we assigned Euler angles to the 30 best class averages taken out of a total of 200 characteristic views from the first rounds of alignment and classification. Once a 3D reconstruction is available, some important iterative refinement procedures are used (Fig. 3.2) which gradually improve on the quality of the 3D reconstruction obtained.

ITERATIVE REFINEMENTS

Once a (preliminary) 3D reconstruction is available, a number of iterative refinement procedures can be used which are essentially based on the fact that the 3D reconstruction volume is a self-contained entity and that sets of 2D (re-)projections calculated from the 3D volume are consistent with each other.[55,56] Moreover, the 3D reconstruction of the Ca^{2+}-release channel will also have an imposed 4-fold symmetry which makes the reconstruction volume, and any re-projection calculated from it, consistent with the C4 symmetry of the channel. Re-projections from a 3D volume are perfectly centred with respect to a common 3D origin and the symmetry axes for the given point group symmetry. The in-plane rotations of the reprojections are normally all made identical. Reprojections have less noise than the input projection images (characteristic views) because of the averaging in 3D space that takes place during 3D reconstruction due to the information overlap between different projection images (see Ref. 51).

The first type of iterative refinement concerns the reference images used for the multi-reference alignment of the whole set of molecular images. From the available 3D reconstruction, reprojected images are calculated in directions uniformly covering the entire asymmetric triangle (all possible unique projections) and these reprojections serve as the reference images for a new round of MRA. Being derived from the same 3D volume, the new references share the same 3D origin and therefore share the same (x, y) origin in projection. They also share a common rotational in-plane orientation (α) as was mentioned above. The refinement of the overall alignment of the data set by MRA and a new round of MSA classification are important for identifying rare views which may be significant for attaining a uniform resolution in the 3D reconstruction. Many rare yet unique views of the channel protein were found only in the course of such refinements. The number of reprojections used for MRA purposes is rather high (typically more than 100) and is chosen depending on the point group symmetry of the structure and on the size of the molecule and the desired resolution level.[30,31] The realignment of the full data set with respect to this large number of reference images is one of the most CPU-intensive parts of image analysis and may take many days of CPU on a modern workstation. Interestingly, the earlier 3D refinement schemes[55,56] would stop at this refinement level, since rough Euler angle assignments can already be made based on the correlations with the references. This earlier approach has become very popular in recent years.[57,58]

Another important iterative refinement to be applied once 3D reconstructions are available, is the anchor set Euler angle assignment.[3,52,59] Here one reprojects the 3D reconstruction in various directions again, covering the asymmetric triangle but here typically only 10–30 reprojections are used. The Euler angle assignments of a projection to be added to the reconstruction are then performed exclusively using the CSCFs of the new projection with respect to these reprojections. The advantages of this procedure over the Euler angle assignment with respect to other experimental projection images are, again, the consistency of the reprojections with respect to each other and with respect to the given pointgroup symmetry, and the much lower noise level in the reprojections than in the experimental class averages.

Iterative refinements take place at various levels of the analysis. After the renewed MRA using reprojection of the earlier 3D reconstruction, the MSA classification will lead to better class averages with a higher degree of internal consistency and a higher resolution. Once these refined class-averages have been obtained, a new assignment of Euler angles is performed using an anchor set of reprojections from the previous 3D reconstruction. A further refinement of the Euler angle assignments is then achieved by calculating a new 3D map using only the best classes. This intermediate reconstruction then only serves to generate a refined set of reprojections to be used as anchor for yet another round of Euler angle assignments. During all phases of the analysis, quality criteria play an important role in monitoring progress towards the final results and, along the way, removing poor averages from the data set.

QUALITY CRITERIA

Not only do quality criteria have a significant role is assessing, say, the resolution of the final 3D reconstruction, but are important tools used throughout the iterative procedures to monitor progress and to identify problems in the data or the procedures.

Classification. In early phases of the analysis, MSA classifications are used to find characteristic views (high SNR class averages) of the molecule. Two quality criteria are used mainly to assess their quality: good classes must be "compact" and thus have a low internal variance[34,49] and the classes must have a high resolution level.[36,60] The class averages with poor quality levels may be removed from the set to be used for 3D analysis, although the images contained in these poor class averages remain in the overall data set and continue to participate in the new rounds of iterative MRA refinements.

Euler angle search. When a class-average is fitted to the 3D volume in terms of its Euler orientation, we use the CSCFs with respect to the "anchor set" as discussed above. Ideally, for perfect projection images, the normalised height of all symmetry peaks should be equal to 1.0. In reality, the height of peaks is lower due to noise and other sources of errors such as structural distortions within the particles. The standard deviation among the peak heights of symmetry related peaks is an excellent quality criterion for the determination of Euler orientation of a given projection image.[3,52] This criterion may, again, be used to exclude poor class averages from the 3D reconstruction procedure.

3D reconstruction. Directly after calculating a 3D reconstruction we calculate reprojections of the 3D map in the Euler directions found for the original classes. Comparison of the reprojections with the input class-averages[3,4] immediately reveals wrong Euler angle assignments, often directly upon visual inspection and even before scrutinising the numerical similarity measures.[60,61]

Resolution in 3D reconstruction. The final quality criterion - often reflected in the title of the resulting papers - is the resolution achieved in the 3D reconstruction experiment. Typically, two 3D reconstructions are performed each using one half of the available data. The two 3D reconstructions are then compared to each other, by calculating the normalised cross correlation coefficient between two corresponding shells in Fourier space.[60,61] Recently, it was realised that the threshold used should be adapted to the pointgroup symmetry of the molecule under investigation.[53] The Fourier Shell Correlation (FSC) is currently one of the most popular 3D resolution criteria[62,63] albeit that in recent papers various authors have started to adopt a fixed value ("0.5") threshold which we consider unjustified. The FSC comparison in both our 3D reconstructions indicated a reproducible resolution of ~28 Å.

It is obvious that there is some overlap between the various quality criteria we use. For example, class averages may be discarded because of their poor internal statistics, or because they do not fit the current anchor set very well. It will require more experience with such criteria to render the angular reconstitution procedures completely automatic. Currently, the experience of the "image processor" still plays an important role in achieving convergence in the data analysis.

OVERALL SHAPE OF THE CA^{2+}-RELEASE CHANNEL

The final structure of the channel in the open state was obtained from the best 165 characteristic views which, together, contained ~3,700 of the 5,700 original molecular images.[4] The structure of the channel in the closed state[3] was calculated from 150 class averages (~2,000 images from 3,000). The Euler angles distributions of the classes used for both reconstructions are shown in Fig. 3.4, and the final 3D structures (at ~28 Å resolution) of the Ca^{2+}-release channel in the open and closed states are shown in Fig. 3.6. The surface of the open state Ca^{2+}-release channel is shown in dark grey, that of the closed state in light grey. The overall features of the open state channel are very similar to those of the closed state channel. Both structures,[3,4] as well as the structure obtained using the conical tilt approach,[64] are relatively empty when compared to familiar soluble proteins. The rather hollow and pretzel appearance of the putative cytoplasmic portion of the structure may be associated with efficient diffusion pathways for the Ca^{2+} into the cytoplasm and may also facilitate the interaction of the channel with different ligands and modulators. An earlier 3D structure of a negatively stained Ca^{2+}-release channel[65], however, did not reflect this open property, probably due to the limitations of the techniques applied, as was discussed by the original authors in Ref. 64. When compared to the newer cryo-microscopical results, this early 3D reconstruction appears to have suffered excessive flattening, an artefact often observed for negatively stained specimens (see Ref. 24).

The side view (Figs. 3.6a and 3.6b) of the channel protein reveals its mushroom-shaped appearance. The "cap" of the mushroom is formed by the cytoplasmic (CY) region of molecule, a square shape of 270 Å × 270 Å × 110 Å. This CY region is connected to the much smaller, yet also square-shaped putative transmembrane (TM) domain by four columns. These columns and the TM domain form the "stem" of the mushroom-shape with a height of ~65 Å, part of which is thought to correspond to the membrane-spanning domain of the channel protein. The lowest part of the mushroom stem is formed by 4 bean-shaped sub-domains which can be seen in the bottom view (Fig. 3.6e and 3.6f). The sides of this

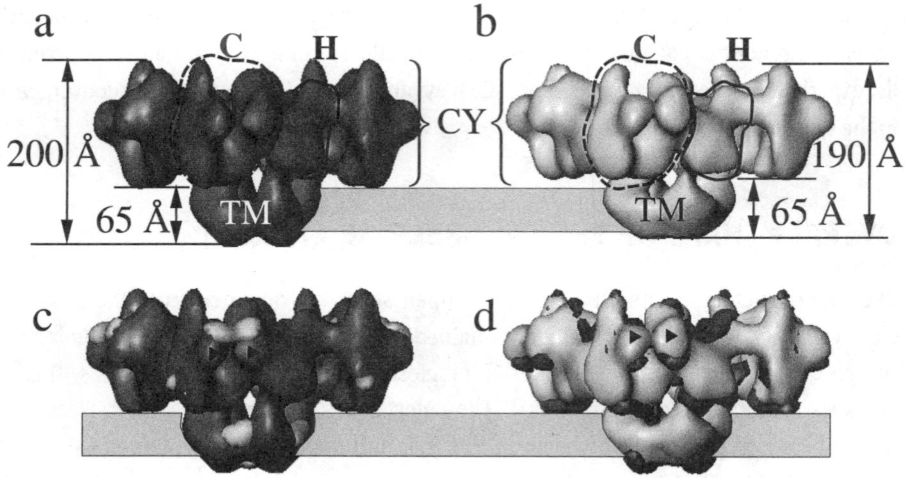

Fig. 3.6 Surface representation of the 3D structure of the Ca^{2+}-release channel in the open (Fig. 3.6 left column) and closed states (Fig. 3.6 right column). The volume of the reconstruction at ths chosen density threshold level corresponds to a mass of ~2.4 million Dalton, assuming a protein density of 1.35 g/cm^3. The 3D structure of the channel in open state is shown in dark grey, the channel in the closed state is shown in light grey. The difference map between closed and open states (C-O) is shown in light grey and superimposed upon the structure of the channel in the open state. The difference map between open and closed states (O-C) is shown in dark grey and superimposed upon the channel structure in the closed state. Fig. 3.6(**a**): Side view of the 3D reconstruction of the channel in the open state. Dashed and solid lines (C, and H) denote the clamp domain (C) and handle domain (H). Figure 3.6(**b**): Side view of the 3D reconstruction of the channel in the closed state. Figure 3.6(**c**): The C-O difference map superimposed upon the open-state structure. Figure 3.6(**d**): The O-C difference map superimposed upon the closed-state structure. Figure 3.6(**e**): View from SR lumen towards the cytoplasm of the 3D reconstructions of the Ca^{2+}-release channel in its open state. The transmembrane (TM) region of the open-state reconstruction shows a central opening. Figure 3.6(**f**): Oblique view of the structure of the Ca^{2+}-release channel in the closed state. Figure 3.6(**g**): The C-O difference map superimposed upon the open-state structure. The light grey spots of the C-O difference map indicate changes in, for example, the transmembrane domain. Figure 3.6(**h**): The O-C difference map imposed upon the closed-state structure. The dark grey spots of the difference map reveal changes in the angle of the bean-like sub-domains of the transmembrane domain and reveal an elongation of the clamp domain.

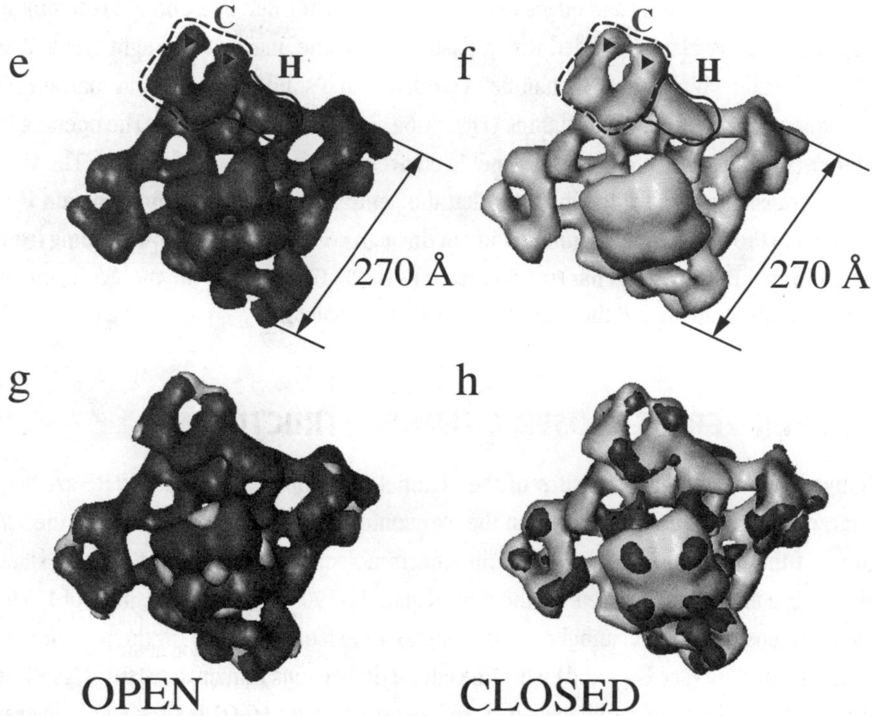

Fig. 6 (*Contineud*)

square-shaped TM domain measure ~120 Å. The number of residues on the luminal side of the SR membrane is thought to be small.[11,12] The membrane-spanning region of the channel is thus likely to be close to the bottom part of the mushroom stem. It has been suggested that the membrane-spanning domain contains between 4 and 10 transmembrane a helices per subunit.[11,12] Taking into account a typical membrane thickness of ~45 Å, the volume of the membrane-spanning domain calculated from the map is ~120,000 Å^3 per subunit. Such a molecular volume could accommodate a bundle of up to 10 nearly parallel α-helices.

The CY and the TM squares are rotated with respect to each other by an angle of ~ 40° (Figs. 3.6e and 3.6f). The side of the protein (CY) facing the cytoplasm shows a central opening ~50 Å in diameter (Fig. 3b in Ref. 3 and Fig. 3 in Ref. 4). At the current resolution level, the CY part of the molecule appears to form a network of densities consisting of 3 "domains": **1**. The corners and edges of the CY region are formed by a domain resembling a closed laboratory clamp (circled with a dashed black line, marked C in light grey). **2**. A bridging domain which we term "handle" (circled with a solid black line and marked H) interconnects two neighbouring clamps (Figs. 3.6a, 3.6b, 3.6e and 3.6f). **3**. The opening in the inner part of the CY site of the channel is encircled by four central domains. The view from the transmembrane side illustrates that the central part of the bridging domain H is connected to the membrane-spanning domain through columns of density protruding from the membrane. Each column has two connections in the CY region of the molecule, one is with the handle-domain and the other with the central domain.

THE OPEN VERSUS CLOSED CHANNEL STRUCTURES

Although the overall characteristics of the channel in its open and closed state are very similar, clear differences exist between the two conformations in both the CY and the TM regions of the molecule. The height of the entire molecule is ~200 Å in the open state, slightly more than that observed for the closed state (~190 Å) (Figs. 3.6a and 3.6b). The vertical dimension of the channel protein appears to expand by ~10 Å in the presence of Ca^{2+} and ryanodine (see Fig. 3.6d) while the lateral dimensions remain constant (Figs. 3.6e and 3.6f). The difference map between closed and open states (C-O) is shown in light grey on the structure in the open state (Fig. 3.6g), and the difference map between open and closed states (O-C) is shown in dark grey on the structure on the closed state (Fig. 3.6h). Figures 3.6c and 3.6g show the difference map C-O superimposed on 3D-map of open state, while in Figs. 3.6d and Fig. 3.6h the O-C difference map is superimposed on the closed structure. The threshold level used for displaying the difference maps is the same as that used for the original 3D maps.

The clamp-shaped domains at the four corners of the square shaped CY region show significant rearrangements of density in the presence of ryanodine and Ca^{2+} compared to the closed state (Figs. 3.6e and 3.6f). In the open state, these domains are elongated vertically as can be seen in the O-C difference map (Fig. 3.6d) in which dark grey densities are located above and below the clamp domain region. Additionally, two finger-like sub-domains of the clamp domain (marked with solid triangles in Figs. 3.6e and 3.6f) are separated in the open state but become joined in the closed state. This rearrangement is associated with the movement of both fingers away from the SR membrane region (Figs. 3.6c and 3.6d) as illustrated by the light grey density on the upper side of the fingers (Fig. 3.6c) and the dark grey density on the lower side of the fingers (Fig. 3.6d).

Apart from the changes in the CY domain described above, closing of the channel is also associated with structural changes in the TM domain. In the open state, the four bean-like sub-domains which form the TM domain are elongated in the vertical direction and have a pronounced central aperture ~20 Å in diameter. Additionally, indentures on the sides of the TM domain appear in the open state. The transition to the closed state appears to be associated with the tilting of the bean-like sub-domains, leading to both the closure of the central aperture and the indentures on the sides of the square being filled. These changes, which can be seen most clearly where the C-O difference map is superimposed upon the open structure in Fig. 3.6g, make the TM domain look more compact in the closed state than in the open state. In addition, the tilting of the bean-like sub-domains in the closed state reduces the vertical height of the TM domain - this is best appreciated in Fig. 3.6d in which the O-C difference map is superimposed upon the closed structure. Associated with these changes is a clockwise ~4° rotation of the TM domain in the open state with respect to its position in the closed state.

FUNCTIONAL INTERPRETATIONS OF THE CHANNEL STRUCTURES

The observed structural changes in the central aperture of the TM domain can be related to functional measurements of the channel. The diameter of the narrowest part of the conduction pathway in the open state has been estimated to be ~7 Å, based on organic cation permeability measurements.[66] Moreover, the length of the voltage drop in the conduction pathway has been estimated, to be ~10 Å by using trimethylammonium blocking cations.[67] The opening in the transmembrane domain in the open-state reconstruction has a diameter of 18 ± 7 Å (in its narrowest region) and a length of ~20 Å and is located at ~20 Å from the lumen face of the channel. The parameters determined in the electrophysiological investigations appear quite compatible to the cryo-microscopical structural data at the present level of resolution.

We have observed structural changes throughout the molecule, associated with the opening/closing of the channel. These changes are seen in both the putative membrane spanning portion and in the cytoplasmic region of the protein. Other types of measurements have already suggested widespread rather than local conformational changes in the Ca^{2+}-release channel. Mutations in the channel associated with malignant hyperthermia and with central core disease are all in the putative cytoplasmic regions of the protein and alter the function of the channel.[68,69] Fluorescent probe studies by Ikemoto and co-workers[70] also support widespread conformational changes in response to activators. Widespread conformational changes upon ligand binding have also been shown for other types of channels such as the GAP junction and the nicotinic acetylcholine receptor.[71,72]

The structures of the Ca^{2+}-release channel shown here may provide some new insights into both the physiological path followed by calcium ions when released from the SR, and the association of the Ca^{2+}-release channel with the voltage sensor transverse-tubules in the triad junction. The newly identified "hole" in the putative membrane spanning region of the protein[4] may represent the exit route for Ca^{2+} from the SR lumen when the channel opens. The cytoplasmic region contains one or more sites which interact either directly or indirectly with the voltage sensor in the T-tubule membrane. Meissnėr and colleagues (see Ref. 10) have demonstrated that the II-III cytoplasmic loop of the "a" subunit of the dihydropyridine receptor appears to preferentially interact with the open state of the Ca^{2+}-release channel during excitation-contraction coupling. The vertical extension of the channel may facilitate the interactions with the dihydropyridine receptor, and the observed large conformational difference between the open and closed states of the Ca^{2+}-release channel may be essential for regulating such interactions with other molecular components involved in excitation-contraction coupling in muscle.

Outlook

The angular reconstitution approach applied to the 3D structure of the Ca^{2+}-release channel has already elucidated structural differences associated with differences in functional states at a resolution level of ~25 Å. Independent groups have achieved similar 3D reconstructions which speaks for the reproducibility of the "single particle" data processing methodologies.[64] To achieve a higher level of resolution for these structures, much of the research emphasis will focus on improving the electron microscopical sample preparation (see Ref. 53) and on controlling the biochemical parameters precisely so as to lock the channel in a definite conformational state. An excellent example for such a functional-states study is the study on the 70S E. coli ribosome in its pre- and post-translocational states[38] and with elongation factors attached.[37] Resolution level of ~10 Å have already been achieved for highly symmetrical icosahedral particles[62,63] or helical assemblies.[72] Given these impressive recent successes in electron cryo-microscopy, it now seems conceivable that the "pipe dream" of obtaining atomic resolution structures of the Ca^{2+}-release channel in its various states and isoforms may be indeed realised.

ACKNOWLEDGEMENTS

We would like to thank Michael Schatz and Ralf Schmidt of Image Science Software GmbH for support, and Ed Morris for critically reading early versions of the manuscript. This work was supported by grants from the EU-BIO4-CT96-0592 to M. van Heel, from the NCRR of National Institutes of Health RP02250 to W. Chiu, from MDA and the National Institutes of Health to S.L. Hamilton.

REFERENCES

1. Dubochet J, Chang J-J, Freeman R *et al.* Frozen aqueous suspensions. *Ultramicroscopy* 1982; **10**:55–62.
2. van Heel M. A posteriory assignment of projection directions for 3D reconstruction. *Ultramicroscopy* 1987; **21**:111–124.
3. Serysheva II, Orlova EV, Chiu W *et al.* Electron cryomicroscopy and angular reconstitution used to visualize the skeletal muscle calcium release channel. *Nature Struct. Biol.* 1995; **2**:18–24.
4. Orlova EV, Serysheva II, van Heel M *et al.* Two structural configurations of the skeletal muscle calcium release channel. *Nature Struct. Biol.* 1996; **3**:547–552.
5. Heilbrunn LV, Wiercinski FV. The action of various cations on muscle protoplasm. *J. Cell. Comp. Physiol.* 1947; **29**:15–32.
6. Mathews CK, van Holde KE. *Biochemistry, 2d ed.* The Bendjamin/Cummings Pub. Comp. Inc., 1996.
7. Ebashi, S, Endo M. Calcium ion and muscle contraction. *Biophys. Mol. Biol.* 1968; **18**:123–183.
8. Franzini-Armstrong C. Studies of the triad. 1. Structure of the junction in frog twitch fibers. *J. Cell Biol.* 1970; **47**:488–499.
9. Saito A, Inui M,. Radermacher M *et al.* Ultrastructure of the calcium release channel of the sarcoplasmic reticulum. *J. Cell. Biol.* 1988; **107**:211–219.
10. Lu X, Xu L, Meissner G. Activation of the skeletal muscle calcium release channel by a cytoplasmic loop of the dihidropyridine receptor. *J.Biol.Chem.* 1994; **269**:6511–6516.
11. Takeshima H, Nishimura S, Matsumoto T *et al.* Primary structure and expression from complementary DNA of skeletal muscle ryanodine receptor. *Nature* 1989; **339**:439–445.
12. Zorzato F, Fujii J, Otsu K. Molecular cloning of cDNA encoding human and rabbit forms of the Ca^{2+} release channel (ryanodine receptor) of skeletal muscle sarcoplasmic reticulum. *J. Biol. Chem.* 1990; **265**:2244–2256.
13. Jayaraman T, Brillantes AM, Timerman AP. FK506 binding protein associated with the calcium release channel (ryanodine receptor). *J. Biol. Chem.* 1992; **267**:9474–9477.
14. Lai FA, Misra M, Xu L. The ryanodine receptor Ca^{2+} release channel complex of skeletal muscle sarcoplasmic reticulum: evidence for a cooperatively charged homotetramer. *J. Biol. Chem.* 1989; **264**:16776–16785.

15. Smith JS, Coronado R, Meissner G. Single channel measurements of the calcium release channel from skeletal muscle sarcoplasmic reticulum. *J. Gen. Physiol.* 1986; **88**:573–588.
16. Smith JS, Imagawa T, Ma J *et al.* Purified ryanodine receptor from rabbit skeletal muscle is the calcium release channel of sarcoplasmic reticulum. *J. Gen. Physiol.* 1988; **92**:1–26.
17. Wang JP, Needleman DH, Hamilton SL. Relationship of low affinity [^{3}H] ryanodine binding sites to high affinity sites on the skeletal muscle Ca^{2+} release channel. *J. Biol. Chem.* 1993; **268**: 20974–20982.
18. Cifuentes ME, Ronjat M, Ikemoto N. Polylysine induces a rapid Ca^{++} release from sarcoplasmic reticulum vesicles by meditation of its binding to the foot protein. *Arch. Biochem. Biophys.* 1989; **273**:554–561.
19. Pessah IN, Stambu RA, Casida JE. Ca^{++} -activated ryanodine binding: mechanisms of sensitivity and intensity modulation by Mg^{++}, caffeine, and adenine nucleotides. *Mol. Pharmacol.* 1990; **31**:232–238.
20. Chu A, Diaz-Munoz M, Hawkes MJ *et al.* Ryanodine as a probe for the functional state of the skeletal muscle sarcoplasmic calcium release channel. *Mol. Pharmacol* 1990; **37**:735–741.
21. Chiu, W. What does electron cryomicroscopy provide that X-ray crystallography and NMR spectroscopy cannot? *Ann. Rev. Biophys. Biomol. Struct.* 1993; **22**:233–255.
22. Harris JR. *Negative staining and cryoelectron microscopy: the thin film techniques.* 1t ed. Oxford, UK: BIOS Scientific Publishers Limited, 1997.
23. Kellenberger E, Haener M, Wurtz M. The wrapping phenomenon in air-dried and negatively stained preparations. *Ultramicroscopy* 1982; **9**:139–150.
24. Cejka Z, Santini C, Tognon G *et al.* The molecular architecture of the extracellular hemoglobin of *Ophelia bicornis*: Analysis of two-dimensional crystalline arrays. *J. Struct. Biol.* 1991; **107**: 259–267.
25. Bremer A, Henn C, Engel A *et al.* Has negative staining still a place in biomolecular electron microscopy? *Ultramicroscopy* 1992; **46**:85–111.
26. Harris JR, Horn RW. Negative staining: A brief assessment of current technical benefits, limitations and future possibilities. *Micron* 1994; **25**:5–13.
27. Dubochet J, Adrian M, Chang J-J *et al.* Cryo-electron microscopy of vitrified specimens. *Q. Rev. Biophys.* 1988; **21**:129–228.
28. Hoppe W, Gassmann J, Hunsmann N *et al.* Three-dimensional reconstruction of individual negatively stained yeast fatty-acid synthetase molecules from tilt series in the electron microscope. *Hoppe-Seyler's Z. Physiol. Chem* 1974; **355**:1483–1487.
29. Radermacher, M. Three-dimensional reconstruction of single particles from random and nonrandom tilt series. *J. Electr. Microsc. Tech.* 1988; **9**:359–394.
30. Schatz M, Orlova EV, Dube P *et al.* Angular reconstitution in 3D electron microscopy: practical and technical aspects. *Scanning Microscopy* 1998. (in press)
31. van Heel M, Orlova EV, Harauz G *et al.* Angular reconstitution in 3D electron microscopy: high resolution structures from uncrystallised macromolecules. *Scanning Microscopy* 1998. (in press)
32. Frank J. *Three-dimensional electron microscopy of macromolecular assemblies.* 1t ed. New York: Academic Press 1995.

33. Crowther RA, DeRosier DJ, Klug A. The reconstruction of a three-dimensional structure from projections and its application to electron microscopy. *Proc. Roy. Soc. London* 1970; **317**: 319–340.
34. van Heel M. Classification of very large electron microscopical image data sets. *Optik* 1989; **8**:114–126.
35. van Heel M, Frank J. Use of multivariate statistics in analysing the images of biological macromolecules. *Ultramicroscopy* 1981; **6**:187–194.
36. van Heel M, Stöffler-Meilicke M. Characteristic views of E. Coli and B. stearothermophilus 30S ribosomal subunits in the electron microscope. *EMBO J.* 1985; **4**:2389–2395.
37. Stark H, Orlova EV, Rinke-Appel J *et al.* Arrangement of tRNAs in Pre and Posttranslocational ribosomes revealed by electron cryomicroscopy. *Cell* 1997; **88**:19–28.
38. Stark H, Rodnina M, Rinke-Appel J *et al.* Visualization of elongation factor on the Esherichia coli ribosome. *Nature* 1997; **389**:403–406.
39. Stewart PL, Chiu CY, Huang S *et al.* Cryo-EM visualization of an exposed RGD epitope on adenovirus that escapes antibody neutralization. *EMBO J.* 1997; **16**:1189–1198
40. Hawkes MJ, Diaz-Munoz M, Hamilton SL. A procedure for purification of the ryanodine receptor from skeletal muscle. *Memb. Biochem.* 1989; **8**:133–145.
41. Zhou ZH., Hardt S, Wang B *et al.* CTF determination of images of ice-embedded single particles using a graphics interface. *J. Struct. Biol.* 1996; **116**:216–222.
42. Hawkes PW. Image processing based on linear theory of image formation, in: *Computer Processing of Electron Microscope Images.* ed. Hawkes PW. Berlin, Heidelberg, New York *Springer Verlag.* 1980:1–33.
43. van Heel M, Keegstra W. IMAGIC: a fast flexible and friendly image analysis software system. *Ultramicroscopy* 1981; **7**:113–130.
44. van Heel M, Harauz G, Orlova E *et al.* A new generation of the IMAGIC image processing system. *J. Struct. Biol.* 1996; **116**:17–24.
45. Frank J, Verschoor A, Boublik M. Computer averaging of electron micrographs of 40S ribosomal subunits. *Science* 1981; **214**:1353–1355.
46. Steinkilberg M, Schramm HJ. Eine verbesserte Drehkorrelationsmethode für die Strukturbestimmung biologischer Macromoleküle durch Mittelung elektronenmikro-skopischer Bilder. Hoppe-Seyler's Zeitschr. f. Phys. Chem. 1980; **361**:1363–1369.
47. van Heel M, Schatz M, Orlova EV. Correlation functions revisited, *Ultramicroscopy* 1992; **46**: 307–316.
48. Dube P, Tavares P, Lurz R, *et al.* The portal protein of bacteriophage SPP1: a DNA pump with 13-fold symmetry. *EMBO J.* 1993;**12**:1303–1309.
49. van Heel M. Multivariate statistical classification of noisy images (randomly oriented biological macromolecules). *Ultramicroscopy* 1984; **13**:165–184.
50. Hall, E. L. Computer image processing and recognition.New York Academic Press: 1979.
51. Harauz, G, van Heel M. Exact filters for general geometry three-dimensional reconstruction. *Optik* 1986; **73**:146–156.

64. Radermacher M, Rao V, Grassucci, et R, *et al.* Cryo-electron microscopy and three-dimensional reconstruction of the calcium release channel ryanodine receptor from skeletal muscle. *J. Cell Biol.* 1994; **127**:411–423.
65. Wagenknecht T, Grassucci R, Frank J, *et al.* 3-Dimensional architecture of the calcium release channel/foot structure of sarcoplasmic reticulum. *Nature 1989;***338**:167–170.
66. Tinker A, Wiliams AJ. Probing the structure of the conduction pathway of the sheep cardac sarcoplasmic reticulum calicium-release channel with permanent and inpermament organic cations. *J. Gen. Physiol.* 1993; **102**:1107–1129.
67. Tinker A, Williams AJ. Measuring the length of the pore of the sheep cardiac sarcoplasmic reticulum calicium-release channel using related trimethylammonium ions as molecular calipers. *Biophys. J.* 1995; **68** 111–120.
68. Fujii J, Otsu K, Zorzoto F. Identification of a mutation in porcine ryanodine receptor associated with malignant hyperthemia. *Science 1991;* **253** 448–451.
69. Zhang Y, Chen HS, Khanna VK. A mutation in the human ryanodine receptor gene associated with central core disease. *Nature Genetics* 1993; **5**:46–50.
70. Ikemoto N, Antoniuu B, Meszaros LG. Rapid flow chemical quench studies of calcium release from isolated sarcoplasmic reticulum. *J. Biol. Chem* 1985; **260**:14096–14100.
71. Unwin PNT, Ennis PD. Calcium-mediated changes in gap junction structure: Evidence from isolated sarcoplasmic reticulum. *J. Cell Biol.* 1983; **97**:1459–1466.
72. Unwin PNT. Acetylcholine receptor channel imaged in the open state. *Nature* 1995; **373**: 37–43.

Chapter 4

Regulation of Ryanodine Receptor Channel Gating

Rebecca Sitsapesan

INTRODUCTION

Changes in intracellular Ca^{2+} levels regulate a variety of diverse processes including muscle contraction, neurotransmitter release, cell development and fertilisation. The sarcoplasmic reticulum (SR) Ca^{2+} release /ryanodine receptor (RyR) channel is one of two main types of channels through which Ca^{2+} is released from intracellular stores. The RyR is an intracellular membrane channel and therefore the gating and ion handling properties of the channel have always been investigated after reconstitution into planar phospholipid bilayers. Single channel studies have demonstrated that the gating of RyR can be regulated by a bewildering number of physiological and pharmacological agents. Furthermore, the effects of many of these agents appear to depend on the particular isoform of RyR involved. The mammalian cardiac (RyR2) and skeletal (RyR1) isoforms are the most frequently studied at the single channel level. This is usually because other isoforms of RyR (for example, RyR3) may only be expressed in low amounts or may be co-expressed with other isoforms. There appear to be certain universal features of gating which can be observed in all types of RyR incorporated into bilayers. While these properties have been observed in RyR1 and RyR2, some have been observed in other isoforms but may not have been investigated in detail.

1. The first and perhaps the most important feature of the gating of RyR is that cytosolic Ca^{2+} activates (opens) the channels. The mechanisms involved in Ca^{2+}-activation vary subtly for each isoform of RyR.

2. Many other ligands affect channel gating but virtually all ligands (whose action is readily reversible) appear to be Ca^{2+}-dependent. The list of agents shown to affect RyR gating is a continually growing one and many of the most important ligands have been listed in previous reviews. [1–4]

3. The basic gating modes observed when RyR are activated by Ca^{2+} alone or by Ca^{2+} plus a second ligand have been shown to be modified by the following:

A. The environment of the channel including ionic composition and strength, pH, temperature and holding potential

B. The phosphorylation state of RyR

C. Agents present at the luminal side of RyR

D. Junctional proteins closely associated with RyR

E. Covalent modifiers of the channel

F. Agents interacting with thiol groups

4. Ryanodine binds to all isoforms of RyR which have been studied at the single channel level.

Cytosolic Ca^{2+} Activation and Inhibition

All ryanodine receptors so far studied after incorporation into bilayers are activated by cytosolic Ca^{2+}. The effects of Ca^{2+} on the gating of RyR have been studied in more detail than the effects of any other ligand and yet there is very little consensus of opinion concerning the mechanisms involved in Ca^{2+} activation and inhibition of RyR. Part of this problem stems from the wide variation in the effects of cytosolic Ca^{2+} that appears to occur between RyR isoforms and between species. In addition, different investigators often perform experiments under widely varying experimental conditions including different ionic composition of recording solutions, temperature, pH, membrane potential, the use of purified or native RyR and the method used to elevate cytosolic $[Ca^{2+}]$; all conditions shown to affect the response of RyRs to changes in $[Ca^{2+}]$. Ca^{2+} activation of RyR also appears to be a very variable process even within channels from the same membrane preparation[5,6] and this can, in part, be attributed to changes in gating to low or high open probability (Po) modes.[7–9] Understanding mechanisms of cytosolic Ca^{2+} regulation of RyR is crucial because the effects of most other ligands which interact with the channels are Ca^{2+}-dependent.

In the absence of activating ligands on the cytosolic channel side, the Po of RyR channels is zero. Micromolar levels of Ca^{2+} open the channels in the absence of other activating ligands but in general cannot alone fully activate the channels.[5–8,10–12] Reported EC_{50} values for activation of RyR by Ca^{2+} as the sole ligand vary from approximately 1 μM to 40 μM depending on the isoform of RyR and the method of raising the free $[Ca^{2+}]$.[5–8,10–13] Some authors report a Hill coefficient of approximately 1 for Ca^{2+}-activation of RyR indicating only a single cytosolic Ca^{2+} activation site[7,10,12,14] while others report Hill coefficients >1[5,6] indicating that multiple Ca^{2+} ions bind in a positively cooperative manner to increase Po. Different isoforms of RyR exhibit subtle differences in the mechanisms of activation and inactivation by cytosolic Ca^{2+}. Single point mutations, as occur in RyR from malignant hypothermic skeletal muscle, also lead to marked changes in the effects of cytosolic Ca^{2+} on the gating of the channel [15–17] (see Chapter 12).

High levels of cytosolic Ca^{2+} lead to a reduction in Po. EC_{50} values for inhibition of channel opening vary widely from approximately 400 μM in RyR1[10] to 1–15 mM in RyR2 and lobster skeletal RyR.[6,12,13,18] Other types of RyR, including chicken and frog skeletal and a recombinant RyR3 from mammalian smooth muscle, inactivate with intermediate levels of cytosolic Ca^{2+}.[8,19,20] These effects of Ca^{2+} have led to the suggestion that high affinity activation and low affinity inactivation Ca^{2+} binding sites are present on the cytosolic face of all isoforms of RyR.

Experiments using flash photolysis of DM-nitrophen, a caged Ca^{2+} compound, have suggested that the rate of change of $[Ca^{2+}]$ at the cytosolic face of the RyR determines the final Po that the channel reaches.[21] However this result has not been reproduced using methods that cause a rapid step change in $[Ca^{2+}]$ to a constant $[Ca^{2+}]$.[22–24] Rapid changes in cytosolic $[Ca^{2+}]$ using the flash photolysis method causes rapid increases in Po followed by a gradual fade in Po with a time constant of 1.3 s.[21] This fade in Po has been termed adaptation and is characterised by the ability of the channels to be activated further with subsequent increases in $[Ca^{2+}]$. In contrast, RyR which have been activated by a step change to a constant $[Ca^{2+}]$ by rapidly switching the perfusing solutions and which involve slower rises in $[Ca^{2+}]$ (between 1–50 ms) at the bilayer, do not exhibit adaptation[22–24] Under certain conditions, no reduction in Po is observed but by varying the holding potential, by increasing cytosolic $[Ca^{2+}]$ to high levels or by adding secondary activating ligands, inactivation of the channels can occur which appears to be quite distinct from the adaptation observed in the flash photolysis experiments.[22–24] Chapters 13 and 14 discuss these issues in more detail.

Other cations can also interact with the Ca^{2+} binding sites on the cytosolic face of RyR channels. Strontium can activate mammalian skeletal RyR[25,26] and appears to be of similar or slightly higher potency as Ca^{2+}. Cytosolic barium has either been reported to reduce Po or have little effect.[25,27] It has also been suggested that monovalent cations may bind to the Ca^{2+} activation sites.[28] Millimolar Mg^{2+} causes a reduction in the Po of all isoforms of RyR where the effect has been investigated.[7,19,29,30,31] It is generally thought that the mechanism for Mg^{2+} inhibition is competition with Ca^{2+} for the cytosolic Ca^{2+} activation and inhibition binding sites.[7,30,31] The affinity and efficacy of Mg^{2+} for the Ca^{2+}-activation sites are much lower than for Ca^{2+} but are reported to be similar to Ca^{2+} for the Ca^{2+}-inhibition sites.[7,30,31] As inactivation by cytosolic Ca^{2+} occurs at much lower concentrations in skeletal than in cardiac RyR it is expected that Mg^{2+} binding to inhibition sites will have a much greater effect on skeletal RyR than cardiac RyR at physiological $[Mg^{2+}]$ *in situ*.[6,12,30,31] The mechanism for Mg^{2+} inhibition is discussed in Chapter 14. The phosphorylation state of RyR has been reported to affect the response of the channel to Mg^{2+} [34,35] and this is discussed in more detail in Chapter 8.

Activation by Secondary Ligands

Many agents have been demonstrated to affect the gating of RyR incorporated into bilayers. Virtually all agents appear to act from the cytosolic channel side but the exact location of the ligand binding sites is not known. Figure 4.1 illustrates a cartoon of RyR and some modulators of mammalian cardiac and skeletal RyR channel gating.

An important aspect of the action of RyR activators is that they require the presence of activating cytosolic Ca^{2+} for maximum effect. Some agents can activate the channel in the absence of cytosolic Ca^{2+} but the effect on Po is small. The sum of the individual effects of Ca^{2+} alone plus the effects of a secondary ligand (for example ATP) alone, is less than the simultaneous effect of Ca^{2+} plus the secondary activating ligand. This can be observed in Fig. 4.2 where ATP is the secondary activating ligand. Clearly the synergistic action of Ca^{2+} and ATP results in a massive increase in Po that does not occur in the presence of either ligand alone. ATP is not merely sensitising the channel to Ca^{2+}. In mammalian isoforms of RyR this synergistic mechanism of RyR activation has been demonstrated for agonists which bind to the adenine nucleotide/nucleoside sites, to the caffeine sites and to the suramin sites.[36–38]

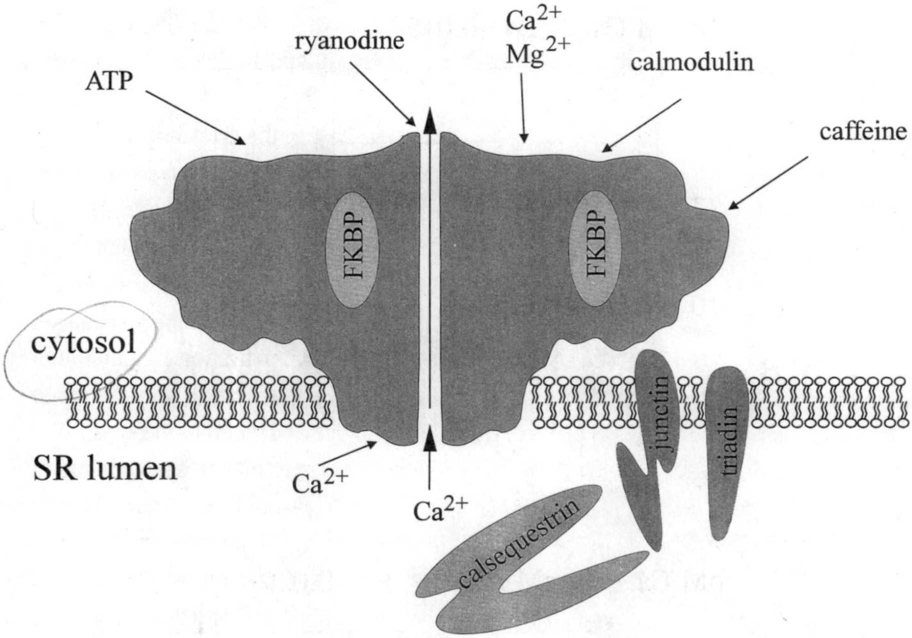

Fig. 4.1 Cartoon representation of the RyR-channel in the junctional SR membrane together with associated proteins, FK-506 binding protein (FKBP), junctin, triadin and calsequestrin. Most ligands regulate the gating of RyR by binding to the cytosolic portion of the channel but more specific localization of the various binding sites has not yet been achieved. Ca^{2+} may bind to luminal sites on RyR.

Some ligands have no effect in the absence of Ca^{2+} and are unable to fully activate the channel even in the presence of activating concentrations of Ca^{2+}. These agents appear to sensitise the channel to Ca^{2+} as the mechanism for channel activation is similar to that of Ca^{2+} alone. In the presence of a maximally activating level of cytosolic Ca^{2+}, no further increase in Po can be attained by adding these agents which include inorganic phosphate (Pi)[37] and cardiac glycosides.[39] Figure 4.3 demonstrates the effect of Pi on the gating of a typical sheep cardiac RyR. Caffeine is widely known as a sensitizer of RyR to Ca^{2+} but this is only strictly true at low caffeine concentrations.[40] Millimolar levels of caffeine and related agents act synergistically with Ca^{2+} to cause a different mode of channel gating whereby the channels can be fully opened.[36,40]

Agents which act at particular binding sites appear to induce gating characteristics which are similar for all agents binding to that site.

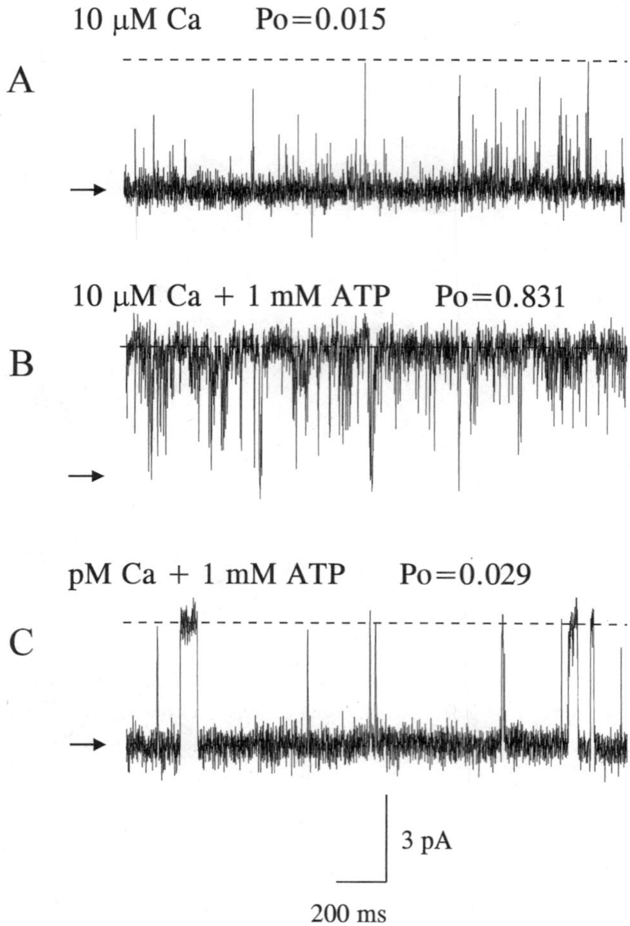

Fig. 4.2 Different gating mechanisms of RyR. Current fluctuations through a typical single sheep cardiac RyR incorporated into a phosphatidylethanolamine lipid bilayer. Ca^{2+} is the permeant ion (luminal [Ca^{2+}] 50 mM) and current flows through the channel from the luminal to the cytosolic side of the bilayer. The holding potential is 0 mV. The dotted lines indicate the fully open channel level and the arrows indicate the closed channel level. Po values are shown above each trace. In (A) the channel was activated solely by 10 μM cytosolic Ca^{2+}. In (B) the channel was activated by 10 μM Ca^{2+} plus 1 mM ATP and Po increased to 0.831. In (C) the channel was activated by 1 mM ATP in the absence of activating cytosolic Ca^{2+} (pM Ca^{2+} after the addition of 12 mM cytosolic EGTA). Note the long open and closed events when the channel is activated by ATP alone. In (A), (B) and (C), three apparently distinct gating mechanisms are operating. In B, the simultaneous presence of Ca^{2+} plus ATP causes almost maximal activation of the channel and the degree of activation is much greater than the sum of the individual actions of Ca^{2+} alone (A) and ATP alone (C).

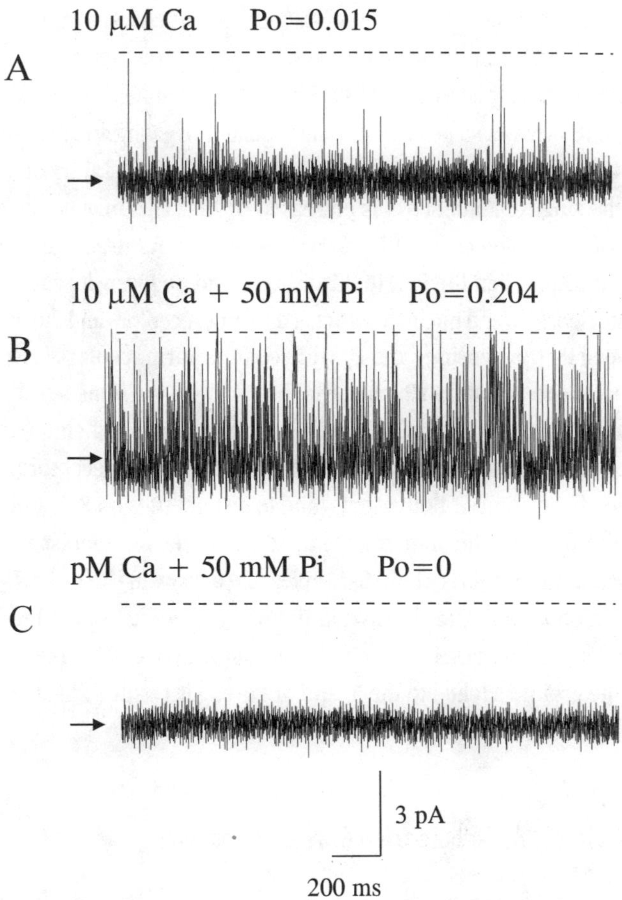

Fig. 4.3 Apparent sensitisation of RyR to Ca^{2+} by Pi. Current fluctuations through a representative single sheep cardiac RyR with 50 mM luminal Ca^{2+} as the permeant ion. Current flow is in the luminal to the cytosolic direction with a holding potential of 0 mV. The dotted lines indicate the fully open channel level and the arrows indicate the closed channel level. The channel is activated by 10 μM Ca^{2+} in (A), by 10 μM Ca^{2+} plus 50 mM Pi in (B) and by 50 mM Pi in the absence of activating cytosolic [Ca^{2+}] in (C).

ATP sites

Experiments in skinned cardiac cells indicated that the SR contained a Ca^{2+} efflux channel sensitive to adenine nucleotides.[41] Ca^{2+} efflux studies from isolated SR vescles and activation of single RyR incorporated into bilayers confirmed this.[11,30,37,42,43] Non-hydrolysable analogues of ATP exert similar effects on SR Ca^{2+}efflux and channel gating indicating that the observed effects of adenine nucleotides result from a direct interaction with receptor sites on the channel rather than by phosphorylation of the channel.[11,30,43] (Phosphorylation of ryanodine receptors may also be an important regulator of gating and this will be discussed briefly later in this chapter and in depth in Chapter 8). The Po of all ryanodine receptors investigated at the single channel level is increased by ATP. This may not be a general rule and in fact the pharmacological profile of Ca^{2+}-release from intracellular membranes of sea urchin eggs indicates that the RyR in this tissue could be insensitive to ATP.[44]

In general, nucleosides and nucleotides act at different receptors and the overall sequence homology between P_1 (nucleoside) and P_2 (nucleotide) purinoceptors is less than 20%.[45] On RyR however, adenosine and ATP appear to interact with the same sites[43] and therefore the pharmacology of the ATP site on RyR appears to be unique. Caffeine (an antagonist at certain P_1 receptors) and suramin (an antagonist of certain P_2 receptors) do not act as antagonists of adenosine or ATP. Both agents bind to distinct sites on RyR and act as agonists to increase Po.[38,43] The adenine moiety appears to be crucial for agonist activity at the ATP site as GTP, UTP and CTP have very little Ca^{2+} releasing activity in skeletal SR vesicles.[42] ATP and adenosine can both fully activate RyR in the presence of activating levels of cytosolic Ca^{2+}. However ADP is less potent than ATP and can only partially activate the channel indicating that the group attached to the 5′ end of the sugar residue affects the efficacy and affinity of the adenine compounds.[37]

Is there a distinct cyclic ADP-ribose (cADPR) receptor on RyR?

cADPR can release Ca^{2+} from intracellular stores in sea urchin eggs and it has been proposed as a novel second messenger which acts to open ryanodine receptors in a manner analogous to the effect of $InsP_3$ on $InsP_3$ receptors. In sea urchin eggs, cADPR induced Ca^{2+} release appears to be dependent on the presence of other agents including Ca^{2+},[46] calmodulin,[47] calmodulin dependent protein kinase[48] and $[Mg^{2+}]$;[44] all agents known to affect the gating of skeletal and cardiac isoforms of RyR. Ryanodine was also shown to cause Ca^{2+} release in sea urchin eggs.[49] cADPR has been reported to release intracellular Ca^{2+} in numerous other cell types including cardiac muscle cells however there are large discrepancies in the observed results from different laboratories.[50,51] Most evidence suggests that if cADPR does trigger the opening of RyR in sea urchin eggs, the sea urchin isoform of RyR must be modulated by cADPR in a manner different to that of RyR1 and RyR2.[52] cADPR has been reported to have a small effect or no significant effect on cardiac and skeletal RyR gating or on ryanodine binding to and Ca^{2+} efflux from cardiac and skeletal SR.[52–57] In cardiac and skeletal RyR, cADPR appears to increase Po by binding to the same site as ATP and physiological concentrations of cADPR have no effect in the presence of physiological levels of ATP.[53,54] It has been suggested that cADPR may have an additional effect by binding to closely associated proteins and that this cannot be observed during the single channel experiments because the proteins have not incorporated into the bilayer along with the RyR.[52] This is a possibility but does not explain why cADPR (1μM) did not induce SR Ca^{2+} release from isolated skeletal or cardiac SR or cause an increase in ryanodine binding to the same SR membrane fraction.[56] Sea urchin RyR may possess novel cADPR binding sites on the channels or on closely associated proteins and may either lack the presence of or have very different ATP sites. Two studies, one from skeletal SR[57] and the other from cardiac SR,[58] suggest that cADPR-induced release of Ca^{2+} from intracellular stores should not necessarily be attributed to the activation of RyR channels. Both studies demonstrate that cADPR can release SR Ca^{2+} by a mechanism independent of RyR channel activation. Therefore it is possible that a novel channel may be the pathway for the cADPR-induced release of intracellular Ca^{2+} in certain tissues.

Caffeine sites

As mentioned earlier, caffeine and other methylxanthines are commonly used as antagonists at adenosine (P_1) receptors[45] but appear not to interact with adenine nucleotide/nucleoside sites on RyR.[14,40,42,43,60] RyR have distinct caffeine binding sites and caffeine acts as an agonist increasing the Po of the channel. Caffeine potentiates the effect of adenosine and other agents acting at the adenine nucleotide binding sites.[43] Low concentrations of caffeine analogues cause only an increase in the frequency of channel opening (as does Ca^{2+} as the sole ligand) and therefore appear to sensitise the sheep cardiac RyR to Ca^{2+}. High concentrations act by a different mechanism and can fully activate the channel by increasing both the frequency and the duration of channel openings.[36,40] More potent agents have been proposed to activate RyR via the caffeine binding sites. The EC_{50} for activation of cardiac RyR by caffeine is between 2 and 10 mM[14,40] and for sulmazole is approximately 1 mM.[36,60] The analogues of sulmazole, EMD 46512 and EMD 41000 exhibited EC_{50} values of approximately 40 μM and are approximately 100 times more potent than caffeine at increasing Po.[60] 9-Me-7-Br-Eudistomin-D (MBED) is reported to be more than 1000 times more potent than caffeine at releasing Ca^{2+} from skeletal SR although the single channel properties of this compound have not been investigated.[61]

There is evidence that the caffeine sites may be different in different isoforms of RyR. Ryanodine binding studies indicate that the channel from thoracic muscle of the insect *Heliothis virescens* either does not have caffeine binding sites or that the sites have very low affinity for caffeine.[62] Interestingly, caffeine is not very effective at increasing the Po of RyR from lobster skeletal muscle and cannot fully open the channel in the presence of activating Ca^{2+}.[13] Thus more primitive isoforms of RyR may lack caffeine binding sites or may possess caffeine sites with lower affinity for caffeine.

Calmodulin sites

Calmodulin affects the gating of RyR at concentrations expected to be present physiologically (nanomolar-micromolar range). Data from sequence analysis, [^{125}I]calmodulin binding studies and fluorescence anisotropy measurements predict the presence of multiple calmodulin binding sites (up to 24 high affinity sites[63–66]). [^{125}I] calmodulin binding to purified RyR indicate that at low free [Ca^{2+}] (<0.1 μM), the rabbit skeletal RyR binds 16 calmodulins and at 100 μM Ca^{2+} binds 4 calmodulins. Single channel, $^{45}Ca^{2+}$ efflux and ryanodine binding studies indicated that calmodulin increased the Po of RyR at <0.2 μM Ca^{2+} but decreased Po at micromolar [Ca^{2+}].[66] Understanding the physiological requirement for all the calmodulin binding sites is difficult given that the Hill coefficients for activation and inhibition of ryanodine binding were both 1.3 indicating very little cooperativity.[66] In contrast to these results, Fuentes *et al.* 1994[67] observed only inhibition of ryanodine binding and channel activity at low and high [Ca^{2+}] in preparations containing pig skeletal RyR. Activation of calmodulin dependent protein kinase (CAM kinase) has been reported to reverse the inhibition caused by the direct interaction of calmodulin with RyR[68] (see Chapter 8). The complex summation of the direct and indirect effects of calmodulin on RyR gating is predicted to influence the activity of RyR channels significantly *in situ* and is another area of RyR function which must be explored in more detail. It has also been suggested that direct interactions of calmodulin with RyR may contribute to inactivation mechanisms leading to the termination of SR Ca^{2+} release.[66]

Lipid derivatives

Fatty acid metabolites such as palmitoyl carnitine or palmitoyl coenzyme A (CoA) are found normally in muscle cells and occur in increased levels during metabolic stress. It is therefore interesting that these agents can increase the open probability of cardiac and skeletal RyR at micromolar concentrations.[69–71] Palmitoyl CoA is also found in millimolar concentrations in sea urchin eggs and the levels decrease after fertilisation.[72,73] Palmitoyl CoA was shown to potentiate cADPR-induced release of Ca^{2+} from intracellular stores in sea urchin eggs and led to the idea that palmitoyl CoA was acting via ryanodine receptors.[74] It has been suggested that palmitoyl CoA may be a physiological regulator of ryanodine receptors and serves to overcome Mg^{2+}-inhibition.[71]

Sphingosine and related compounds have been shown to stimulate Ca^{2+}-release from intracellular stores in non-muscle cells and from smooth muscle cells.[2,75–77] In contrast, sphingosine causes a reduction in skeletal muscle SR Ca^{2+}-release[78] and a decrease in the Po of RyR1 incorporated into planar lipid bilayers. Sphingosine appears to interact directly with RyR1 although the mechanism of action is not known and therefore it may play a role as an endogenous modulator of RyR function.

Isoform specific sites

Most ligands do have slightly different effects on the gating of different isoforms of RyR although there are some agents which appear to interact with only one isoform. The differences may be due to alterations in the ligand binding sites or, since most effectors are Ca^{2+}-sensitive, may be the result of differences in the mechanisms of Ca^{2+}-activation. The specificity of ligands has usually been tested by comparing the gating of two isoforms of RyR, commonly RyR1 and RyR2. Ligands reported to affect the gating of certain isoforms but not others include digoxin,[39] cADPR (although this is controversial[52]) and certain anions, for example perchlorate.[8,79] Agents reported to have significantly different effects on different isoforms include:

Mg^{2+}, which is reported to inhibit the opening of RyR1 more than RyR2.[30,31]

Caffeine and related compounds which activate most RyR isoforms studied including RyR1, RyR2 and RyR3[14,19,40,59] but do not appear to be very effective on insect thoracic muscle RyR[62] or lobster abdominal muscle RyR.[13]

Suramin is more effective at increasing the open lifetime duration of RyR2 than RyR1.[38]

Ryanodine binding studies have suggested that inorganic phosphate (Pi) and other anions have a much greater activating effect on the gating of RyR1 than on that of RyR2.[79–81]

Peptide toxins, for example, imperatoxin activates RyR1 but not RyR2.[82,83] The effects of certain peptide toxins on RyR gating are detailed in Chapter 10.

Modification of Gating

Environment of RyR

Single channel studies together with $[^3H]$ryanodine binding and Ca^{2+}-efflux studies from isolated SR vesicles indicate that the response of RyR to activating and inactivating ligands depends very strongly on the ionic environment and that different isoforms of RyR are not always affected to the same degree by changes in the ionic environment.

(a) Effects of pH

SR Ca^{2+}-release and the Po of RyR incorporated into planar lipid bilayers is very dependent on pH. Reductions in pH from pH 7.5 to 6.0 cause $[H^+]$-dependent reductions in Po and in $[^3H]$ryanodine binding to and Ca^{2+}-efflux from SR vesicles.[84,85] The effect of changing pH on the gating of RyR appears to be related to the mechanism of channel activation and the Po of the channels.[85]

(b) Effects of ionic strength and composition

An increase in ionic strength has been shown to increase the Po of RyR1 and RyR2.[28,79] This has been suggested to occur by: 1. an interaction of monovalent and divalent cations with the cytosolic Ca^{2+} binding sites on RyR, 2. by the binding of anions to specific sites[28] on the channels or 3. by the chaotropic properties of the anions involved and therefore does not involve binding to discrete sites on the channel.[79] The increases in Po observed in skeletal RyR with increasing concentrations of Cl^- and other ions such as perchlorate and nitrate are not observed in cardiac RyR.[79–81] Perchlorate also activates the chicken skeletal α but not the β isoform.[8] Inorganic phosphate (Pi) has been suggested to interact with RyR1 by a different mechanism to that of other anions and it has been suggested that specific sites for Pi exist on the channels.[79]

(c) Effects of temperature

Changes in temperature between 5 and 30°C have been demonstrated to cause dramatic changes in the Po and the duration of open lifetimes of sheep cardiac RyR.[86] Fig. 4.4 illustrates

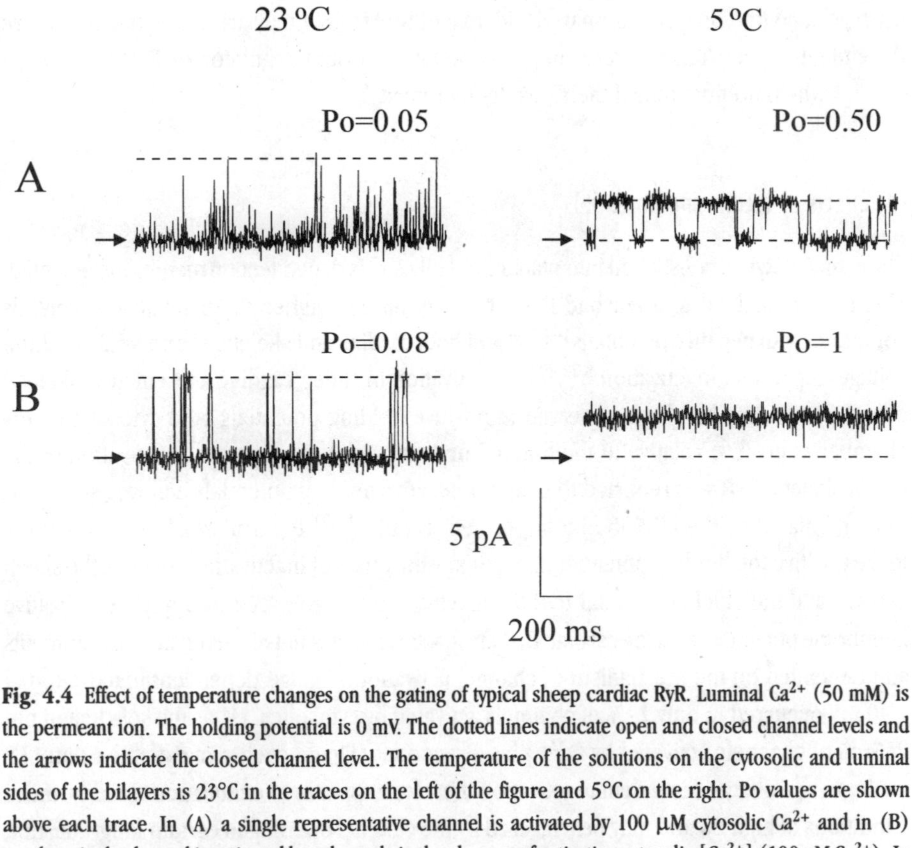

Fig. 4.4 Effect of temperature changes on the gating of typical sheep cardiac RyR. Luminal Ca^{2+} (50 mM) is the permeant ion. The holding potential is 0 mV. The dotted lines indicate open and closed channel levels and the arrows indicate the closed channel level. The temperature of the solutions on the cytosolic and luminal sides of the bilayers is 23°C in the traces on the left of the figure and 5°C on the right. Po values are shown above each trace. In (A) a single representative channel is activated by 100 μM cytosolic Ca^{2+} and in (B) another single channel is activated by sulmazole in the absence of activating cytosolic [Ca^{2+}] (100 pM Ca^{2+}). In both experiments cooling the channels causes a reduction in current amplitude and a large increase in Po, although the Po changes are more marked with sulmazole-activated channels than with Ca^{2+}-activated channels (Modified from Sitsapesan *et al.*[86]).

how changing the temperature can modify RyR channel function. It is interesting that the cooling-induced increase in Po of the channels was more marked with sulmazole than with cytosolic Ca^{2+} as the channel activator. The effects of temperature on the function of other isoforms of RyR have not been investigated and therefore it is not known if differences in the temperature regulation of gating exists between isoforms. Studies involving reconstitution of RyR into planar phospholipid bilayers are generally performed at room temperature, however the physiological temperature of mammalian tissue is more than 10°C higher than that of most bilayer experiments and therefore significant differences in the gating of channels incorporated into bilayers compared with that of RyR *in situ* may arise due to temperature dependent effects. Temperature may also be an important regulator of RyR function in species which do not control their own temperature.

(d) Effects of membrane potential

The gating of RyR reconstituted into planar lipid bilayers is dependent on membrane potential. Mammalian skeletal and cardiac RyR channels have a higher Po at positive potentials compared with negative potentials[14,87,88] and both cardiac and skeletal channels also exhibit voltage-dependent inactivation.[8,22,89,90] Inactivation of sheep cardiac and chicken skeletal α and β isoforms has been observed at positive holding potentials (the cytosolic or *cis* chamber is positive relative to the luminal or *trans* side). In contrast, inactivation of the rabbit skeletal RyR was reported to occur at negative holding potentials and was suggested to be mediated by the FK-506 binding protein (FKBP).[89,91] Percival *et al.*[8] found that the more positive the holding potential, the quicker the onset of inactivation. In both the sheep cardiac and the chicken skeletal RyR the inactivation was reversed by applying a negative membrane potential. With the cardiac RyR the inactivation was not observed in all experiments and depended on the mechanism of channel activation. Voltage-dependent inactivation at +40 mV occurred in only 17% of channels activated by cytosolic Ca^{2+} as the sole ligand but in 56% of channels activated by Ca^{2+} plus a secondary ligand such as a caffeine or an ATP analogue.[22] Voltage-dependent inactivation therefore appears to occur more readily under conditions of high Po and may be triggered by the channel opening to certain long-duration open states. Once inactivation has occurred, it cannot be reversed by removing and re-applying the agonists. It is only be reversed by changing the voltage.[22] It is interesting that the luminal $[Ca^{2+}]$ can regulate the relationship between Po and membrane potential[87,92] and therefore the voltage at which inactivation occurs may be regulated by the luminal $[Ca^{2+}]$. If such a mechanism existed *in situ,* it could act as a negative control mechanism leading to inactivation of SR Ca^{2+} release. It is not known if there is any change in potential across the SR membrane during normal excitation-contraction (E-C) coupling although it is generally assumed that very little change occurs. It has been suggested that the current caused by Ca^{2+}-release from skeletal SR is unlikely to change the SR membrane potential more than 7 mV.[93] In addition, isolated SR vesicle studies indicate that the SR K^+ channel can rapidly shunt K^+ across the SR and would be expected to maintain the SR membrane potential near to zero.[94] Under physiological conditions, it is not certain what changes in SR potential would occur or what the time course of any such changes would be.

Phosphorylation of RyR

Evidence suggests that phosphorylation of RyR modifies the gating characteristics of the channel and the manner in which the channel responds to certain ligands. The reported effects of phosphorylation are diverse and often conflicting and this is an important area for future study. It appears that different isoforms of RyR are not controlled in the same way by phosphorylation. A detailed review of this subject can be found in Chapter 8.

Agents which may act from the luminal side of RyR

Very few agents have been shown to have an effect if they are added to the luminal side of the bilayer. In most cases, even if an effect is observed, unequivocal demonstration that the site of action is the luminal side of the channel or associated protein has not been demonstrated. Agents suggested to act from the luminal side are annexin,[95] chlorocresol[96] and Ca^{2+}, either directly or via calsequestrin.[97–101]

(a) Luminal Ca^{2+}

In cardiac cells, the amount of Ca^{2+} in the SR lumen has been shown to markedly affect E-C coupling.[102–107] Changes in Ca^{2+}-loading of the SR under pathophysiological conditions such as ischaemia and hypertrophy are linked to the abnormal functioning of the muscle cells.[108] Greater loading of the SR with Ca^{2+} causes a greater amount of Ca^{2+} to be released for a given Ca^{2+} trigger.[102–107] Chapter 15 describes detailed experiments on isolated cardiac myocytes which illustrate the marked effect that SR Ca^{2+} load has on E-C coupling. Measurements of Ca^{2+} release from isolated skeletal SR vesicles also indicate that more Ca^{2+} is released for a given trigger if the vesicles are loaded with more Ca^{2+}.[97,98,109] Experiments with reconstituted native and purified single channels demonstrate that changes in luminal Ca^{2+} cause not only the expected conductance changes but also lead to alterations in channel gating.[87,92,107,110–112] Increasing the luminal $[Ca^{2+}]$ in the micromolar to millimolar range results in an increase in Po. At present there are two schools of thought concerning the site of action of luminal Ca^{2+}. Two groups report that the increase in Po (with either RyR1 or RyR2) depends on the mechanism of channel activation from the cytosolic side.[87,92,107] If the channels are activated by cytosolic Ca^{2+} as the sole ligand, then luminal Ca^{2+} has no effect. If, however, the channels are activated by Ca^{2+} plus a second ligand or activated solely by a secondary ligand then luminal Ca^{2+} does activate the channels. These results led to the conclusion that luminal Ca^{2+} binds to a luminal site on the channels. Other investigators, working with purified rabbit skeletal RyR, find essentially the same effects of luminal Ca^{2+}, an increase in Po which is characterised by long open events but find that luminal Ca^{2+} increases the Po of channels activated by cytosolic Ca^{2+} alone.[111,112] The voltage-dependence of the effect led the authors to conclude that luminal Ca^{2+} affects Po by flowing through the channel and binding to cytosolic Ca^{2+} activation and inactivation sites. The effects of luminal Ca^{2+} can be observed in both native and purified channels and therefore suggest direct effects on RyR.[87,92,107,111,112] Other experimental techniques indicate that calsequestrin may be involved the effects of luminal Ca^{2+} on channel gating and these results are discussed in the following section.

Junctional proteins closely associated with RyR

In skeletal muscle, the mechanism for E-C coupling is charged coupled or depolarisation-induced Ca^{2+}-release and there is thought to be a direct physical link between some RyR and the L-type Ca^{2+} channels. The nature of the physical link is not known and this could be either a direct RyR/L-type channel link (see Chapter 9 for a review on this subject) or a link via one or more secondary proteins. The glycolytic enzymes, aldolase and glyceraldehyde-3-phosphate dehydrogenase (GADP), have been shown to associate with RyR and it has been suggested that GADP may form a physical link between L-type Ca^{2+} channels and RyR.[113,114] In addition to L-type channels, the RyR at the T-tubule/SR junctions are structurally associated with certain other distinct "triadic" proteins which have been purified and cloned[115,116] and there is increasing evidence that the associated proteins may be involved in the control of SR Ca^{2+} release in skeletal and cardiac muscle E-C coupling.[117] Junctional proteins shown to be directly or indirectly associated with and proposed to modulate the function of cardiac (RyR2) and skeletal (RyR1) channels include calsequestrin, triadin, junctin and FKBP.

Calsequestrin is a high capacity Ca^{2+} binding protein which can exhibit considerable conformational changes when binding divalent cations and it has been located in the lumen of the SR of cardiac and skeletal muscle.[118,119] It has been proposed that calsequestrin may play a role in the regulation of Ca^{2+}-release from SR. Different isoforms of calsequestrin have been identified and are between approximately 55 and 65 kDa in weight. The different isoforms have been shown to undergo different Ca^{2+}-induced conformational changes.[119] Some Ca^{2+} flux studies from skeletal SR vesicles indicate that luminal Ca^{2+} may not act directly on RyR to influence channel gating or that an additional mechanism for regulation by luminal Ca^{2+} exists. Such studies have suggested that calsequestrin may mediate some or all of the effects of luminal Ca^{2+}.[97–100] Single channel experiments investigating the effects of calsequestrin on RyR1 indicate that changes in luminal Ca^{2+} may affect channel gating indirectly via calsequestrin.[101] Addition of calsequestrin to the luminal side of native RyR1 incorporated into bilayers increased Po in a Ca^{2+}-dependent manner. At present we can only estimate the free $[Ca^{2+}]$ at the luminal face of RyR during contraction and rest and this is a major problem when trying to understand the relationship between SR load and RyR gating in vivo. The exact role of calsequestrin in the regulation of intracellular Ca^{2+} release is therefore unknown and consideration must be given to the possibility that associated proteins which bind calsequestrin to RyR may also be regulated by luminal $[Ca^{2+}]$.[117] These proteins are discussed in the following section.

Isoforms of triadin and junctin are found in both cardiac and skeletal muscle and the proteins co-localise with calsequestrin and ryanodine receptors.[117,120] It has been suggested that RyR, junctin, triadin and calsequestrin may be present in the junctional SR as a functional unit and that the normal physiological regulation of SR Ca^{2+} release may depend on the complete complex of the four proteins.[117,120] In this regard there are emerging studies indicating a role for calsequestrin as a modulator of RyR function and SR Ca^{2+}-release.[101,121] Other studies indicate that triadin[122] and junctin[123] may regulate RyR function and may be involved in calsequestrin-mediated changes in RyR gating.[121]

The immunophilins, FK-506 binding proteins (FKBP12 (skeletal); FKBP12.6 (cardiac)) have been shown to form stable complexes with cardiac and skeletal RyR but their role in modulating channel function is unclear. The skeletal isoform of the ryanodine receptor binds FKBP12 and FKBP12.6.[124] In contrast, only the FKBP12.6 isoform binds to the cardiac RyR.[124] FK-506 displaces both binding proteins from the RyR isoforms.[124] Dissociation of the binding proteins from RyR has been reported to have various effects. Removal of FKBP12 or FKBP12.6 (by FK-506 or rapamycin) has been suggested to: 1. increase Po and induce sub-conductance levels,[125–127] 2. increase Po but exert no effect on conductance,[128] 3. exhibit no effect on Po or conductance[124] or 4. alter the voltage dependence of Po.[89,91] As yet it is not known if these differences are the result of species differences, differences between RyR isoforms or the method by which the channels have been reconstituted into bilayers. The wide range of reported effects observed in single channel experiments makes it difficult to identify the possible physiological roles for these proteins. One suggestion is that the receptor for transforming growth factor-β may be located on FKBP.[129] Another is that FKBP may be the site for cADPR binding.[130] Effects of agents known to displace the binding proteins have also been shown to exert profound effects on E-C coupling in cardiac cells. FK-506 has been shown to increase both the rate of SR Ca^{2+}-release during rest and the fraction of

SR Ca^{2+} released during a normal twitch in rat heart cells.[131,132] McCall *et al.*[131] found an increase in frequency of resting Ca^{2+} sparks but no change in amplitude or duration. In contrast, Xiao *et al.* reported that FK-506 and rapamycin increased the duration of Ca^{2+} sparks in rat heart myocytes and induced sparks of two amplitudes corresponding to full and half normal spark amplitude.[132] Many of these effects could be explained by either an increase in Po of the full conductance channel or the stabilisation of the channel in lower conductance but longer open states. FK-506, however, has been reported to affect other proteins involved in intracellular Ca^{2+} homeostasis, for example Na/Ca exchange.[131] FK-506 also appears to inhibit outward K^+ currents, an effect which causes prolongation of the action potential.[132] Incubation of cardiac RyR with FK-506 slowed the adaptation observed after flash photolysis of caged Ca^{2+} and led to the suggestion that the FKBP proteins may exert an inhibitory action on RyR which leads to channel inactivation and termination of SR Ca^{2+} release.[132] There is evidence that FK-506 and rapamycin may directly affect RyR gating in addition to their action of dissociating FKBP. Ahern *et al.* have demonstrated that rapamycin can reversibly increase the Po of FKBP-stripped channels.[133] Certain pharmacological agents have been proposed to act via FKBP. Bastadins, isolated from the sponge Ianthella basta, have marked effects on RyR gating which can be antagonised by FK-506 and it has been suggested that the site of action is FKBP12.[134]

Covalent modifiers of the channel

Agents which react covalently with RyR lead to a number of varied changes to both the conductance and gating properties of the channels. Some of these effects and the use of covalent modifiers as tools to understanding the structure and function of RyR have been discussed in Chapter 11.

Agents interacting with thiol groups

Sulfhydryl (SH) reagents have been shown to stimulate SR Ca^{2+}-release and are thought to influence RyR gating by reacting with free SH groups on the channel. The binding sites for sulfhydryl agents are probably cysteine residues either on RyR or on closely associated proteins which influence the gating of RyR. A wide variety of chemicals interact with SH groups and have been shown to release Ca^{2+} from the SR and include phthalocyanine dyes, heavy metals, reactive disulfides, reactive oxygen species, anthraquinones, mercaptans.[135–144] The gating effects of the agents are varied and depend on the type of compound used to modify SH groups. The observed effects may reflect the accessibility of the agent to the free SH groups, the type of reaction involved and the specificity of the agent for SH groups. While it has been suggested that thiol oxidation may be physiologically important and may regulate normal SR Ca^{2+}-release[145] it is more readily accepted that oxidation of RyR may become important under pathological conditions or conditions of fatigue when levels of oxidants and free radicals within a cell may be increased.

Effects of Ryanodine

Ryanodine is very selective for RyR at nanomolar concentrations and binds almost irreversibly. It has been used extensively as a tool to purify RyR channels and for detection of RyR in various tissues. Inspection of the effects of ryanodine at the single channel level reveal that ryanodine causes the channel to enter long-lived open states of a modified conductance.[146,147] The mechanisms leading to the distinct alterations in conductance and gating of RyR by ryanodine are complex,[148] however the use of reversible analogues of ryanodine is providing new insight into the mechanisms of ryanodine induced modification of channel function.[149,150] The pharmacology of ryanodine and related compounds has been recently reviewed by Sutko *et al.*[4] and is discussed in Chapter 7.

SUMMARY

All isoforms of RyR so far studied at the single channel level share certain basic mechanisms of gating. All channels are activated by cytosolic Ca^{2+} and inactivated by high $[Ca^{2+}]$. Typically, Ca^{2+}-activated channels exhibit very brief open events. All RyR require the presence of a second activating ligand (for example, ATP) for full, maintained activation of the channels. The gating of all ryanodine receptors is modified by various factors including the ionic environment of the channel, membrane potential and the phosphorylation state of the channel. There are some agents which appear to be specific for a particular isoform of RyR but in general, most agents which affect the gating of one isoform of RyR also modify the gating of other RyR isoforms. The situation is complicated in that all isoforms of RyR respond slightly differently to virtually every ligand or intervention that regulates gating. As there is a huge variety of agents which can modify channel gating this leads to a multitude of possible gating mechanisms. In turn this allows for extensive diversity of the intracellular Ca^{2+}-release mechanisms.

ACKNOWLEDGEMENTS

Supported by the British Heart Foundation.

REFERENCES

1. Coronado R, Morrissette J, Sukhareva M *et al.* Structure and function of ryanodine receptors. *Am J Physiol* 1994;**266**:C1485–C1504.
2. Zucchi R, Ronca-Testoni S. The sarcoplasmic reticulum Ca^{2+} channel/ryanodine receptor: Modulation by endogenous effectors, drugs and disease states. *Pharmacol Rev* 1997;**49**: 1–51.
3. Sutko JL, Airey JA. Ryanodine receptor Ca^{2+} release channels: Does diversity in form equal diversity in function. *Physiol Rev* 1996;**76**:1027–1071.
4. Sutko JL, Airey JA, Welch W *et al.* The pharmacology of ryanodine and related compounds. *Pharmacol Rev* 1997;**49**:53–98.
5. Sitsapesan R, Williams AJ. Gating of the native and purified cardiac SR Ca^{2+}-release channel with monovalent cations as permeant species. *Biophys J* 1994;**67**:1484–1494.
6. Copello JA, Barg S, Onoue H *et al.* Heterogeneity of Ca^{2+} gating of skeletal muscle and cardiac ryanodine receptors. *Biophys J* 1997;**73**:141–156.
7. Ashley RH, Williams AJ. Divalent cation activation and inhibition of single calcium release channels from sheep cardiac sarcoplasmic reticulum. *J Gen Physiol* 1990;**95**:981–1005.
8. Percival AL, Williams AJ, Kenyon JL *et al.* Chicken skeletal muscle ryanodine receptor isoforms: Ion channel properties. *Biophys J* 1994;**67**:1834–1850.
9. Zahradníková A, Zahradník I. Description of modal gating of the cardiac calcium release channel in planar lipid membranes. *Biophys J* 1995;**69**:1780–1788.
10. Chu A, Fill M, Stefani E *et al.* Cytoplasmic Ca^{2+} does not inhibit the cardiac muscle sarcoplasmic reticulum ryanodine receptor Ca^{2+} channel, although Ca^{2+}-induced Ca^{2+} inactivation of Ca^{2+} release is observed in native vesicles. *J Membr Biol* 1993;**135**:49–59.
11. Smith JS, Coronado R, Meissner G. Single channel measurements of the calcium release channel from skeletal muscle sarcoplasmic reticulum. *J Gen Physiol* 1986;**88**:573–588.
12. Laver DR, Roden LD, Ahern GP *et al.* Cytoplasmic Ca^{2+} inhibits the ryanodine receptor from cardiac muscle. *J Membr Biol* 1995;**147**:7–22.
13. Zhang JJ, Sitsapesan R, Williams AJ. Gating properties of the lobster skeletal muscle ryanodine receptor. *J Physiol* 1997;**504.P**:161P–162P.
14. Rousseau E, Meissner G. Single cardiac sarcoplasmic reticulum Ca^{2+}-release channel: activation by caffeine. *Am J Physiol* 1989;**256**:H328–H333.
15. Mickelson JR, Gallant EM, Litterer LA *et al.* Abnormal sarcoplasmic reticulum ryanodine receptor in malignant hyperthermia. *J Biol Chem* 1988;**263**:9310–9315.
16. Fill M, Coronado R, Mickelson JR *et al.* Abnormal ryanodine receptor channels in malignant hyperthermia. *Biophys J* 1990;**57**:471–475.

17. Mickelson JR, Louis CF. Malignant hyperthermia: Excitation-contraction coupling, Ca^{2+} release channel, and cell Ca^{2+} regulation defects. *Physiol Rev* 1996;**76**:537–592.
18. Seok J-H, Xu L, Kramarcy NR *et al.* The 30 S lobster skeletal muscle Ca^{2+} release channel (ryanodine receptor) has functional properties distinct from the mammalian channel proteins. *J Biol Chem* 1992;**267**:15893–15901.
19. Chen SRW, Li X, Ebisawa K *et al.* Functional Characterization of the Recombinant Type 3 Ca^{2+} Release Channel (Ryanodine Receptor) Expressed in HEK293 Cells. *J Biol Chem* 1997;**272**:24234–24246.
20. Bull R, Marengo JJ. Sarcoplasmic reticulum release channels from frog skeletal muscle display two types of calcium dependence. *FEBS Lett* 1993;**331**:223–227.
21. Györke S, Fill M. Ryanodine receptor adaptation: Control mechanism of Ca^{2+}-induced Ca^{2+} release in heart. *Science* 1993;**260**:807–809.
22. Sitsapesan R, Montgomery RAP, Williams AJ. New insights into the gating mechanisms of cardiac ryanodine receptors revealed by rapid changes in ligand concentration. *Circ Res* 1995;**77**: 765–772.
23. Schiefer A, Meissner G, Isenberg G. Ca^{2+} activation and Ca^{2+} inactivation of canine reconstituted cardiac sarcoplasmic reticulum Ca^{2+}-release channels. *J Physiol* 1995;**489**:337–348.
24. Laver DR, Curtis BA. Response of ryanodine receptor channels to Ca^{2+} steps produced by rapid solution. *Biophys J* 1996;**71**:732–741.
25. Rousseau E, Pinkos J, Savaria D. Functional sensitivity of the native skeletal Ca^{2+}-release channel to divalent cations and the Mg-ATP complex. *Can J Physiol Pharmacol* 1992;**70**:394–402.
26. Nagasaki K, Kasai M. Channel selectivity and gating specificity of calcium-induced calcium release channel in isolated sarcoplasmic reticulum. *J Biochem* (Tokyo) 1984;**96**:1769–1775.
27. Meissner G, El-Hashem A. Ryanodine as a functional probe of the skeletal muscle sarcoplasmic reticulum Ca^{2+} release channel. *Mol Cell Biochem* 1992;**114**:119–123.
28. Meissner G, Rios E, Tripathy A *et al.* Regulation of skeletal muscle Ca^{2+} release channel (ryanodine receptor) by Ca^{2+} and monovalent cations and anions. *J Biol Chem* 1997;**272**:1628–1638.
29. Percival A, Williams AJ, Airey JA *et al.* Ion channel properties of the α and β ryanodine receptor isoforms co-expressed in avian skeletal muscle. *Biophys J* 1994;**66(2)**:A415. (Abstract)
30. Meissner G, Darling E, Eveleth J. Kinetics of rapid Ca^{2+} release by sarcoplasmic reticulum. Effects of Ca^{2+}, Mg^{2+} and adenine nucleotides. *Biochemistry* 1986;**25**:236–244.
31. Laver DR, Baynes TM, Dulhunty AF. Magnesium inhibition of ryanodine-receptor calcium channels: Evidence for two independent mechanisms. *J Membr Biol* 1997;**156**:213–229.
32. Rousseau E, Smith JS, Henderson JS *et al.* Single channel and $^{45}Ca^{2+}$ flux measurements of the cardiac sarcoplasmic reticulum calcium channel. *Biophys J* 1986;**50**:1009–1014.
33. Meissner G, Henderson JS. Rapid Ca release from cardiac sarcoplasmic reticulum vesicles is dependent on Ca^{2+} and is modulated by Mg^{2+}, adenine nucleotide and calmodulin. *J Biol Chem* 1987;**262**:3065–3073.
34. Hain J, Nath S, Mayrleitner M *et al.* Phosphorylation modulates the function of the calcium release channel of sarcoplasmic reticulum from skeletal muscle. *Biophys J* 1994;**67**: 1823–1833.

35. Hain J, Onoue H, Mayrleitner M *et al.* Phosphorylation modulates the function of the calcium release channel of sarcoplasmic reticulum from cardiac muscle. *J Biol Chem* 1995;**270**: 2074–2081.
36. Williams AJ, Holmberg SRM. Sulmazole (AR-L 115BS) activates the sheep cardiac muscle sarcoplasmic reticulum calcium-release channel in the presence and absence of calcium. *J Membr Biol* 1990;**115**:167–178.
37. Kermode H, Sitsapesan R, Williams AJ. ADP and inorganic phosphate activate the sheep cardiac sarcoplasmic reticulum Ca^{2+}-release channel. *Biophys J* 1998; **74**: 1296–1304.
38. Sitsapesan R, Williams AJ. Modification of the conductance and gating properties of ryanodine receptors by suramin. *J Membr Biol* 1996;**153**:93–103.
39. McGarry SJ, Williams AJ. Digoxin activates sarcoplasmic reticulum Ca^{2+}-release channels: a possible role in cardiac inotropy. *Br J Pharmacol* 1993;**108**:1043–1050.
40. Sitsapesan R, Williams AJ. Mechanisms of caffeine activation of single calcium-release channels of sheep cardiac sarcoplasmic reticulum. *J Physiol* 1990;**423**:425–439.
41. Fabiato A. Calcium-induced release of calcium from the cardiac sarcoplasmic reticulum. *Am J Physiol* 1983;**245**:C1–C14.
42. Meissner G. Adenine nucleotide stimulation of Ca^{2+}-induced Ca^{2+} release in sarcoplasmic reticulum. *J Biol Chem* 1984;**259**:2365–2374.
43. McGarry SJ, Williams AJ. Adenosine discriminates between the caffeine and adenine nucleotide sites on the sheep cardiac sarcoplasmic reticulum calcium-release channel. *J Membr Biol* 1994;**137**:169–177.
44. Graeff RM, Podein RJ, Aarhus R *et al.* Magnesium ions but not ATP inhibit cyclic ADP-ribose-induced calcium release. Biochem Biophys Res Commun 1995;**206**:786–791.
45. Fredholm BB, Abbracchio MP, Burnstock G *et al.* Nomenclature and classification of purinoceptors. *Pharmacol Rev* 1994;**46**:143–156.
46. Guo XQ, Becker PL. Cyclic ADP-ribose-gated Ca^{2+} release in sea urchin eggs requires an elevated $[Ca^{2+}]$. *J Biol Chem* 1997;**272**:16984–16989.
47. Lee HC, Aarhus R, Graeff R *et al.* Cyclic ADP ribose activation of the ryanodine receptor is mediated by calmodulin. *Nature* 1994;**370**:307–309.
48. Takasawa S, Ishida A, Nata K *et al.* Requirement of calmodulin-dependent protein kinase II in cyclic ADP-ribose-mediated intracellular Ca^{2+} mobilization. *J Biol Chem* 1995;**270**: 30257–30259.
49. Galione A, Lee HC, Busa WB. Ca^{2+}-induced Ca^{2+} release in sea urchin egg homogenates: Modulation by cyclic ADP-ribose. *Science* 1991;**253**:1143–1146.
50. Rakovic S, Galione A, Ashamu GA *et al.* A specific cyclic ADP-ribose antagonist inhibits cardiac excitation-contraction coupling. *Curr Biol* 1996;**6**:989–996.
51. Guo XQ, Laflamme MA, Becker PL. Cyclic ADP-ribose does not regulate sarcoplasmic reticulum Ca^{2+} release in intact cardiac myocytes. *Circ Res* 1996;**79**:147–151.
52. Sitsapesan R, McGarry SJ, Williams AJ. Cyclic ADP-ribose, the ryanodine receptor and Ca^{2+}-release. *TIPS* 1995;**16**:386–391.

53. Sitsapesan R, McGarry SJ, Williams AJ. Cyclic ADP-ribose competes with ATP for the adenine nucleotide binding site on the cardiac ryanodine receptor Ca^{2+}-release channel. *Circ Res* 1994;**75(3)**:596–600.
54. Sitsapesan R, Williams AJ. Cyclic ADP-ribose and related compounds activate sheep skeletal sarcoplasmic reticulum Ca^{2+} release channel. *Am J Physiol* 1995;**268**:C1235–C1240.
55. Meszáros LG, Bak J, Chu A. Cyclic ADP-ribose as an endogenous regulator of the non-skeletal type ryanodine receptor Ca^{2+} channel. *Nature* 1993;**364**:76–79.
56. Fruen BR, Mickelson JR, Shomer NH *et al.* Cyclic ADP-ribose does not affect cardiac or skeletal muscle ryanodine receptors. *FEBS Lett* 1994;**352**:123–126.
57. Morrissette J, Heisermann G, Cleary J, *et al.* Cyclic ADP-ribose induced Ca^{2+} release in rabbit skeletal muscle sarcoplasmic reticulum. *FEBS Lett* 1993;**330**:270–274.
58. Lahouratate P, Guibert J, Faivre JF. cADP-ribose releases Ca^{2+} from cardiac sarcoplasmic reticulum independently of ryanodine receptor. *Am J Physiol Heart Circ Physiol* 1997;**273**:H1082–H1089.
59. Rousseau E, LaDine J, Liu Q-Y *et al.* Activation of the Ca^{2+} release channel of skeletal muscle sarcoplasmic reticulum by caffeine and related compounds. *Arch Biochem Biophys* 1988;**267**:75–86.
60. McGarry SJ, Williams AJ. Activation of the sheep cardiac sarcoplasmic reticulum Ca^{2+}-release channel by analogues of sulmazole. *Br J Pharmacol* 1994;**111 (4)**:1212–1220.
61. Kobayashi J, Ishibashi M, Nagai U *et al.* 9-Methyl-7-bromoeudistomin D, a potent inducer of calcium release from sarcoplasmic reticulum of skeletal muscle. Experientia 1989;**45**: 782–783.
62. Scott-Ward T, Windass JD, Dunbar SJ *et al.* Preliminary characterisations of a ryanodine receptor-calcium release channel from the flight muscle of Heliothis virescens. *Biophys J* 1997;**72**:A158.
63. Takeshima H, Nishimura S, Matsumoto T *et al.* Primary structure and expression from complementary DNA of skeletal muscle ryanodine receptor. *Nature* 1989;**339**:439–445.
64. Zorzato F, Fujii J, Otsu K *et al.* Molecular cloning of cDNA encoding human and rabbit forms of the Ca^{2+} release channel (ryanodine receptor) of skeletal muscle sarcoplasmic reticulum. *J Biol Chem* 1990;**265**:2244–2256.
65. Menegazzi P, Larini F, Treves S *et al.* Identification and characterization of three calmodulin binding sites of the skeletal muscle ryanodine receptor. *Biochemistry* 1994;**33**:9078–9084.
66. Tripathy A, Xu l, Mann G *et al.* Calmodulin activation and inhibition of skeletal muscle Ca^{2+} release channel (ryanodine receptor). *Biophys J* 995;**69**:106–119.
67. Fuentes O, Valdivia C, Vaughan D *et al.* Calcium-dependent block of ryanodine receptor channel of swine skeletal muscle by direct binding of calmodulin. *Cell Calcium* 1994;**15**:305–316.
68. Witcher DR, Kovacs RJ, Schulman H *et al.* Unique phosphorylation site on the cardiac ryanodine receptor regulates calcium channel activity. *J Biol Chem* 1991;**266**:11144–11152.
69. El-Hayek R, Valdivia C, Valdivia HH *et al.* Activation of the Ca^{2+} release channel of skeletal muscle sarcoplasmic reticulum by palmitoyl carnitine. *Biophys J* 1993;**65**:779–789.
70. Fulceri R, Nori A, Gamberucci A *et al.* Fatty acyl-CoA esters induce calcium release from terminal cisternae of skeletal muscle. *Cell Calcium* 1994;**15**:109–116.

71. Connelly T, Ahern C, Sukhareva M *et al.* Removal of Mg^{2+} inhibition of cardiac ryanodine receptor by palmitoyl coenzyme A. *FEBS Lett* 1994;**352**:285–290.
72. Mita M, Yasumasu I. Inhibition of glucose 6-phosphate dehydrogenase and 6-phosphogluconate dehydrogenase in sea urchin eggs by palmitoyl-coenzyme A reversal by polyamines. *Arch Biochem Biophys* 1980;**201**:322–329.
73. Mita M, Yasumasu I. Regulation of the tricarboxylic acid cycle in sea urchin eggs and embryos. *J Exp Zool* 1983;**228**:71–77.
74. Chini EN, Dousa TP. Palmitoyl-CoA potentiates the Ca^{2+} release elicited by cyclic ADP-ribose. *Am J Physiol* 1996;**270**:C530–C537.
75. Ghosh TK, Bian J, Gill DL. Intracellular calcium release mediated by sphingosine derivatives generated in cells. *Science* 1990;**248**:1653–1656.
76. Martonosi AN. Mechanisms of Ca^{2+} release from sarcoplasmic reticulum of skeletal muscle. *Physiol Rev* 1984;**64**:1240–1320.
77. Olivera A, Zhang H, Carlson RO *et al.* Stereospecificity of sphingosine-induced intracellular calcium mobilization and cellular proliferation. *J Biol Chem* 1994;**269**:17924–17930.
78. Needleman DH, Aghdasi B, Seryshev AB *et al.* Modulation of skeletal muscle Ca^{2+}-release channel activity by sphingosine. *Am J Physiol* 1997;**272**:C1465–C1474.
79. Fruen BR, Kane PK, Mickelson JR *et al.* Chloride-dependent sarcoplasmic reticulum Ca^{2+} release correlates with increased Ca^{2+} activation of ryanodine receptors. *Biophys J* 1996;**71**:2522–2530.
80. Fruen BR, Mickelson JR, Roghair TJ *et al.* Anions that potentiate excitation-contraction coupling may mimic effect of phosphate on Ca^{2+} release channel. *Am J Physiol* 1994;**266**: C1729–C1735.
81. Fruen BR, Mickelson JR, Shomer NH *et al.* Regulation of the sarcoplasmic reticulum ryanodine receptor by inorganic phosphate. *J Biol Chem* 1994;**269**:192–198.
82. Valdivia HH, Kirby MS, Lederer WJ *et al.* Scorpion toxins targeted against the sarcoplasmic reticulum Ca^{2+}-release channel of skeletal and cardiac muscle. *Proc Natl Acad Sci USA* 1992;**89**:12185–12189.
83. El-Hayek R, Lokuta AJ, Ar_valo C *et al.* Peptide probe of ryanodine receptor function - Imperatoxin A, a peptide from the venom of the scorpion *Pandinus imperator*, selectively activates skeletal-type ryanodine receptor isoforms. *J Biol Chem* 1995;**270**:28696–28704.
84. Rousseau E, Pinkos J. pH modulates conducting and gating behaviour of single calcium release channels. *Pflugers Arch* 1990;**415**:645–647.
85. Xu l, Mann G, Meissner G. Regulation of Cardiac Ca^{2+} Release Channel (Ryanodine Receptor) by Ca^{2+},H^{+},Mg^{2+} and Adenine Nucleotides Under Normal and Simulated Ischemic Conditions. *Circ Res* 1996;**79**:1100–1109.
86. Sitsapesan R, Montgomery RAP, MacLeod KT *et al.* Sheep cardiac sarcoplasmic reticulum calcium release channels: modification of conductance and gating by temperature. *J Physiol* 1991;**434**:469–488.
87. Sitsapesan R, Williams AJ. Regulation of the gating of the sheep cardiac sarcoplasmic reticulum Ca^{2+}-release channel by luminal Ca^{2+}. *J Membr Biol 1*994;**137**:215–226.

88. Ma JJ, Fill M, Knudson M *et al.* Ryanodine receptor of skeletal muscle is a gap junction-type channel. *Science* 1988;**242**:99–102.
89. Ma JJ, Bhat MB, Zhao JY. Rectification of skeletal muscle ryanodine receptor mediated by FK506 binding protein. *Biophys J* 1995;**69:**2398–2404.
90. Ma J. Desensitization of the skeletal muscle ryanodine receptor: evidence for heterogeneity of calcium release channels. *Biophys J* 1995;**68**:893–899.
91. Chen SRW, Zhang L, MacLennan DH. Asymmetrical blockade of the Ca^{2+} release channel (ryanodine receptor) by 12-kDa FK506 binding protein. *Proc Natl Acad Sci USA* 1994;**91**:11953–11957.
92. Sitsapesan R, Williams AJ. The gating of the sheep skeletal sarcoplasmic reticulum Ca^{2+}-release channel is regulated by luminal Ca^{2+}. *J Membr Biol* 1995;**146**:133–144.
93. Oetliker H. An appraisal of the evidence for a sarcoplasmic reticulum membrane potential and its relation to calcium release in skeletal muscle. *J Muscle Res Cell Mot* 1982;**3**:247–272.
94. Garcia AM, Miller C. Channel mediated monovalent cation-fluxes in isolated sarcoplasmic reticulum vesicles. *J Gen Physiol 1*984;**83**:819–839.
95. Diaz-Munoz M, Hamilton SL, Kaetzel MA *et al.* Modulation of Ca2+ release channel activity from sarcoplasmic reticulum by Annexin VI (67-kDalton Calcimedin). *J Biol Chem* 1990;**265**:15894–15899.
96. Herrmann-Frank A, Richter M, Sarközi S *et al.* 4-chloro-*m*-cresol, a potent and specific activator of the skeletal muscle ryanodine receptor. *Biochim Biophys Acta* 1996;**1289**:31–40.
97. Gilchrist JSC, Belcastro AN, Katz S. Intraluminal Ca^{2+} dependence of Ca^{2+} and ryanodine-mediated regulation of skeletal muscle sarcoplasmic reticulum Ca^{2+} release. *J Biol Chem* 1992;**267**:20850–20856.
98. Donoso P, Prieto H, Hidalgo C. Luminal calcium regulates calcium release in triads isolated from frog and rabbit skeletal muscle. *Biophys J* 1995;**68**:507–515.
99. Ikemoto N, Ronjat M, Meszáros LG, *et al.* Postulated role of calsequestrin in the regulation of calcium release from the sarcoplasmic reticulum. *Biochemistry* 1989;**28**:6764–6771.
100. Ikemoto N, Antoniu B, Kang J-J *et al.* Intravesicular calcium transient during calcium release from sarcoplasmic reticulum. *Biochemistry* 1991;**30**:5230–5237.
101. Kawasaki T, Kasai M. Regulation of calcium channels in sarcoplasmic reticulum by calsequestrin. *Biochem Biophys Res Commun* 1994;**199**:1120–1127.
102. Fabiato A, Fabiato F. Calcium and cardiac excitation-contraction coupling. *Annu Rev Physiol* 1979;**41**:473–484.
103. Fabiato A, Fabiato F. Use of chlorotetracycline fluorescence to demonstrate calcium-induced release of calcium from the sarcoplasmic reticulum of skinned cardiac cells. *Nature* 1979;**281**:146–148.
104. Bassani JWM, Yuan W, Bers DM. Fractional SR Ca release is regulated by trigger Ca and SR Ca content in cardiac myocytes. *Am J Physiol* 1995;**268**:C1313–C1319.
105. Han S, Schiefer A, Isenberg G. Ca^{2+} load of guinea pig ventricular myocytes determines efficacy of brief Ca^{2+} currents as trigger for Ca^{2+} release. *J Physiol* 1994;**480**:411–421.
106. Spencer CI, Berlin JR. Control of sarcoplasmic reticulum calcium release during calcium loading in isolated rat ventricular myocytes. *J Physiol* 1995;**488**:267–279.

107. Lukyanenko V, Györke I, Györke S. Regulation of calcium release by calcium inside the sarcoplasmic reticulum in ventricular myocytes. *Pflugers Arch* 1996;**432**:1047–1054.
108. Niggli E, Lipp P. Subcellular features of calcium signalling in heart muscle: what do we learn? *Cardiovas Res* 1995;**29**:441–448.
109. Nelson TE, Nelson KE. Intra- and extraluminal sarcoplasmic reticulum membrane regulatory sites for Ca^{2+}-induced Ca^{2+} release. *FEBS Lett* 1990;**263**:292–294.
110. Tinker A, Lindsay ARG, Williams AJ. Cation conduction in the cardiac sarcoplasmic reticulum calcium-release channel under physiological and pathophysiological conditions. *Cardiovas Res* 1993;**27**:1820–1825.
111. Tripathy A, Meissner G. Sarcoplasmic reticulum lumenal Ca^{2+} has access to cytosolic activation and inactivation sites of skeletal muscle Ca^{2+} release channel. *Biophys J* 1996;**70**:2600–2615.
112. Herrmann-Frank A, Lehmann-Horn F. Regulation of the purified Ca^{2+} release channel/ryanodine receptor complex of skeletal muscle sarcoplasmic reticulum by luminal calcium. *Pflugers Arch* 1996;**432**:155–157.
113. Cadwell JJS, Caswell AH. Identification of a constituent of the junctional feet linking terminal cisternae to transverse tubules in skeletal muscle. *J Cell Biol* 1982;**93**:543–550.
114. Caswell AH, Brunschwig JP. Identification and extraction of proteins that compose the triad junction of skeletal muscle. *J Cell Biol* 1984;**99**:929–939.
115. Knudson CM, Stang KK, Moomaw CR *et al.* Primary structure and topological analysis of a skeletal muscle-specific junctional sarcoplasmic reticulum glycoprotein (Triadin). *J Biol Chem* 1993;**268**:12646–12654.
116. Jones LR, Zhang L, Sanborn K *et al.* Purification, primary structure, and immunological characterization of the 26-kDa calsequestrin binding protein (junctin) from cardiac junctional sarcoplasmic reticulum. *J Biol Chem* 1995;**270**:30787–30796.
117. Zhang L, Kelley J, Schmeisser G *et al.* Complex formation between junction, triadin, calsequestrin, and the ryanodine receptor - Proteins of the cardiac junctional sarcoplasmic reticulum membrane. *J Biol Chem* 1997;**272**:23389–23397.
118. Yano K, Zarain-Herzberg A. Sarcoplasmic reticulum calsequestrins: Structural and functional properties. *Mol Cell Biochem* 1994;**135**:61–70.
119. Damiani E, Margreth A. Characterization study of the ryanodine receptor and of calsequestrin isoforms of mammalian skeletal muscles in relation to fibre types. *J Muscle Res Cell Mot* 1994;**15**:86–101.
120. Guo W, Campbell KP. Association of triadin with the ryanodine receptor and calsequestrin in the lumen of the sarcoplasmic reticulum. *J Biol Chem* 1995;**270**:9027–9030.
121. Ohkura M, Ide T, Furukawa K *et al.* Calsequestrin is essential for the Ca^{2+} release induced by myotoxin a in skeletal muscle sarcoplasmic reticulum. *Can J Physiol Pharmacol* 1995;**73**:1181–1185.
122. Liu G, Pessah IN. Molecular interaction between ryanodine receptor and glycoprotein triadin involves redox cycling of functionally important hyperreactive sulfhydryls. *J Biol Chem* 1994;**269**:33028–33034.

123. Yamaguchi N, Kawasaki T, Kasai M. DIDS binding 30-kDa protein regulates the calcium release channel in the sarcoplasmic reticulum. *Biochem Biophys Res Commun* 1995;**210**:648–653.
124. Timerman AP, Onoue H, Xin HB *et al.* Selective binding of FKBP12.6 by the cardiac ryanodine receptor. *J Biol Chem* 1996;**271**:20385–20391.
125. Ahern GP, Junankar PR, Dulhunty AF. Subconductance States in Single-channel Activity of Skeletal Muscle Ryanodine Receptors After Removal of FKBP12. *Biophys J* 1997;**72**:146–162.
126. Brillantes A-MB, Ondrias K, Scott A *et al.* Stabilization of calcium release channel (ryanodine receptor) function by FK506-binding protein. Cell 1994;**77**:513–523.
127. Kaftan E, Marks AR, Ehrlich BE. Effects of rapamycin on ryanodine receptor Ca^{2+}-release channels from cardiac muscle. *Circ Res* 1996;**78**:990–997.
128. Mayrleitner M, Timerman AP, Wiederrecht G *et al.* The calcium release channel of sarcoplasmic reticulum is modulated by FK-506 binding protein: Effect of FKBP-12 on single channel activity of the skeletal muscle ryanodine receptor. *Cell Calcium* 1994;**15**:99–108.
129. Wang T, Donahoe PK, Zervos AS. Specific interaction of type I receptors of the TGF-beta family with the immunophilin FKBP-12. *Science* 1994;**265**:674–676.
130. Noguchi N, Takasawa S, Nata K *et al.* Cyclic ADP-ribose binds to FK506-binding protein 12.6 to release Ca^{2+} from islet microsomes. *J Biol Chem* 1997;**272**:3133–3136.
131. McCall E, Li L, Satoh H *et al.* Effects of FK-506 on Contraction and Ca^{2+} Transient in Rat Cardiac Myocytes. *Circ Res* 1996;**79**:1110–1121.
132. Xiao R-P, Valdivia HH, Bogdanov K *et al.* The immunophilin FK506-binding protein modulates calcium release channel closure in rat heart. *J Physiol* 1997;**500**:343–354.
133. Ahern GP, Junankar PR, Dulhunty AF. Ryanodine receptors from rabbit skeletal muscle are reversibly activated by rapamycin. *Neurosci Lett* 1997;**225**:81–84.
134. Mack MM, Molinski TF, Buck ED *et al.* Novel modulators of skeletal muscle FKBP12/calcium channel complex from *Ianthella basta*. Role of FKBP12 in channel gating. *J Biol Chem* 1994;**269**:23236–23249.
135. Oba T, Koshita M, Yamaguchi M. H_2O_2 modulates twitch tension and increases P_o of Ca_{2+} release channel in frog skeletal muscle. *Am J Physiol* 1996;**271**:C810–C818.
136. Trimm JL, Salama G, Abramson JJ. Sulfhydryl oxidation induces rapid calcium release from sarcoplasmic reticulum vesicles. *J Biol Chem* 1986;**261**:16092–16098.
137. Salama G, Abramson JJ. Silver ions trigger Ca2+ release by acting at the apparent physiological release site in sarcoplasmic reticulum. *J Biol Chem* 1984;**259**:13363–13369.
138. Stuart J, Pessah IN, Favero TG *et al.* Photooxidation of skeletal muscle sarcoplasmic reticulum induces rapid calcium release. *Arch Biochem Biophys* 1992;**292**:512–521.
139. Favero TG, Henry R, Abramson JJ. Reduced oxygen species stimulate calcium release from skeletal muscle sarcoplasmic reticulum. *Biophys J* 1993;**64**:A303. (Abstract)
140. Zaidi NF, Lagenaur CF, Abramson JJ *et al.* Reactive disulfides trigger Ca^{2+} release from sarcoplasmic reticulum via an oxidation reaction. *J Biol Chem* 1989;**264**:21725–21736.
141. Liu G, Abramson JJ, Zable AC *et al.* Direct evidence for the existence and functional role of hyperreactive sulfhydryls on the ryanodine receptor-triadin complex selectively labeled by the coumarin maleimide 7-diethylamino-3-(4'-maleimidylphenyl)-4- methylcoumarin. *Mol Pharmacol 1*994;**45**:189–200.

142. Boraso A, Williams AJ. Modification of the gating of the cardiac sarcoplasmic reticulum calcium-release channel by hydrogen peroxide and dithiothreitol. *Am J Physiol* 1994;**267**:H1010–H1016.

143. Eager KR, Roden LD, Dulhunty AF. Actions of sulfhydryl reagents on single ryanodine receptor Ca^{2+}-release channels from sheep myocardium. *Am J Physiol Cell Physiol* 1997;**272**: C1908–C1918.

144. Ondrias K, Borgatta L, Kim DH *et al.* Biphasic effects of doxorubicin on the calcium release channel from sarcoplasmic reticulum of cardiac muscle. *Circ Res* 1990;**67**:1167–1174.

145. Koshita M, Miwa K, Oba T. Sulfhydryl oxidation induces calcium release from fragmented sarcoplasmic reticulum even in the presence of glutathione. *Experientia* 1993;**49**:282–284.

146. Meissner G. Ryanodine activation and inhibition of the Ca^{2+} release channel of sarcoplasmic reticulum. *J Biol Chem* 1986;**261**:6300–6306.

147. Williams AJ. Ion conduction and discrimination in the sarcoplasmic reticulum ryanodine receptor/calcium-release channel. *J Muscle Res Cell Mot* 1992;**13**:7–26.

148. Lindsay ARG, Tinker A, Williams AJ. How does ryanodine modify ion-handling in the sheep cardiac sarcoplasmic reticulum Ca^{2+}-release channel? *J Gen Physiol* 1994;**104(3)**:425–447.

149. Tinker A, Sutko JL, Ruest L *et al.* Electrophysiological effects of ryanodine derivatives on the sheep cardiac sarcoplasmic reticulum calcium-release channel. *Biophys J* 1996;**70**: 2110–2119.

150. Tanna B, Ruest L, Welch W *et al.* The interaction of 21-amino-9alpha-hydroxy-ryanodine with the sheep cardiac sarcoplasmic reticulum (SR) Ca^{2+}-release channel is influenced by voltage. *J Physiol* 1997;**504**.P:163P–164P.

Chapter 5

Ryanodine Receptor Ion Conduction and Selectivity

Alan J. Williams

WHAT DOES THE RYANODINE RECEPTOR (RyR) CHANNEL DO?

The functional role of the ryanodine receptor is that of an intracellular membrane Ca^{2+}-release channel, elevating cytosolic Ca^{2+} in response to stimuli. The effectiveness of the channel as a Ca^{2+}-release pathway will depend upon two major functional aspects of the protein. Firstly, it will depend upon how the probability of channel opening is modulated or controlled by protein and chemical regulators (these issues are discussed in many chapters of this book). Secondly, and equally important, it will depend upon how the channel discriminates between and conducts permeant ions; in other words, the mechanisms that make it a Ca^{2+} channel.

An efficient Ca^{2+}-release pathway should (a) be selective for Ca^{2+}, (b) be able to translocate large quantities of Ca^{2+} from the storage organelle to the cytosol on a physiologically relevant (millisecond) time scale; at the single-channel level this is equivalent to a high single-channel conductance and (c) be able to maintain this high Ca^{2+} flux at the relatively low concentrations of free Ca^{2+} present within the storage organelle (~1 mM); the release pathway should have a high affinity for Ca^{2+}. Does the RyR channel satisfy these criteria and if it does, what are the mechanisms underlying its function?

CHARACTERIZATION OF ION HANDLING BY THE RyR CHANNEL

As a channel located in an intracellular membrane system, the properties of RyR cannot be determined by conventional electrophysiological techniques such as patch-clamp. However the function of individual RyR channels can be monitored, under voltage clamp conditions, following the incorporation of the channel into an artificial membrane system such as a planar phospholipid bilayer. Initial "reconstitution" studies involved the incorporation of purified vesicles of native junctional sarcoplasmic reticulum (SR) isolated from mammalian skeletal or cardiac muscle.[1–4] While the incorporation of intact SR vesicles permits the examination of the regulation of channel gating, the scope for characterizing ion handling in the channel is severely restricted by the co-incorporation of two other species of SR ion channel, one permeable to monovalent cations and the other to monovalent anions. The characterization of ion handling by a channel requires the investigation of ion translocation under a wide range of ionic conditions and this is not possible if the system contains more than one species of channel. However, such studies are possible using functional, purified RyR channels.

The ion handling properties of ion channels are determined from measurements of the relative permeability, relative conductance and relative affinity of the channel for a range of permeant ions, together with information revealed from block of permeant ion current by impermeant species. A systematic investigation of these properties has been carried out in this laboratory for the sheep cardiac muscle RyR channel and the data from these studies will be described here. In broad terms, information on the ion handling properties of other RyR channels, where it is available, is in agreement with that determined for the sheep cardiac channel. Specific differences will be highlighted as appropriate.

IS THE RyR CHANNEL A Ca^{2+} CHANNEL?

This may seem a strange question to ask, as there is no doubt that in the cell Ca^{2+} is the charge carrying species in this channel. However studies using purified RyR channels have demonstrated that Ca^{2+} is certainly not the only ion that is permeant in this channel. In practice, the RyR channel is permeable to a wide range of divalent and monovalent inorganic cations and to a large number of organic monovalent cations. In the remaining sections of this chapter I will describe the characteristics of translocation of these various classes of cation and consider how the mechanisms underlying these processes contribute to the efficiency of RyR as a Ca^{2+}-release channel.

BASIC ION HANDLING PROPERTIES OF RyR - WHAT IS PERMEANT, WHAT IS IMPERMEANT?

RyR is a Cation Channel

Determinations of single channel current-voltage relationships in asymmetric solutions of KCl yield reversal potentials indistinguishable from those predicted from the Nernst equation for an ideally K^+-selective system.[5] These experiments indicate that RyR is impermeable to anions.

Monovalent Inorganic Cations

In symmetrical solutions of the group 1a monovalent cations, RyR single-channel current varies linearly with holding potential.[6] Slope conductances in solutions containing 210 mM X^+ are shown in Table 5.1 as are permeability ratios (pX^+/pK^+) determined under bi-ionic

Tabel 5.1 Conduction and Permeability of Group Ia Monovalent Cations in RyR

Cation	G_{210} (pS)	pX^+/pK^+	K_D (mM)	G_{max} (pS)
K^+	723	1.00	19.9	900
Na^+	446	1.15	17.8	516
Cs^+	460	0.61	34.0	588
Li^+	215	0.99	9.1	248
Rb^+	621	0.87	ND	ND

ND = not determined.

conditions with K^+ at the luminal side, and X^+ at the cytosolic side of the channel. The table also gives details of the maximal conductance (G_{max}) and the ionic activity at which 50% maximal conductance occurs (K_D), obtained from single-channel conductance-activity relationships.

The striking feature arising from these data is that RyR, nominally a Ca^{2+} channel, is permeable to the group 1a monovalent cations; the channel shows relatively little discrimination between these ions in terms of permeability and, while there is some variation, all these cations carry exceedingly large currents in RyR.

Divalent Inorganic Cations

RyR does, however, discriminate between inorganic monovalent and divalent cations. The relative permeabilities of these two classes of cation have been established under bi-ionic conditions with K^+ as the charge carrier at the cytosolic face of the channel and various alkaline earth divalents as charge carriers on the luminal side of the channel. Under these conditions, the relative permeabilities of these cations (pX^{2+}/pK^+) can be calculated from the observed reversal potential.[7] Similarly, the permeabilities of Ca^{2+}, Mg^{2+} and Sr^{2+} to Ba^{2+} (pX^{2+}/pBa^{2+}) have been determined under bi-ionic conditions with Ba^{2+} at the cytosolic face of the channel and the other divalent at the luminal face of the channel. Data derived from these experiments are given in Table 5.2. Single channel conductance information monitored either as the slope conductance in symmetrical 210 mM solutions for Ba^{2+} and Sr^{2+} or from the slope of the appropriate limb of a current-voltage relationship monitored under bi-ionic conditions with K^+ at the cytosolic face of the channel and the divalent cation at the luminal face of the channel, are given in Table 5.3.[7]

These experiments demonstrate that RyR does show some discrimination between divalent and monovalent inorganic cations ($pX^{2+}/pK^+ \sim 6$), however all the alkaline earth divalents examined are essentially equally permeant.

Table 5.2 Relative Permeability of Divalent Cations in RyR

Cation	E_{rev} vs K^+ (mV)	pX^{2+}/pK^+	E_{rev} vs Ba^{2+} (mV)	pX^{2+}/pBa^{2+}
Ba^{2+}	37.1 ± 0.5	5.8	–	–
Sr^{2+}	39.0 ± 1.2	6.7	2.0 ± 0.4	1.1
Ca^{2+}	38.5 ± 0.6	6.5	1.5 ± 0.6	1.1
Mg^{2+}	37.3 ± 0.4	5.9	2.0	1.1

Table 5.3 Conduction of Divalent Cations in RyR

Cation	Conductance in 210 mM X^{2+}, symmetric (pS)	Conductance in 210 mM X^{2+}, bi-ionic (pS)
Ba^{2+}	199 ± 7	202 ± 5
Sr^{2+}	183 ± 5	166 ± 4
Ca^{2+}	–	135 ± 5
Mg^{2+}	–	89 ± 4

It is not possible to determine conductance in symmetric 210 mM Ca^{2+} or Mg^{2+}. With these concentrations of these cations at the cytosolic face of the channel Po is zero.

As is the case with the monovalent inorganic cations, there is some variation in conductance within the group of divalent cations. In all cases the single channel conductance is high. The demonstration that Mg^{2+}, which has a particularly large dehydration energy, has a permeability equivalent to the other divalents, indicates that dehydration of the permeant ions is unlikely to contribute significantly to discrimination in RyR.

RyR has a high affinity for divalent inorganic cations. The dependence of single channel conductance on Ba^{2+} activity has been determined for the sheep cardiac RyR;[7] yielding a K_D of approximately 400 μM. However, the true value is probably even lower.[8]

Organic Monovalent Cations

In addition to the inorganic cations, RyR is permeable to some organic cations. We have monitored the single channel conductance, determined as slope conductance in symmetrical 210 mM solutions and the relative permeability to K^+ (K^+ at the luminal side, and the organic cation at the cytosolic face of the channel) of a wide range of these cations. Data from some representative cations are given in Table 5.4. A full listing of the organic cations that have been examined is given in Tinker & Williams (1993).[9] Table 5.4 also contains details of the minimum circular radius through which the cation can pass.[9]

Unlike the monovalent and divalent inorganic cations which show little variation in permeability within the group, the organic monovalents display a wide range of relative permeabilities. This is not surprising given the more heterogeneous nature of this group of cations. Relative permeability declines as the radius of the cation increases and cations with a minimum circular radius of ≥ 3.5 Å are impermeant. It would appear that the

Table 5.4 Conduction and Permeability of Organic Monovalent Cations in RyR

Cation	E_{rev} vs K^+ (mV)	pX^+/pK^+	Cond. (pS)	R (Å)
ammonia	-8.8 ± 0.2	1.42	594 ± 4.0	1.7
methylamine	10.1 ± 0.2	0.67	286 ± 5.0	1.9
dimethylamine	28.5 ± 0.3	0.32	132 ± 4.0	2.3
trimethylamine	48.9 ± 0.4	0.14	59.0 ± 2.0	2.8
ethylamine	17.0 ± 0.6	0.51	105 ± 3.0	2.2
propylamine	22.0 ± 0.6	0.42	37.0 ± 1.4	2.3
dimethylamine	59.9 ± 1.2	0.09	10.4 ± 0.6	3.1
ethylethanolamine	38.3 ± 0.6	0.22	23.0 ± 0.1	2.6
diethanolamine	40.3 ± 0.8	0.20	31.0 ± 1.0	2.8
dimethylethanolamine	50.3 ± 0.9	0.14	29.5 ± 0.8	2.9
ethyldiethanolamine	≤ 80	≤ 0.04	≤ 10	3.5
triethylamine	≤ 80	≤ 0.04	≤ 10	3.6
triethanolamine	≤ 80	≤ 0.04	≤ 10	3.6
diethylethanolamine	≤ 80	≤ 0.04	≤ 10	3.6

permeability of these cations is determined by a sieving mechanism at a selectivity filter located towards the luminal end of the conduction pathway.[9] Application of the excluded area hypothesis to this data yields a minimum radius of the sheep cardiac RyR conduction pathway of ~3.4 Å.[9]

At first sight it might appear that the conductance of the organic cations in RyR is controlled by a similar sieving mechanism. However a closer inspection of the data reveals some inconsistencies in this simple picture; cations with similar relative permeability do not necessarily have similar conductance. As an example, consider trimethylamine and diethylmethylamine; cations with similar relative permeability but very different conductance. An inspection of the other ion handling properties of these two cations revealed that the lower single channel conductance of diethylmethylamine reflects a higher relative affinity of the conduction pathway for this ion. Overall, transport of the organic cations is governed by a selection process based on size at a selectivity filter and binding to a hydrophobic binding site.[9]

BLOCK OF RyR

Organic cations of low permeability in RyR (e.g. trimethylamine and diethylmethylamine), will reduce the rates of translocation of more permeant cations; i.e. they act as permeant blockers, by competing for occupancy of the conduction pathway.[9]

Many larger organic cations, although impermeant in RyR, do interact with the conduction pathway of the channel and influence the conduction of permeant cations. The addition of millimolar concentrations of small tetraalkyl ammonium (TAA) ions (e.g. tetramethyl ammonium (TMA), tetraethyl ammonium (TEA), tetrapropyl ammonium (TPrA)) and charged local anesthetics (QX222 & procaine) bring about a reduction of monovalent cation conduction. In all cases, the blocking cation is only effective when added to the cytosolic face of the channel and the reduction in single channel current is dependent upon voltage; the degree of block increases as holding potential is made more positive.[6,10,11] Quantitative analysis of voltage-dependent block of K^+ current by these cations indicates the presence of two sites of interaction within the voltage drop across the channel. The smallest TAA, TMA, interacts with a site approximately 50% across the voltage drop from the cytosolic face of the channel. Access to this site appears to be limited by steric factors; increasing the dimensions of TMA by the substitution of either an ethyl or a propyl group for one of the methyl groups of TMA increases the voltage-dependence of block. The site of interaction of these substituted TMAs, together with the larger, TAAs, (TEA and TPrA) and the local anesthetics, is located approximately 90% across the voltage drop from the cytosolic face of the channel.[10] Determination of the zero-voltage dissociation constant ($K_b(0)$) of the TAAs interacting with the 90% site indicates that the affinity of the site for the TAA increases as the number of methyl and methylene groups increases suggesting the involvement of hydrophobic interactions at this site.[10]

Still larger TAAs, tetrabutyl ammonium (TBA) and tetrapentyl ammonium (TPeA) also reduce the conduction of permeant cations in RyR.[12] As with the smaller members of this group, block is dependent upon voltage and the blocking cations are effective only from the cytosolic face of the channel. However these cations do not produce a straightforward reduction in single channel conductance, rather, in the presence of TBA and TPeA channel opening events occur as a mixture of normal, full amplitude, events and opening events to a conductance level considerably less than normal. The relative conductance of the reduced state does not vary with holding potential but the probability of occurrence of the state rises as the holding potential is made more positive and as the concentration of the TAA is increased. It seems probable that these cations induce a form of partial block which results from the interaction of the cations with sites at the cytosolic extremity of the voltage drop across the channel. Again, hydrophobic interactions appear to play a role in the stabilisation of these cations at their binding sites.[12]

THE DIMENSIONS OF THE CONDUCTION PATHWAY AND MECHANISMS OF ION HANDLING IN RyR

At the outset I should define what is meant by "conduction pathway"; in this chapter this is the region of the channel protein in which ions leave an environment in which their movement is governed by aqueous diffusion and fall under the influence of the trans-membrane electric field.[13] The physical dimensions of the conduction pathway of an ion channel will clearly be a major factor in the regulation of ion translocation. As outlined above, measurements of the relationship between the relative permeability and the minimum circular radius in which an organic monovalent cation can be enclosed have indicated that the conduction pathway of RyR has a minimum radius of approximately 3.4 Å. Equally important in determining ion handling will be the length of the conduction pathway; the distance over which the trans-membrane potential falls. Sites within the conduction pathway are commonly identified in terms of electrical distance; for example, measurements of the variation in block of K^+ current by TEA with holding potential in RyR indicate that this cation interacts with a site ~90% into the voltage drop from the cytosolic side of the channel. How is electrical distance related to physical distance in RyR? An estimate of the length of the voltage drop in RyR has been obtained by investigating the blocking parameters of a range of monovalent and divalent trimethyl ammonium derivatives of varying length (monovalent: $CH_3\text{-}(CH_2)_{n-1}\text{-}N^+(CH_3)_3$; divalent: $(CH_3)_3N^+\text{-}(CH_2)_n\text{-}N^+(CH_3)_3$); a technique used previously to monitor the lengths of the voltage drops in the SR K^+ channel[14] and the Ca^{2+}-activated K^+ channel.[15] Both sets of cations are voltage-dependent blockers of K^+ current when added to the cytosolic face of RyR.[16] Quantitative analysis of block by monovalent derivatives indicates that the effective valence of block is not influenced by varying chain length. In all cases a value of 0.9 was obtained indicating that the trimethyl ammonium group of each derivative interacts with a site located 90% of the way across the voltage drop; the site of interaction of TEA. A different picture was obtained with the divalent derivatives; increasing chain length from n=2 to n=7 produced a linear decrease in effective valence from a value of approximately 1.5 until at n=7 this parameter approached the value determined for the monovalent derivatives. Beyond this chain length (n=8 and n=9), effective valence did not change. To interpret these findings it was assumed that on entering the conduction pathway one of the trimethyl ammonium groups of the divalent derivatives interacts with the site located 90% of the way across the voltage drop, i.e. the site of interaction of the trimethyl ammonium group of the monovalent derivatives. The second trimethyl ammonium group will then act a sensor of the trans-membrane potential. With very short chain derivatives this second group will be positioned very close to the bound group and the monitored effective valence will reflect this. As chain length increases, the location of the second trimethyl ammonium group will approach the cytosolic limit of the voltage drop and effective valence will decrease accordingly. Eventually, with increasing chain length, the second trimethyl ammonium group should extend beyond the voltage drop and the monitored effective valence should plateau. As described above, this occurs in RyR when n=7. By measuring the N^+-N^+ distance for the divalent derivatives it is then possible to provide an estimate of the physical length of the voltage drop. Assuming that the voltage drop in RyR is linear, this calculation reveals that the electrical potential across the channel falls over a distance of 10.4 Å.[16]

Combining the findings of the two approaches outlined in this section it would appear that RyR contains a region of approximately 10 Å in length within which ions will be influenced by trans-membrane voltage, this is presumably equivalent to the region of the channel in which permeant ions come into contact with polar residues of the channel protein and the bulk of "selection" takes place. In agreement with this observation, the measurement of streaming potentials in the canine cardiac RyR indicates that the region of the channel in which permeant ions and water are constrained to diffuse in single file is likely to have a length of approximately 10 Å.[17] The minimum radius of the pathway within this region is approximately 3.4 Å. Studies with voltage-dependent blockers of ion conduction in RyR establish that this region may be considered as a constriction or "filter" located at the luminal extremity of the conduction pathway. A number of impermeant TAAs interact with a site 90% into the voltage drop from the cytosolic face of the voltage drop. These TAAs are ineffective when added to the luminal face of the channel, presumably as they are prevented from reaching their binding site by the filter. The RyR selectivity filter is large in comparison with the minimum pore radii determined for voltage-dependent Na^+ and K^+ channels and given the essentially equal permeability within the group 1a monovalent cations and within the alkaline earth divalent cations it seems probable that physical sieving plays little or no part in selection of these inorganic ions within RyR. The translocation of these ions may not require the substitution of all hydrating water molecules by polar residues of the conduction pathway. This would be consistent with the observation that Mg^{2+}, an ion with a high hydration energy, has a high conductance in RyR.

These estimates of the physical dimensions of the RyR conduction pathway also indicate that this region is likely to be only a very a small portion of the overall complex visualised in three-dimensional reconstructions of RyR.[18,19] The location of the conduction pathway within the RyR complex is not known, however recent studies in which a carboxyl-terminal portion of RyR1, containing the putative trans-membrane regions of the protein, has been expressed in CHO cells have established that this protein functions as an ion channel displaying ion handling properties similar to those of the full length RyR.[20]

The structural features discussed above are summarised in Fig. 5.1. How might these features contribute to the rather unusual ion handling properties of RyR? In keeping with its role as an effective Ca^{2+}-release pathway, rates of ion translocation in RyR are very high; in fact for several cations saturating rates are above maximal theoretical values calculated for selective systems (Hille, (1992) Chapter 12 page 333).[21] As suggested by Latorre and Miller (1983),[13] the conductance of any channel could be maximised if the length of the pore in which restricted diffusion takes place is short; this will optimize the rate of exit from the channel. Similarly, a structure with a large capture radius; effectively the difference between the radius of the opening of the conduction pathway and the radius of the ion, could maximize ion entry into the channel by overcoming diffusion-limited access to the conduction pathway. In theory, ion entry could be further enhanced by dipole layers at the mouth of the conduction pathway which would provide a means of preliminary selection of ions and an enrichment of ions above the concentration of the bulk solution.[22] Based on the evidence set out above it is probable that such physical features underlie the ion handling properties of RyR, contributing to the observed, enormous, single channel conductance. The conduction pathway is short and the capture radius, at least at the cytosolic face of the channel, appears to be large. Recent studies in which it has been demonstrated that K^+ channel N-type inactivation peptides interact with the conduction pathway of RyR; blocking K^+ conduction in a voltage-dependent manner from the cytosolic face of the channel, indicate that the rate of entry of the inactivation peptide to its blocking site is enhanced by modifications in peptide structure that result in an increase in the net positive charge of the peptide. This result indicates that the cytosolic mouth of the conduction pathway of RyR is likely to contain significant negative charge.[23,24] In addition, experiments in which carboxyl neutralizing agents have been used to modify ion handling in canine cardiac RyR channels have been interpreted as suggesting that negative surface charge is present in the vicinity of the luminal mouth of the channel but outside the electric field of the membrane.[25] The rate of ion translocation through RyR might also be enhanced if the functional channel complex contained multiple conductance pathways gating in a co-ordinated fashion (see later).

The ion handling properties of a channel will be determined by the interaction of ions with the conduction pathway of the protein and the possible interactions of ions within this region. Channels are classified as either single-ion or multi-ion. In the former, the conduction pathway of the channel can be occupied by only one ion at a time. In the latter group the

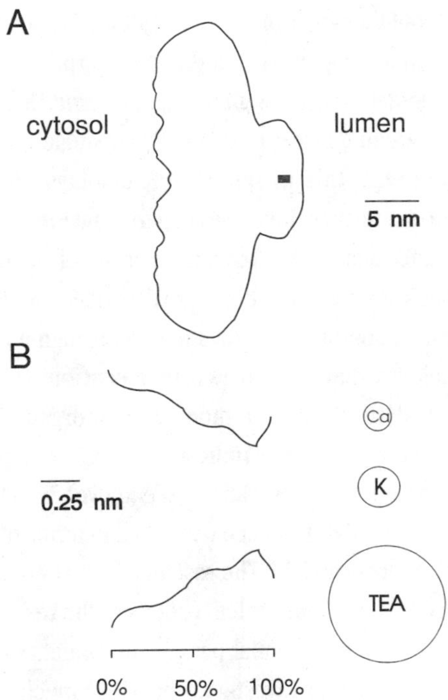

Fig. 5.1 (A) The relative dimensions of RyR (as determined in 3-D reconstruction) and the conduction pathway (as determined using methods described in the chapter). The long axis of the shaded oblong represents the distance over which voltage drops and the short axis represents the minimum diameter of the pathway. The exact location of the pathway within the trans-membrane region of the RyR complex is unknown. It is possible that the functional channel consists of multiple pathways with identical ion handling properties. (B) A representation of the conduction pathway of RyR. As in (A) the long axis corresponds to the length of the voltage drop. The minimum pore diameter is represented by the constriction located in the last 10% of the voltage drop at the luminal end of the pathway; this is the selectivity filter. The fine structure of the pathway is entirely fanciful and in any case should probably not be thought of as a rigid structure, however it is probable that the pathway widens somewhat from the filter. This would be required for large blocking cations such as TEA to gain access to the site of interaction located 90% in to the voltage drop from the cytosolic side of the channel; hydrophobic interactions play a role in stabilising blocking cations at this site. Blocking studies have also identified a site 50% into the voltage drop at which TMA binds. This is the location of the high affinity site for monovalent and divalent cations which is a feature of the rate theory model of conduction in RyR. The cytosolic mouth of the pathway is likely to be the region in which large organic cation interactions result in partial block of conduction and in which K^+-channel N-type inactivation peptides bind. Again hydrophobic interactions play a role here. We have proposed that the conduction pathway contains a high density of negative charge; such an arrangement would favour the passage of divalent rather than monovalent cations. The dimensions of Ca^{2+}, K^+ and TEA^+ are shown for reference.

conduction pathway can contain two or more ions at a time, a situation which can result in interactions between ions and complex ion handling properties.[21] As discussed by Latorre and Miller (1983),[13] a channel in which high unit conductance is achieved as the result of a short conduction pathway would be expected to display single-ion occupancy due to ion-ion repulsion. Consistent with this proposal RyR displays ion handling properties characteristic of single-ion occupancy. Increases in permeant ion activity result in saturation of single channel conductance that can be described in terms of Michaelis-Menten kinetics.[6] The relative permeability of RyR for various ions is not dependent upon the total concentration of the ions.[7,9] Single channel conductance does not go through a minimum or maximum value with variations in mole-fraction of mixtures of permeant ions (anomalous mole-fraction dependence).[6,7] Finally, analysis of block of monovalent inorganic cation conductance in RyR by small tetraalkyl ammonium cations indicates that block results from occupation of the conduction pathway by one blocking cation (see above).[6,10] Consistent with this, the translocation of cations through RyR has been described in terms of a conventional single-ion occupancy Eyring rate theory model.[8] The essential features of this model are that RyR discriminates between divalent and monovalent cations as the result of differential binding to a central binding site and the preferential passage of divalent cations between binding sites. This model has been used to investigate possible mechanisms involved in the translocation of various classes of permeant ion in RyR.[8,9,26–28] When considering barrier models such as this it is important to remember that they do not provide information on the physical structures involved in translocation. As highlighted by Moss and Moczydlowski,[29] they are "primarily useful in determining whether a particular mechanism can account for a particular phenomenon." Potentially more physically realistic descriptions of translocation are provided by models based on diffusion theory and one such model has recently been used to describe K^+ translocation in RyR.[30]

FACTORS THAT MODIFY ION HANDLING IN RyR

As outlined above, the translocation of permeant ions in RyR can be modified by less permeant cations such as trimethylamine or by impermeant cations such as TEA. These ions behave as conventional, voltage-dependent, blockers. However, the ion handling properties of RyR can also be affected by processes involving mechanisms that are inconsistent with classical block.

Ryanodine and Ryanodine Derivatives

As is discussed elsewhere in this book, the binding of ryanodine to its high affinity site on the channel protein results in altered function; open probability rises dramatically and single channel conductance changes.[27,31,32] Investigations in which the ion handling properties of the ryanodine-modified channel were assessed have revealed that altered conductance involves changes in various aspects of ion translocation.[27] In these studies, the conductance of the ryanodine-modified channel was expressed as a fraction of the conductance of the channel in the absence of ryanodine (fractional conductance - FC). FC is not constant; its value varies with the nature of the permeant ion. Values of FC for the group 1a monovalent cations are essentially the same, ranging from 0.60 to 0.66; while values for divalent alkaline earth cations are significantly lower (Ba^{2+}, 0.22 and Sr^{2+}, 0.28). With organic monovalent cations as the permeant species, FC varied considerably (ammonia, 0.66; ethylamine, 0.76; propanolamine, 0.65, diethanolamine, 0.92; diethylamine, 1.2). In other words, with some organic monovalent cations the interaction of ryanodine results in an increased single channel conductance rather than a reduced conductance state; ryanodine does not simply block the channel. In an attempt to uncover the mechanisms underlying these variations in FC, experiments were carried out to determine various parameters of ion handling in the ryanodine-modified channel and these were compared with values obtained for the channel in the absence of ryanodine.

When the high affinity binding site is occupied by ryanodine the affinity of the conduction pathway for the group 1a monovalent cations is increased (e.g. K_D for K^+ is 20 mM in the unmodified channel and 11.6 mM in the modified channel). The relative permeability of the group 1a monovalent cations is essentially unaffected by the interaction of ryanodine. The affinity of the conduction pathway of RyR for divalent inorganic cations increases following the interaction of ryanodine and this is accompanied by a decrease in the permeability of the channel for this class of cations relative to K^+, in comparison to that monitored in the absence of ryanodine. The influence of ryanodine binding on the translocation of organic monovalent cations is influenced by both the size and the properties of the cation. As with the group 1a monovalent cations, the permeability, relative to K^+, of the smaller organic monovalent cations (e.g. ammonia and ethylamine) is essentially unaffected by ryanodine. Larger cations (e.g. diethanolamine and diethylamine) have significantly greater relative permeabilities in the ryanodine-modified channel. The FC of these cations is also influenced by the relative importance of interactions with both hydrophobic and hydrophilic sites within the conduction pathway of the channel.[27]

These findings were interpreted as indicating that the modification in ion handling that occurs on the interaction of ryanodine with its high affinity binding site on the channel protein results from an alteration in conformation. It was suggested that in physical terms the observed alterations in ion handling might arise if, with ryanodine bound, the conduction pathway of RyR adopted a conformation in which the field strength of the principal cation binding site was increased, the density of negative charge in the conduction pathway was decreased and the minimum pore radius was increased.[27] Consistent with this last proposal, measurements of streaming potentials indicate that ryanodine-modification results in an increase in the radius of the narrow region of the conduction pathway of RyR.[17]

Changing the structure of ryanodine changes FC. With K^+ as charge carrier, FC was determined for a number of ryanodine derivatives; in all cases the interaction of the ryanoid with the channel resulted in the occurrence of a high open probability state of reduced conductance. However, the amplitude of the ryanoid-induced conductance state varied widely with FC values of ~0.06 for guanidino propionyl ryanodine to ~0.7 with ryanodol.[33] Comparative molecular field analysis was used to analyze the relationship between the structure of these ryanoids and FC and these studies revealed strong correlations between FC and specific structural loci on the ryanoid. In addition, these investigations suggest that the modification of ion handling in RyR observed on the binding of ligand to the high affinity site on the channel protein involves an allosteric mechanism in which the ryanoid induces or stabilizes a conformational change in the protein. The analysis revealed no evidence for a direct interaction between the ryanoids and the translocated cations.[34]

Removal of the FK-506 Binding Protein (FKBP)

In situ, RyR exists as a member of a complex of proteins involved in Ca^{2+} release.[35] The FK-506 binding proteins are accessory proteins that co-purify with RyR at a stoichiometry of one FKBP per RyR monomer.[36] RyR1 and RyR2 associate with different isoforms of FKBP. FKBP12 is the predominant isoform in the cytosol of both skeletal and cardiac muscle cells, however RyR1 preferentially associates with FKBP12 while RyR2 associates only with FKBP12.6.[37] The possibility that these proteins may play a role in regulating the ion handling properties of RyR was highlighted by the observation that on reconstitution, RyR1, purified following expression in SF9 cells, formed ion channels displaying large numbers of sub-conductance events. Co-expression of FKBP12 markedly increased the incidence of channels displaying full conductance events. The displacement of the FKBP by FK506 or rapamycin from this latter group of channels increased the likelihood of the occurrence of sub-conductance states. Brillantes *et al.*[38] concluded that FKBP stabilizes the function of RyR by in some way co-ordinating the gating of the monomers in each functional homotetramer. In such a scheme, the sub-conductance states observed in the absence of FKBP would represent ion flow through uncoordinated monomers of the tetramer.[36]

In subsequent investigations it was demonstrated that the addition of FK506[39] or rapamycin[40] to native rabbit skeletal RyR channels increased the occurrence of sub-conductance states. Similarly, sub-conductance states have been reported to be induced by rapamycin in native canine cardiac RyR channels[41] and by FK506 in native rat cardiac channels.[42] In all of these investigations FK506 and/or rapamycin also increased single channel Po.

The removal of FKBPs from RyR channels is not always associated with the occurrence of sub-conductance states; Timerman *et al.*[43] found that removal of FKBP from canine cardiac RyR with FK590 had no effect on channel function. Barg *et al.*[37] have described a series of experiments in which rabbit RyR1 and canine RyR2 channels were depleted of FKBP with FK590. This procedure resulted in an increase in Po of RyR1 channels which could be reversed by the addition of either FKBP12 or FKBP12.6. No subconductance states were seen in these experiments. These authors reported that FKBP depletion of RyR2 produced no modification of channel function. Similarly, the subsequent addition of exogenous FKBP12 or 12.6 was without effect.

While there is a widely held opinion that FKBPs influence ion handling in RyR by co-ordinating gating, presumably of four individual conductance pathways associated with each of the RyR monomers,[40] the mechanisms involved in the production of sub-conductance states on the removal of these accessory proteins are still far from clear.

How Many Conduction Pathways?

The possibility that the ion handling properties of RyR monitored in single channel experiments represents the co-ordinated gating of multiple conductance pathways is a recurrent theme in this chapter. Some preparations of purified RyR show multiple conductance states;[22] some studies indicate that the removal of FKBP results in the occurrence of multiple conductance states.[36,40] Underlying these observations, the structural features revealed by 3-D reconstructions of cryo-electron micrographs of purified RyRs, have provided hints that the structure may contain more than one trans-membrane pathway. This possibility of multiple conductance pathways in RyR was addressed in experiments carried out by Quinn and Ehrlich.[44] These authors observed a time-dependent, step-wise reduction in single channel current amplitude when RyR channels were exposed to sulfhydryl-reactive compounds. The likelihood of the transition from one conductance state to the next was not random; the time between transitions increased during continued exposure to the reagent suggesting that the successive modifications of ion handling were not independent events, as would be expected if the step-wise reduction in current resulted from the chemical modification and block of separate conduction pathways. Rather, Quinn and Ehrlich argued that the progressive reduction in single channel current reflected the successive chemical modification of residues within a single conduction pathway.

SUMMARY

RyR is a Ca^{2+}-release channel; its function is to provide a pathway for the rapid translocation of large amounts of Ca^{2+} from an intracellular storage organelle to the cytosol. To fulfil this task, RyR has an unusually high single channel conductance. The investigations outlined in this chapter have provided us with some information on the structure and properties of the conduction pathway of RyR that underlie this process.

ACKNOWLEDGEMENTS

The work from my laboratory described in this chapter has been funded by The British Heart Foundation, The Medical Research Council and the Wellcome Trust. I am grateful to Andrew Tinker, Allan Lindsay, Rebecca Sitsapesan, Fiona Mead and Andrew Griffin for their contributions to these investigations.

REFERENCES

1. Smith JS, Coronado R, Meissner G. Sarcoplasmic reticulum contains adenine nucleotide-activated calcium channels. *Nature* 1985;**316**:446–449.
2. Rousseau E, Smith JS, Henderson JS, *et al.* Single channel and $^{45}Ca^{2+}$ flux measurements of the cardiac sarcoplasmic reticulum calcium channel. *Biophys J* 1986;**50**:1009–1014.
3. Nagasaki K, Fleischer S. Ryanodine sensitivity of the calcium release channel of sarcoplasmic reticulum. *Cell Calcium* 1988;**9**:1–7.
4. Holmberg SRM, Williams AJ. Single channel recordings from human cardiac sarcoplasmic reticulum. *Circ Res* 1989;**65**:1445–1449.
5. Lindsay ARG, Williams AJ. Functional characterisation of the ryanodine receptor purified from sheep cardiac muscle sarcoplasmic reticulum. *Biochim Biophys Acta* 1991;**1064**:89–102.

6. Lindsay ARG, Manning SD, Williams AJ. Monovalent cation conductance in the ryanodine receptor-channel of sheep cardiac muscle sarcoplasmic reticulum. *J Physiol* 1991;**439**: 463–480.
7. Tinker A, Williams AJ. Divalent cation conduction in the ryanodine receptor-channel of sheep cardiac muscle sarcoplasmic reticulum. *J Gen Physiol* 1992;**100**:479–493.
8. Tinker A, Lindsay ARG, Williams AJ. A model for ionic conduction in the ryanodine receptor-channel of sheep cardiac muscle sarcoplasmic reticulum. *J Gen Physiol* 1992;**100**: 495–517.
9. Tinker A, Williams AJ. Probing the structure of the conduction pathway of the sheep cardiac sarcoplasmic reticulum calcium-release channel with permeant and impermeant organic cations. *J Gen Physiol* 1993;**102**:1107–1129.
10. Tinker A, Lindsay ARG, Williams AJ. Block of the sheep cardiac sarcoplasmic reticulum Ca^{2+}-release channel by tetraalkyl ammonium cations. *J Membr Biol 1992*;**127**:149–159.
11. Tinker A, Williams AJ. Charged local anesthetics block ionic conduction in the sheep cardiac sarcoplasmic reticulum calcium-release channel. *Biophys J* 1993;65:852-864.
12. Tinker A, Lindsay ARG, Williams AJ. Large tetraalkyl ammonium cations produce a reduced conductance state in the sheep cardiac sarcoplasmic reticulum Ca^{2+}-release channel. *Biophys J* 1992;**61**:1122–1132.
13. Latorre R, Miller C. Conduction and selectivity in in potassium channels. *J Membr Biol* 1983;**71**:11–30.
14. Miller C. Bis-Quaternary ammonium blockers as structural probes of the sarcoplasmic reticulum K^+ channel. *J Gen Physiol* 1982;**79**:869–891.
15. Villarroel A, Alvarez O, Oberhauser A, *et al.* Probing a Ca^{2+}-activated K^+ channel with quaternary ammonium ions. *Pflugers Arch* 1988;**413**:118–126.
16. Tinker A, Williams AJ. Measuring the length of the pore of the sheep cardiac sarcoplasmic reticulum calcium-release channel using trimethylammonium ions as molecular calipers. *Biophys J* 1995;**68**:111–120.
17. Tu Q, Velez P, Brodwick MS, *et al.* Streaming potentials reveal a short ryanodine-sensitive selectivity filter in cardiac Ca^{2+} release channel. *Biophys J* 1994;**67**:2280–2285.
18. Radermacher M, Rao V, Grassucci R, *et al.* Cryo-electron microscopy and three-dimensional reconstruction of the calcium release channel/ryanodine receptor from skeletal muscle. *J Cell Biol* 1994;**127**:411–423.
19. Orlova EV, Serysheva II, Van Heel M, *et al.* Two structural configurations of the skeletal muscle calcium release channel. *Nature-Str Biol* 1997;**3**:547–551.
20. Bhat MB, Zhao JY, Takeshima H, *et al.* Functional calcium release channel formed by the carboxyl-terminal portion of ryanodine receptor. *Biophys J* 1997;**73**:1329–1336.
21. Hille B. *Ionic channels of excitable membranes*. 2nd ed. Sunderland, Mass.: Sinauer, 1992.
22. Williams AJ. Ion conduction and discrimination in the sarcoplasmic reticulum ryanodine receptor/calcium-release channel. *J Muscle Res Cell Mot* 1992;**13**:7–26.
23. Mead FC, Tinker A, Sullivan D, *et al.* Interaction of the shaker B inactivation peptide with the ryanodine-modified SR Ca^{2+}-release channel. *Biophys J* 1995;**68(2)**:A376. (Abstract)

24. Mead FC, Sullivan D, Williams AJ. Block of the ryanodine-modified cardiac sarcoplasmic reticulum (SR) Ca^{2+} release channel by shakerB K^+ channel N-type inactivation peptides is influenced by net charge of the peptide. *J Physiol* 1997;**504P**:162P. (Abstract)
25. Tu Q, Velez P, Cortes-Gutierrez M, *et al.* Surface charge potentiates conduction through the cardiac ryanodine receptor channel. *J Gen Physiol* 1994;**103**:853–867.
26. Tinker A, Lindsay ARG, Williams AJ. Cation conduction in the cardiac sarcoplasmic reticulum calcium-release channel under physiological and pathophysiological conditions. *Cardiovas Res* 1993;**27**:1820–1825.
27. Lindsay ARG, Tinker A, Williams AJ. How does ryanodine modify ion-handling in the sheep cardiac sarcoplasmic reticulum Ca^{2+}-release channel? J *Gen Physiol* 1994;**104(3)**:425–447.
28. Tripathy A, Meissner G. Sarcoplasmic reticulum lumenal Ca^{2+} has access to cytosolic activation and inactivation sites of skeletal muscle Ca^{2+} release channel. *Biophys J* 1996;**70**:2600–2615.
29. Moss GWJ, Moczydlowski E. *Concepts of single-channel analysis: inferring function from fluctuations.* In: Ashley RH., ed.Ion Channels. A Practical Approach. Oxford, New York, Tokyo: IRL Press at Oxford University Press, **1995**:69–112.
30. Chen D, Xu l, Tripathy A, *et al.* Permeation through the calcium release channel of cardiac muscle. *Biophys J* 1997;**73**:1337–1354.
31. Rousseau E, Smith JS, Meissner G. Ryanodine modifies conductance and gating behavious of single Ca^{2+} release channel. *Am J Physiol* 1987;**253**:C364–368.
32. Ashley RH, Williams AJ. Divalent cation activation and inhibition of single calcium release channels from sheep cardiac sarcoplasmic reticulum. *J Gen Physiol* 1990;**95**:981–1005.
33. Tinker A, Sutko JL, Ruest L, *et al.* Electrophysiological effects of ryanodine derivatives on the sheep cardiac sarcoplasmic reticulum calcium-release channel. *Biophys J* 1996;**70**:2110–2119.
34. Welch W, Williams AJ, Tinker A, *et al.* Structural components of ryanodine responsible for modulation of sarcoplasmic reticulum calcium channel function. *Biochemistry* 1997;**36**: 2939–2950.
35. Zhang L, Kelley J, Schmeisser G, *et al.* Complex formation between junction, triadin, calsequestrin, and the ryanodine receptor - Proteins of the cardiac junctional sarcoplasmic reticulum membrane. *J Biol Chem* 1997;**272**:23389–23397.
36. Marks AR. Cellular functions of immunophilins. *Physiol Rev* 1996;**76**:631–649.
37. Barg S, Copello JA, Fleischer S. Different interactions of cardiac and skeletal muscle ryanodine receptors with FK-506 binding protein isoforms. *Am J Physiol* 1997;**272**:C1726–C1733.
38. Brillantes A-MB, Ondrias K, Scott A, *et al.* Stabilization of calcium release channel (ryanodine receptor) function by FK506-binding protein. *Cell* 1994;**77**:513–523.
39. Ahern GP, Junankar PR, Dulhunty AF. Single channel activity of the ryanodine receptor calcium release channel is modulated by FK-506. *FEBS Lett* 1994;**352**:369–374.
40. Ahern GP, Junankar PR, Dulhunty AF. Subconductance States in Single-channel Activity of Skeletal Muscle Ryanodine Receptors After Removal of FKBP12. *Biophys J* 1997;**72**:146–162.
41. Kaftan E, Marks AR, Ehrlich BE. Effects of rapamycin on ryanodine receptor Ca^{2+}-release channels from cardiac muscle. *Circ Res* 1996;**78**:990–997.
42. Xiao R-P, Valdivia HH, Bogdanov K, *et al.* The immunophilin FK506-binding protein modulates calcium release channel closure in rat heart. *J Physiol* 1997;**500**:343–354.

43. Timerman AP, Onoue H, Xin HB, *et al.* Selective binding of FKBP12.6 by the cardiac ryanodine receptor. *J Biol Chem* 1996;**271**:20385–20391.
44. Quinn KE, Ehrlich BE. Methanethiosulfonate derivatives inhibit current through the ryanodine receptor/channel. *J Gen Physiol* 1997;**109**:255–264.

Chapter 6

Ryanodine Binding Sites on the Sarcoplasmic Reticulum Ca^{2+} Release Channel

Irina I. Serysheva and Susan L. Hamilton

INTRODUCTION

One of the most important contributions to our ability to study the structure and function of the sarcoplasmic reticulum (SR) Ca^{2+} release channels was the discovery of the plant alkaloid ryanodine (Fig. 6.1). Ryanodine, initially discovered in the search for a better insecticide, is found in the stem and root of *Ryania speciosa*[1] and induces muscle paralysis in insects. Ryanodine alters Ca^{2+} release from the sarcoplasmic reticulum[2,3] by chonging the conductance and gating properties of the sarcoplasmic reticulum Ca^{2+} channels.[4] Ryanodine also binds to the channel with high specificity and affinity.[5,6] These findings led to the hypothesis that ryanodine was a highly specific modulator of the skeletal muscle Ca^{2+} release channel.

Skeletal muscle contraction is initiated by depolarization of the transverse (T) tubule membrane and the subsequent release of Ca^{2+} from the SR. The release of Ca^{2+} occurs

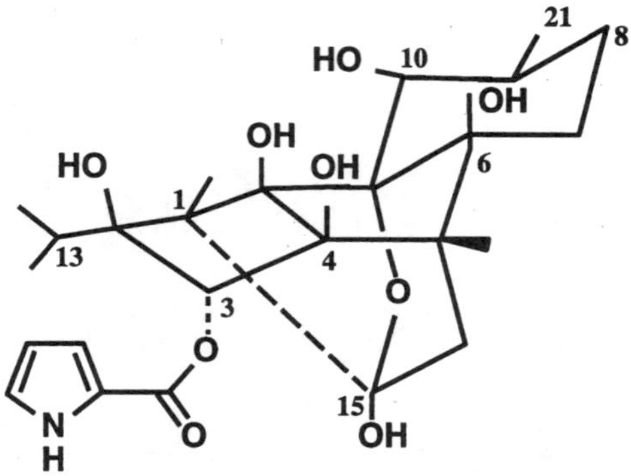

Fig. 6.1 Ryanodine.

through channels located in the specialized junctional regions of the sarcoplasmic reticulum. The purified ryanodine receptor forms a large conductance cation channel upon reconstitution into planar lipid bilayers and its pharmacological properties are similar to those of the native Ca^{2+} release channel.[7] The purified protein also has a morphology closely resembling the "foot" structure[6] observed originally *in situ* by Franzini-Armstrong[8] in the gap between the junctional face membrane of the terminal cisternae SR and the transverse tubule. These studies established that the ryanodine receptor, the Ca^{2+} release channel, and the foot structure are the same protein.

Several groups utilized [^{3}H]ryanodine binding to identify and purify the ryanodine receptor (RyR) from the junctional sarcoplasmic reticulum.[6,9–11] Sequence from the purified protein then led to the cloning of the cDNA and the elucidation of the primary sequence of this protein.[12,13] The ryanodine receptor (RyR) from the skeletal muscle sarcoplasmic reticulum is a homotetramer with 4-fold symmetry composed of subunits with molecular mass of 565,000 Da.[12]

USE OF [^{3}H]RYANODINE BINDING TO PROBE CHANGES IN THE FUNCTIONAL STATE OF RyR

Because of its specificity and high affinity for interacting with and modulating RyR, ryanodine's interaction with RyR has been extensively studied. It has served as a powerful biochemical tool to study Ca^{2+} release channel structure and function. The binding of [^{3}H]ryanodine to the Ca^{2+} release channel is, however, quite complex. Equilibrium and kinetic analyses of the binding of [^{3}H]ryanodine to skeletal and cardiac SR have demonstrated the existence of two classes of sites: a low affinity and a high affinity site.[9,14–23] There is apparently only a single high affinity site for [^{3}H]ryanodine per receptor tetramer,[9,22] but depending on the binding conditions, the apparent dissociation constant (K_d) of this site can vary from 2 nM to 200 nM. One possible explanation for the range of apparent K_d values is that the measured value represents a weighted average of the K_ds for [^{3}H]ryanodine binding to different, interconvertible states of the channel.[20,21,24] A model for RyR in which there are 2 or more interconvertible states with different affinities for [^{3}H]ryanodine would predict linear Scatchard plots with the measured K_d reflecting the proportion of RyR which are in the conformation favored for binding. The favored conformation appears to be the open state of the channel.[22,24,25] If this model is correct, then modulators which open the channel should increase the apparent affinity of RyR for [^{3}H]ryanodine while inhibitors, by decreasing channel opening, should lower the apparent affinity. Indeed, channel activators have been shown to enhance [^{3}H]ryanodine binding while inhibitors decrease binding.[18,24–27] Channel activators increase the affinity of the channel for [^{3}H]ryanodine by increasing its rate of association with the binding site.[24] Inhibitors decrease the rate of association[18,24,27–29] and, as will be discussed below, can also influence the rate at which [^{3}H]ryanodine dissociates from its binding site.

An additional complication in interpreting [^{3}H]ryanodine binding is the presence of the low affinity binding sites.[14,21–23] These sites have K_ds for [^{3}H]ryanodine which are greater than 1 μM. The low affinity sites require high concentrations of [^{3}H]ryanodine to reach saturation and the bound ligand may dissociate rapidly,[23] making it difficult to accurately quantitate the number of low affinity sites. As a result, there is considerable controversy concerning the relationship and the ratio of high to low affinity ryanodine binding sites. Two groups[14,21] reported a ratio of 1:3 for high affinity to low affinity binding sites and have suggested that the low affinity sites arise from the negatively co-operative binding of [^{3}H]ryanodine at one of four initially identical sites.[14,19–21] Negative co-operativity in binding of a radioligand requires that the presence of excess unlabeled ligand increase the dissociation rate of the bound radioligand and/or that the association rate of the radioligand be dependent on the fractional occupancy of the binding sites. Neither of these kinetic phenomena have been observed for [^{3}H]ryanodine binding to the Ca^{2+} release channel. In fact, the presence of excess unlabeled ryanodine slows the dissociation rate of bound [^{3}H]ryanodine,[22,24] suggesting that the two classes of binding sites are allosterically coupled and do not arise from negatively co-operative binding at one of four initially identical sites.[23] The dramatic slowing of [^{3}H]ryanodine dissociation from the high affinity site by the addition of excess unlabeled ryanodine could involve a closing of the channel around the ligand bound at the high affinity sites. This trapping or slowing of the dissociation of bound [^{3}H]ryanodine can also be accomplished by the addition of either neomycin or ruthenium red to the [^{3}H]ryanodine bound receptor.[27,29] Such an effect is difficult to explain by negatively co-operative binding of ryanodine. Another finding inconsistent with negative co-operative binding of ryanodine is that a ryanodine derivative,b-alanyl ryanodine, activates the channel but does not, at higher concentrations, produce inhibition.[30]

The effects of other channel modulators on [^{3}H]ryanodine binding can provide important clues about the ryanodine binding sites. These studies provide additional evidence for the usefulness of [^{3}H]ryanodine binding to probe changes in the functional state of the channel. The affinity of RyR for [^{3}H]ryanodine is modulated by the same ligands that influence the activity of the Ca^{2+} release channel reconstituted into planar lipid bilayers.[7,15,24,29,31–33] [^{3}H]ryanodine binding to the high affinity site on the Ca^{2+} release channel is altered by ionic strength, pH, and both physiologically and pharmacologically relevant chemical effectors of Ca^{2+} release from the SR (e.g., Ca^{2+}, Mg^{2+}, sphingosine, caffeine, ruthenium red, neomycin and adenine nucleotides),[10,18,19,24,26,29,34,35] with Ca^{2+} as the most important modulator. Both channel activity and [^{3}H]ryanodine binding to RyR show a bell shaped dependence on Ca^{2+} concentration.[16,24,36,37] Micromolar Ca^{2+} enhances [^{3}H]ryanodine millimolar Ca^{2+} inhibits binding. At least two Ca^{2+} binding proteins, calmodulin[36,38,39] and S100A[40,41] bind to RyR. However, it is not yet known whether the Ca^{2+} dependence of channel activity and [^{3}H]ryanodine binding reflect a contribution from Ca^{2+} binding to these modulators.

Ligands which are known to open the channel and stimulate Ca^{2+} release from the SR vesicles (e.g. mM Ca^{2+}, mM caffeine, and mM ATP) stimulate [^{3}H]ryanodine binding to the high-affinity site by increasing its rate of association.[24] There appears, however, to be a quantitative discrepancy between the K_d for [^{3}H]ryanodine to SR vesicles and the concentration of ryanodine required to modify the activity of RyR reconstituted in planar lipid bilayers. Much higher concentrations of ryanodine are required to detect the functional effects of ryanodine on single channel behavior than would be predicted from the affinity of the channel for [^{3}H]ryanodine. This discrepancy has been attributed to the slow rate at which ryanodine binds to RyR.[7] The rate at which [^{3}H]ryanodine binds to its high affinity site on RyR is much slower than expected if the binding represented a diffusion limited interaction. Since channel activators enhance the rate of association and channel inhibitors slow it, the rate is, in some way, controlled by channel opening. However, even under maximally activating conditions, the rate is still slower than diffusion limited. The reasons for the slow rate of association have yet to be determined.

All known inhibitors of the channel also inhibit [^{3}H]ryanodine binding. These inhibitors can be divided into three classes based on differences in their effects on the kinetics of

[^{3}H]ryanodine binding to RyR. Millimolar concentrations of Ca^{2+} and Mg^{2+} inhibit the channel and slightly increase the rate of dissociation of [^{3}H]ryanodine bound at the high affinity site.[24,26,36] These agents also appear to inhibit binding primarily by decreasing the rate at which [^{3}H]ryanodine binds to the high affinity site.[24,26] When added after [^{3}H]ryanodine is bound, Ca^{2+} and Mg^{2+} do not trap ryanodine in its high affinity site. Mg^{2+} at low concentrations may also inhibit RyR activity by preventing Ca^{2+} from binding to its activating site, while at higher concentrations, it appears to inhibit in the same manner as mM Ca^{2+}.[16,36] Neomycin and ruthenium red also slow the rate at which [^{3}H]ryanodine binds to the high affinity site. In addition, these compounds and high concentrations of ryanodine greatly slow the dissociation of [^{3}H]ryanodine from its high affinity binding site. These findings suggest that these ligands are either noncompetitive with ryanodine at the high affinity site[27] or that neomycin and ruthenium red also have multiple binding sites on RyR1.[20,42] One possible explanation of their ability to slow [^{3}H]ryanodine dissociation is that they cause the channel to close, trapping [^{3}H]ryanodine in its binding site.

Sphingosine represents a third type of inhibitor of the channel. Sphingosine inhibits caffeine-induced Ca^{2+} release from skinned skeletal muscle fibers[34] and the activity of RyR1 incorporated into planar lipid bilayers.[35] Sphingosine has, however, been suggested to be a channel activator at high concentrations.[34] Sphingosine inhibits [^{3}H]ryanodine binding[34,35] by a noncompetitive mechanism[35] and causes both a dramatic increase in the rate of dissociation of bound [^{3}H]ryanodine from SR membranes and a decrease in the association rate of [^{3}H]ryanodine to the high affinity-site.[35] These studies suggest that the mechanism of sphingosine inhibition of [^{3}H]ryanodine binding is completely distinct from that of Ca^{2+}/Mg^{2+} or neomycin/ruthenium red. Elucidation of the locations of the binding sites for these ligands should provide important new information about the regulation of the channel and the nature of the ryanodine binding site. Both the activator and the inhibitor studies support a model in which the binding of [^{3}H]ryanodine depends upon the functional state of the Ca^{2+} release channel. Therefore, its binding can be used to probe the functional state of the channel and to analyze the effects of modulators on the Ca^{2+} release channel function.

In addition to the small molecules discussed above, a number of proteins have been suggested to interact with RyR. These include the DHPR,[43] calmodulin,[36,38,39] FKBP12,[44,45] annexin VI,[46] triadin,[47–49] and calsequestrin.[50,51] Although some of these proteins may play purely structural roles, others have been shown to be important modulators of channel function. Similar to the small molecule modulators, the protein modulators also alter the binding of [^{3}H]ryanodine.

FUNCTIONAL EFFECTS OF RYANODINE

The functional effects of ryanodine on the Ca^{2+} release channel have been demonstrated both with $^{45}Ca^{2+}$ flux from SR vesicles and channel behavior upon reconstitution into planar lipid bilayers. Binding of ryanodine to the high affinity site in skeletal and cardiac muscle SR correlates with stimulation of Ca^{2+} efflux from Ca^{2+} passively loaded junctional SR vesicles[7,9,16,27,31,52] and with the formation of reduced conductance channels which have long mean channel open times.[7] In contrast, the occupation of the low affinity ryanodine binding sites correlates with closing of the channels, thereby enhancing Ca^{2+} loading of muscle microsomes.[16,22,53,54] No resolvable open channel events are detected at high concentrations of ryanodine (>100 μM) suggesting that the channels are closed under these conditions.[7,31,56]

ELUCIDATING THE MECHANISM OF CHANNEL MODIFICATION BY RYANODINE

The molecular mechanism by which binding of ryanodine to the Ca^{2+} release channel alters its activity remains unknown. Tetrameric assembly of the Ca^{2+} release channel is apparently required for the binding of ryanodine at its high-affinity site[14] and there is only one high affinity ryanodine binding site per homotetramer. The actual mechanism by which a single molecule of ryanodine can bind to the RyR homotetramer, reduce its conductance, and lock the channel in an open conformation remains a mystery. At least two models have been considered:[23,55] (1) ryanodine binds in the pore of the open channel and, in some way, restricts the movement of Ca^{2+} through the pore, producing the reduced conductance or (2) ryanodine binds to a conformationally sensitive site on the channel which is distal to the pore and drives the conformation of the protein toward a reduced conductance open state. The latter model would require that, in order to have only one site per tetramer, the binding of ryanodine must either be negatively co-operative or the binding site is in a region where all four subunits contribute to the binding of ryanodine. As will be discussed below, there are some data to support each of these models.

Ryanodine may act by either inducing or stabilizing conformational changes in the Ca^{2+} release channel that prevent channel closing and result in the observed alterations in cation conductance.[21] A long-lived reduced conductance state of the Ca^{2+} release channel nas never been observed for the native channel in the absence of ryanodine. Tinker *et al.*[56] reported that ryanodine derivatives could produce different RyR subconductance states, implying a complex interaction between ryanodine and the channel protein rather than stabilization of the open conformation in a two-state process.

Comparative molecular field analysis (CoMFA) of the ryanodine molecule has been employed to correlate the structural features of ryanodine with its binding affinity for the Ca^{2+} release channel[57,58] and its ability to modulate the conductance properties of the channel.[55] A variety of ryanoids with different modifications were shown to cause the transition of the ryanodine receptor to a steady reduced conductance state but the magnitude of the fractional conductance measured as the ratio between the conductance of unmodified and modified RyR, varied depending on the specific structure of the ryanoid.[55] Bulky substitutions at the pyrrole group and the isopropyl region (e.g. ryanodyl nicotinate, anhydroester A) as well as lack of bulk (e.g. ryanodol) markedly decreased the binding affinity of the ryanoid for the ryanodine receptor. In contrast, the presence of large groups (e.g. CBZ glycyl, 21-p-nitro) at the 9-, 10- and 21-positions of ryanodine had relatively minor effects on binding.[57,58] Ryanodine was suggested to bind to the channel with the pyrrole and isopropyl groups (the 3- and 2-positions, respectively) buried within the binding site.[58] The opposite pole of the ryanodine molecule (the 9- and 10-positions) appears to extend out of the binding pocket. Steric and electrostatic replacements in the ryanodine molecule caused differences in the fractional conductance of the Ca^{2+} release channel incorporated into planar lipid bilayers.[55] CoMFA of these results is consistent with an allosteric mechanism of ryanodine action rather than a plug mechanism in which ryanodine binds in the ion conduction pathway and interacts with the translocated ion.[55,57,58] Preferential binding to a conformational state of the channel which the channel assumes only rarely could explain the very slow rate at which ryanodine binds to RyR. The problem with this model is that it does not explain the one ryanodine binding site per tetramer. As discussed above, there is no kinetic evidence for negative co-operativity in $[^3H]$ryanodine binding to RyR and, therefore, an allosteric model for ryanodine regulation of channel activity must somehow be explained by a single ryanodine binding site. One possibility, as suggested by Tinker *et al.*[56] is that ryanodine binds to the outer vestibule of the channel at a site formed by all of the subunits.

CONFORMATIONAL TRANSITIONS OF RyR

Changes in the conformation of RyR have been studied by analyzing the different functional states that the channel can undergo,[59,60] by assessing changes in fluorescence from tags placed in conformationally sensitive sites on RyR[61] and by analyzing the structure of the channel in different functional states.[62] Wyskovsky *et al.*[59] have suggested that the model which best fits their observed rates of Ca^{2+}-release from SR vesicles is the following:

$$O \underset{k_{-1}}{\overset{k_{+1}}{\leftrightarrow}} C \underset{k_{-2}}{\overset{k_{+2}}{\leftrightarrow}} C^*$$

where ***O*** is the open state of the channel, ***C*** is the first closed state and ***C**** is the second closed state. Their data suggests that ruthenium red inhibits the channel by entering the pore and blocking the channel in an open state while neomycin inhibits the channel by increasing k_{+1}. In their model, activators would reduce inactivation by decreasing k_{+1}. Multiple open and closed states of the Ca^{2+}-release channel have also been proposed by Ashley and Williams.[63] These workers analyzed the behavior of the cardiac channel incorporated into planar lipid bilayers and proposed that there are at least 3 closed states and 2 open states of the channel.

Conformational changes of RyR in the presence of modulators have been addressed by Ikemoto *et al.*[64] who used fluorescent derivatives of neomycin and polylysine to place fluorescent tags in their binding sites on RyR. In these elegant studies the crosslinker was cleaved and the modulators were removed from their binding sites, leaving behind the fluorescent tag. These fluorescent tags were sensitive to changes in the conformation of RyR. These investigators found that the conformational change induced by ryanodine involved a local change in the neomycin region of the protein, but the Ca^{2+} induced conformation change involved both the polylysine and neomycin sites.[65] This group[61,66] has also used courmarine derivatives as fluorescent probes to demonstrate that RyR undergoes a major conformational change prior to the opening of the channel and this change was not prevented by Mg^{2+}. These findings support those made by functional analysis; that there are multiple conformational states that this protein can undergo.

Image reconstruction analysis of electron micrographs of the purified skeletal muscle Ca^{2+} release channel[62,67–69] has provided a low resolution three-dimensional structure of the channel complex and supports the hypothesis that ryanodine and Ca^{2+} produce conformational changes in RyR. The channel protein has the characteristic mushroom shape with two distinct portions, one of which (the mushroom stem) contains the trans-membrane-spanning region of the channel and faces to the SR lumen. The other portion is the large square-shaped cytoplasmic region, apparently associated with the T-tubule.[69] The trans-membrane domain undergoes a structural change upon binding both Ca^{2+} and ryanodine to produce what appears to be a pore in the middle of this domain (Fig. 6.2). A pore is also detected in reconstructions with Ca^{2+} alone (unpublished observation). This pore is likely

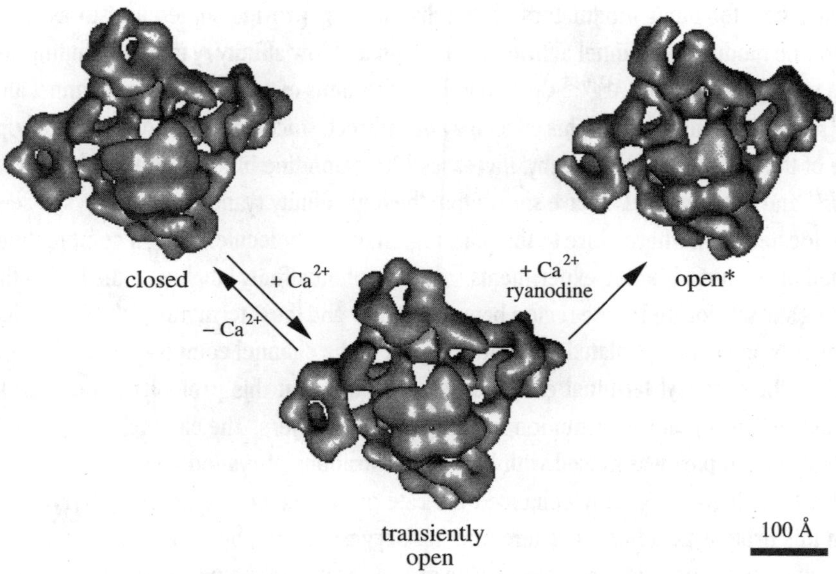

Fig. 6.2 Structural configurations of rabbit skeletal muscle Ca^{2+} release channel in closed, transiently open and ryanodine modified states. The protein is viewed from the lumenal side of the sarcoplasmic reticulum.

to represent the route by which Ca^{2+} exits the lumen during E-C coupling. Another domain of RyR seen in the reconstructions is the large square-shaped cytoplasmic region which is the region which probably interacts with the t-tubules. The clamp-shaped domains of a cytoplasmic region in transiently open and closed channels[69] appear in a more closed conformation than is observed in the ryanodine modified channel.[62] Whether the reconstruction obtained with Ca^{2+} and ryanodine will be the same as a reconstruction obtained with the channel opened with another activator such as ATP remains to be determined. The cytoplasmic domain also has a very hollow appearance which may facilitate the diffusion of Ca^{2+}, after it exits the trans-membrane pore, into the cytoplasm to reach its molecular targets. One such target is likely to be RyR itself since it can be inactivated by high concentrations of Ca^{2+}.

LOCATING THE RYANODINE BINDING SITE

The high and low affinity ryanodine binding sites appear to be either allosterically or sterically coupled.[23] The localization of ryanodine-binding sites within the channel protein as well as binding sites for other modulators of the channel may provide some clues to as to how ryanodine modulates channel activity at both high and low affinity ryanodine binding sites.

A single mutation of (Arg^{615}-Cys) near the N-terminus of the Ca^{2+} release channel alters [^3H]ryanodine binding.[10,70] This effect may be indirect, since this mutation favors the open state of the channel[70] and, thereby, increases [^3H]ryanodine binding. Studies by Callaway *et al.*[71] and Witcher *et al.*[72] have shown that the high affinity ryanodine binding site seems to be located somewhere close to the pore region of the molecule. Using a combination of limited proteolysis, binding experiments,[71] and or photoaffinity labeling,[72] the high affinity binding site was located in the region between Arg^{4475} and the C-terminus of the Ca^{2+} release channel. Wang *et al.*[27] isolated a tryptic complex of the channel composed of amino acids 4476 to the carboxyl terminal end and demonstrated that this proteolytic fragment had channel activity upon reconstitution into planar lipid bilayers. The channel formed by this proteolytic complex was closed with high concentrations of ryanodine and neomycin. In contrast, sphingosine greatly enhanced the rate of dissociation of bound [^3H]ryanodine from this proteolytic complex. There findings suggest that the binding sites for ryanodine (low affinity), neomycin, and sphingosine are all located in the part of the protein which forms the ion channel and has the high affinity ryanodine binding site. If the high affinity ryanodine site is not occupied at the time of the tryptic digestion, the ability of the complex to bind ryanodine at that site is lost. This suggests that either the part of the protein needed to form the binding site is removed or that tryptic cleavage in some other way alters the conformation of the binding site. However, Chen *et al.*[73] suggested that the N-terminal fragment, especially the region between residues 500 and 1300 may form part of the high-affinity ryanodine binding site. Ma and coworkers[74,75] have shown that CHO cells expressing the cDNA encoding a protein consisting of the first 182 amino acids at the N-terminus and the last 1030 amino acids at the C-terminus of skeletal RyR contained the truncated protein. This protein was shown to have ion channel activity in planar lipid bilayers which could be activated by Ca^{2+} and was modified by high concentrations of ryanodine. A truncated protein consisting of the last 656 amino acids, however, showed neither channel activity nor ability to bind [^3H]ryanodine. This raises the possibility that the N-terminal amino acids and/or the amino acids from 4000 to 4600 may contribute to the ryanodine binding site.

SUMMARY

Ryanodine has been an essential tool for characterizing the structure and function of the skeletal muscle Ca^{2+} release channel. This plant alkaloid has enabled workers to identify the Ca^{2+} release channel, purify and clone it, and study the effects of many of the channel modulators. We are, however, still a long way from understanding the molecular mechanisms by which ryanodine modulates the activity of this channel. Elucidation of the mechanism of modulation of the channel by ryanodine will provide us with an important handle with which to assess the mechanism by which other modulators control channel activity and to evaluate their contributions to E-C coupling.

REFERENCES

1. Jenden DJ, Fairhurst AS. The pharmacology of ryanodine. *Pharm Rev* 1969;**21**:1–25.
2. Jones LR, Besch HR, Sutko JL, Willerson JT. Ryanodine-induced stimulation of net Ca^{2+} uptake by cardiac sarcoplasmic reticulum vesicles. *J Pharm Exp Ther* 1979;**209**:48–55.
3. Sutko JL, Willerson JT, Templeton GH *et al.* Ryanodine: Its alterations of cat papillary muscle contractile state and responsiveness to inotropic interventions and suggested mechanism of action. *J Pharm Exp Ther* 1979;**209**:37–47.
4. Sutko J, Ito K, Kenon JL. Ryanodine: a modifier of sarcoplasmic reticulum calcium release channel. *Fed Proc* 1985;**44**:2984–2989.
5. Campbell K, Knudson C, Imagawa T. Identification and characterization of the high affinity [^{3}H]-ryanodine receptor of the junctional sarcoplasmic reticulum Ca^{2+} release channel. *J Biol Chem* 1987;**262**:6460–6463.
6. Inui M., Saito A, Fleischer, S. Purification of the ryanodine receptor and identity with feet structures of junctional terminal cisternae of sarcoplasmic reticulum from fast skeletal muscle. *J Biol Chem* 1987;**262**:1740–1747.
7. Smith JS, Imagawa T, Ma J *et al.* Purified ryanodine receptor from rabbit skeletal muscle is the calcium-release channel of saroplasmic reticulum. *J Gen Physiol* 1988;**92**:1–26.
8. Franzini-Armstrong C. Studies of the triad. I. Structure of the junction in frog twitch fibers. *J Cell Biol* 1970;**47**:488–499.
9. Lai FA, Erickson HP, Rousseau E *et al.* Purifiation and reconstitution of the calcium release channel from skeletal muscle. *Nature* 1988;**331**:315–319.
10. Hawkes MJ, Diaz-Muñoz M, Hamilton SL. A procedure for purification of the ryanodine receptor from skeletal muscle. *Memb Biochem* 1989;**8**:133–145.
11. Imagawa T, Smith J, Coronado R. Purified ryanodine receptor from skeletal muscle sarcoplasmic reticulum is the Ca^{2+}-permeable pore of the Ca^{2+}-release channel. *J Biol Chem* 1987;**262**: 16636–16643.
12. Takeshima H, Nishimura H, Matsumoto T *et al.* Primary structure and expression from complementary DNA of skeletal muscle ryanodine receptor. *Nature* 1989;**339**:439–445.
13. Zorzato F, Fujii J, Otsu K. Molecular cloning of cDNA encoding human and rabbit forms of the Ca^{2+} release channel (ryanodine receptor) of skeletal muscle sarcoplasmic reticulum. *J Biol Chem* 1990;**265**:2244–2256.

14. Lai FA, Misra M, Xu L. The ryanodine receptor Ca^{2+} release channel complex of skeletal muscle sarcoplasmic reticulum: evidence for a cooperatively charged homotetramer. *J Biol Chem.* 1989;**264**:16776–16785.
15. Zimanyi I, Pessah IN. Comparison of [^{3}H] ryanodine receptors and Ca^{2+} release from rat cardiac and rabbit skeletal muscle sarcoplasmic reticulum. *J Pharm Exp Ther* 1991;**256**:938–946.
16. Meissner G, Darling E,.Eveleth J. Kinetics of rapid Ca^{2+} release by sarcoplasmic reticulum. Effects of Ca^{2+}, Mg^{2+}, and adenine nucleotides. *Biochem* 1986;**25**:236–244.
17. Meissner G, Lai FA. Structural and functional correlation of the trypsin-digested Ca^{2+} release channel of skeletal muscle sarcoplasmic reticulum. *J Biol Chem.* 1989;**264**:1715–1722.
18. Meissner G. el-Hashem A. Ryanodine as a functional probe of the skeletal muscle sarcoplasmic reticulum Ca^{2+} release channel. *Mol Cell Biochem* 1992;**114**:119–123.
19. Carrol S, Skarmeta JG, Xiang Yu *et al.* Independence of Ryanodine binding, oligomeric rReceptor interactions, and Ca^{2+} release regulation in junctional sarcoplasmic reticulum. *Arch Biochem Biophys* 1991;**290**:239–247.
20. Buck E, Zimanyi I, Abramson J *et al.* Ryanodine stabilizes multiple conformational states of the skeletal muscle calcium release channel. *J Biol Chem* 1992;**267**:23560–23567.
21. Pessah IN, Zimanyi I. Characterization of multiple [^{3}H] ryanodine sites on the Ca^{2+} release channel of sarcoplasmic reticulum from skeletal and cardiac muscle: Evidence for a sequential mechanism in ryanodine action. *Mol Pharm* 1991;**39**:679–689.
22. McCrew SG, Wolleben C, Siegl P *et al.* Positive cooperativity of ryanodine binding to the calcium release channel of sarcoplasmic reticulum from heart and skeletal muscle. *Biochem* 1989;**28**:1686–1691.
23. Wang JP, Needleman DH, Hamilton SL. Relationships of low affinity [^{3}H] ryanodine binding sites to high affinity sites on the skeletal muscle Ca^{2+} release channel. *J Biol Chem* 1993;**268**: 20974–20982.
24. Chu A, Diaz-Muñoz M, Hawkes MJ *et al.* Ryanodine as a probe for the functional state of the skeletal muscle sarcoplasmic calcium release channel. *Mol Pharmacol* 1990;**37**:735–741.
25. Meissner, G. Ryanodine activation and inhibition of the Ca^{2+} release channel of sarcoplasmic reticulum. *J Biol Chem* 1986;**261**:6300–6306.
26. Pessah IN, Stambuk RA, Casida JE. Ca^{2+}-activated ryanodine binding: Mechanisms of sensitivity and intensity modulation by Mg^{2+}, caffeine, and adenine. *Mol Pharmacol* 1986;**31**: 232–238.
27. Wang, JP Needleman DH, Seryshev AB *et al.* Interaction between ryanodine and neomycin binding sites on Ca^{2+} release channel from skeletal muscle sarcoplasmic reticulum. *J Biol Chem* 1996;**271**:8387–8393.
28. Fleischer S, Inui M. Biochemistry and biophysics of excitation-contraction coupling. *Annu Rev Biophys Biophys Chem* 1989;**18**:333–364.
29. Mack M, Zimanyi I, Pessah I. Discrimination of multiple binding sites for antagonists of the calcium release channel complex of skeletal and cardiac sarcoplasmic reticulum. *J Pharm Exp Ther* 1991;**262**:1028–1037.
30. Humerickhouse RA, Bidasee KR, Gerzon K *et al.* High affinity C10-0eq. ester derivatives of ryanodine. Activator-selective agonists of the sacoplasmic reticulum calcium release channel. *J Biol Chem* 1994;**269**:30243–30253.

31. Rousseau E, Smith JS, Meissner G. Ryanodine modifies conductance and gating behavior of single Ca^{2+} release channel. *Am J Physiol* 1987;**253**:C364–C368.
32. Rousseau E, Meissner G. Single cardiac sarcoplasmic reticulum Ca^{2+}-release channel activation caffeine. *Am J Physiol.* 1989;**256**:H328–H333.
33. Pessah IN, Stambuk RA, Casida, JE. Ca^{2+}-activated ryanodine binding: mechanisms of sensitivity and intensity modulation by Mg^{2+}, caffeine, and adenine nucleotides. *Mol Pharmacol.* 1987;**31**:232–238.
34. Sabbatini RA, Betto R, Teresi A *et al.* The effects of sphingosine on sarcoplasmic reticulum membrane calcium release. *J Biol Chem* 1992;**267**:15475–15484.
35. Needleman DH, Aghdasi B, Seryshev AB *et al.* Modulation of skeletal muscle Ca^{2+}-release channel activity by sphingosine. *Am J Physiol* (Cell Physiol 41) 1997;**272**:C1465–C1474.
36. Meissner G, Henderson JS. Rapid calcium release from cardiac sarcoplasmic reticulum vesicles is dependent on Ca^{2+} and is modulated by Mg^{2+}, adenine nucleotide, and calmodulin. *J Biol Chem* 1987;**262**:3065–3073.
37. Fill M, Coronado R, Mickelson J *et al.* Abnormal ryanodine receptor channels in malignant hyperthermia. *Biophys J* 1990;**50**:471–475.
38. Smith JS, Rousseau E, Meissner G. Calmodulin modulation of single sarcoplasmic reticulum Ca^{2+}-release channels from cardiac and skeletal muscle. *Circ Res* 1989;**64**:352–359.
39. Fuentes O, Valdivia C, Vaughan D *et al.* Calcium-dependent block of ryanodine receptor channel of swine skeletal muscle by direct binding of calmodulin. *Cell Calcium* 1994;**15**:305–316.
40. Marsili V, Mancinelli L, Menchetti G *et al.* S-100ab increases Ca^{2+} release in purified sarcoplasmic reticulum vesicles of frog skeletal muscle. *J Muscle Res Cell Moti* 1992;**13**:511–515.
41. Fano G, Angelelia P, Aisa MC *et al.* S-100ab protein stimulates Ca^{2+} induced Ca^{2+} release from isolated sarcoplasmic reticulum vesicles. *FEBS Lett* 1989;**255**:381–384.
42. Yano M, El-Hayek, R, Ikemoto N. Role of calcium feedback in excitation-contraction coupling in isolated triads. *J Biol Chem* 1995;**270**:19936–19942.
43. Lu X, Xu L,. Meissner G. Activation of the skeletal muscle calcium release channel by a cytoplasmic loop of the dihydropyridine receptor. *J Biol Chem* 1994;**269**:6511–6516.
44. Jayaraman T, Brillantes AM, Timerman AP *et al.* FK506 binding protein associated with the calcium release channel (ryanodine receptor). *J Biol Chem* 1992;**267**:9474–9477.
45. Brilliantes A-M, Ondrias K, Scott A *et al.* Stabilization of calcium release channel (ryanodine receptor) function by FK506-binding protein. *Cell* 1994;**77**:513–523.
46. Diaz-Munoz M, Hamilton SL, Kaetzel MA *et al.* Modulation of Ca^{2+} release channel activity from sarcoplasmic reticulum by annexin VI (67-kDa Calcimedin). *J Biol Chem* 1990; **265**: 15894–15899.
47. Brandt NR, Caswell AH, Wen S-R *et al.* Molecular interactions of the junctional foot protein and dihydropyridine receptor in skeletal muscle triads. *J Membr Biol* 1990;**113**:237–251.
48. Brandt NR, Caswell AH, Brandt T *et al.* Effects of anti-triadin antibody on Ca^{2+} release from sarcoplasmic reticulum. *FEBS Letters* 1992;**299**:57–59.

49. Caswell AH, Brandt NR, Brunschwig J-P *et al.* Localization and partial characterization of the oligomeric disulfide-linked molecular weight 95,000 protein (triadin) which binds the ryanodine and dihydropyridine receptors in skeletal muscle triadic vesicles. *Biochem* 1991; **30**:7507–7513.
50. Collins JH, Tarcsafalvi A, Ikemoto N. Identification of a region of calsequestrin that binds to the junctional face membrane of sarcoplasmic reticulum. *Biochem Biophys Res Commun* 1990;**167**:189–193.
51. Ikemoto NB, Antoniu B, Kang J *et al.* Intravesicular calcium transient during calcium release from sarcoplasmic reticulum. *Biochem* 1991;**30**:5230–5237.
52. Fleischer S, Ogunbunmi T, Dixon MC *et al.* Localization of Ca^{2+} release channels with ryanodine in junctional terminal cisternae of sarcoplasmic reticulum of fast skeletal muscle. *Proc Natl Acad Sci USA* 1985;**82**:7256–7259.
53. Inui M, Wang S, Saito A *et al.* Characterization of junctional and longitudinal sarcoplasmic reticulum from heart muscle. *J Biol Chem* 1988;**263**:10843–10850.
54. Lattanzio FA, Schlatterer RG, Nicar, M *et al.* The effects of ryanodine on passive calcium fluxes across sarcoplasmic reticulum membranes. *J Biol Chem* 1987;**262**:2711–2718.
55. Welch W, Williams A, Tinker A *et al.* Structural components of ryanodine responsible for modulation of sarcoplasmic reticulum calcium channel function. *Biochem* 1997;**36**:2939–2950.
56. Tinker A, Sutko JL, Ruest L *et al.* Electrophysiological effects of ryanodine derivatives on the sheep cardiac sarcoplasmic reticulum calcium-release channel. *Biophys. J* 1996;**70**:2110–2119.
57. Welch W, Ahmad S, Airey JA *et al.* Structural determinants of high-affinity binding of ryanoids to the vertebrate skeletal muscle ryanodine receptor: A comparative molecular field analysis. *Biochem* 1994;**33**:6074–6085.
58. Welch W, Sutko JL, Mitchel KE *et al.* The pyrrole locus is the major orienting factor in ryanodine binding. *Biochem* 1996;**35**:7165–7173.
59. Wyskovsky W, Hohenegger M, Plank B *et al.* Activation and inhibition of the calcium-release channel of isolated skeletal muscle heavy sarcoplasmic reticulum. Models of the calcium-release channel. *Eur J Biochem* 1990;**194**:549–559.
60. Williams AJ. Ion conduction and discrimination in the sarcoplasmic reticulum ryanodine receptor/calcium-release channel. *J Muscle Res Cell Motil.* 1992;**13**:7–26.
61. Ohkusa T, Kang JJ, Morii M *et al.* Conformational change of the foot protein of sarcoplasmic reticulum as an initial event of calcium release. *J Biochem* 1991;**109**:609–615.
62. Orlova EV, Serysheva II, van Heel M *et al.* Two structural configurations of the skeletal muscle calcium release channel. *Nature Struct Biol* 1996;**3**:547–551.
63. Ashley RH, Williams AJ. Divalent cations activation and inhibition of single calcium release channels from sheep cardiac sarcoplasmic reticulum. *J Gen Physiol* 1990;**95**:981–1005.
64. Cifuentes ME, Ronjat M, Ikemoto N. Polylysine induces a rapid Ca^{2+} release from sarcoplasmic reticulum vesicles by mediation of its binding to the foot protein *Arch Biochem Biophys* 1989;**273**:554–561.
65. Kang JJ, Tarcsafalvi A, Carlos AD *et al.* Conformational changes in the foot protein of the sarcoplasmic reticulum assessed by site-directed fluorescent labeling. *Biochem* 1992;**31**:3288–3293.

66. El-Hayek R, Yano M, Ikemoto N. A conformational change in the junctional foot protein is involved in the regulation of Ca^{2+} release from sarcoplasmic reticulum. *J Biol Chem* 1995; **270**: 15634–15638.
67. Wagenknecht T, Grassucci R, Frank J *et al.* 3-dimensional architecture of the calcium release channel/foot structure of sarcoplasmic reticulum. *Nature* 1989;**338**:167–170.
68. Radermacher M, Rao V, Grassucci R *et al.* Cryo-electron microscopy and three-dimensional reconstruction of the calcium channel/ryanodine receptor from skeletal muscle. *J Cell Biol* 1994;**127**:411–423.
69. Serysheva II, Orlova EV, Chiu W *et al.* Electron cryomicroscopy and angular reconstitution to visualize the skeletal muscle calcium-release channel. *Nature Struct Biol.* 1995;**2**:18–24.
70. Mickelson JR, Litterer LA, Jacobson BA *et al.* Stimulation and inhibition of [^{3}H] ryanodine binding to sarcoplasmic reticulum from malignant hyperthermia susceptible pigs. *Arch Biochem Biophys* 1990;**278**:251–257.
71. Callaway C, Seryshev A, Wang JP *et al.* Localization of the high and low affinity [^{3}H] ryanodine binding sites on the skeletal muscle Ca^{2+} release channel. *J Biol Chem* 1994; **269**: 15876–15884.
72. Witcher DR, McPherson PS, Kahl SD *et al.* Photoaffinity labeling of the ryanodine receptor/ Ca^{2+} release channel with an zido derivative of ryanodine. *J Biol Chem* 1994;**269**:13076–13079.
73. Chen SRW, Airey JA, MacLennan DH. Positioning of major tryptic fragments in the Ca^{2+} release channel (ryanodine receptor) resulting from partial digestion of rabbit skeletal muscle sarcoplasmic reticulum. *J Biol Chem* 1993;**268**:22642–22649.
74. Bhat MB, Zhao J, Takeshima H *et al.* Functional calcium release channel formed by the carboxyl-terminal portion of ryanodine receptor. *Biophys J* 1997;**73**:1329–1336.
75. Bhat MB, Zhao J, Hayek S *et al.* Deletion of amino acids 1641-2437 from the foot region of skeletal muscle ryanodine receptor alters the conduction properties of the Ca^{2+} release channel. *Biophys J* 1997;**73**:1320–1328.

Chapter 7

The Ryanodine Receptor: A Report from the Ryanoids

William Welch

> "*When you can measure what you are speaking about, and express it in numbers, you know something about it; but when you cannot measure it, when you cannot express it in numbers, your knowledge is of a meager and unsatisfactory kind: it may be the beginning of knowledge, but you have scarcely, in your thoughts, advanced to the stage of science*".
>
> William Thomson, Lord Kelvin
> Popular Lectures and Addresses

Ryanodine is a peculiar molecule with remarkably complex and often contradictory properties. Although ryanodine has provided the name for what is clearly an important, widespread and diverse set of intracellular calcium ion release channels,[1,2] extremely little is known about the mechanism of ryanodine-receptor interactions. Ryanodine and the ryanodine receptor (RyR) have been the subject of reviews in recent years.[1–13] In this chapter the focus will be on one method of understanding the function of the RyR, that is, on comparative molecular field analysis (CoMFA).

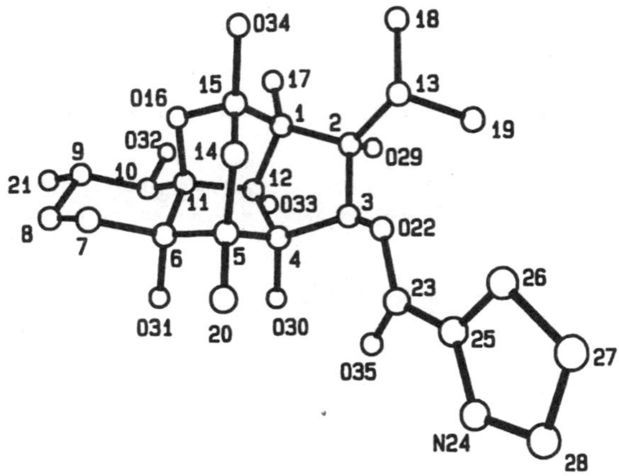

Fig. 7.1 A perspective drawing of ryanodine. The hydrogen atoms have been omitted for clarity and the numbering system used for the ryanoids is illustrated. Reprinted with permission from *Biochemistry* 33:6074-6085. Copyright 1994 American Chemical Society.

A perspective ball and stick rendering of ryanodine is shown in Fig. 7.1. The main body of ryanodine consists of a complicated group of five fused rings. It bears multiple hydroxyl groups and pyrrole carbonyl and isopropyl pendant groups. The figure shows the numbering system that will be used throughout this chapter. The ryanodine molecule is large enough to present distinct hydrophobic and hydrophilic patches to potential receptors.[1]

Binding isotherms of ryanodine to the RyR are complex evidencing both high and low affinity sites.[11,15–19] However, the nature and relationships between these sites is not yet clear. Binding is further complicated by the large number of modulators of ryanodine binding. Modulation of RyR channel function by ryanodine is also complex. At low concentrations ryanodine produces two distinct effects: (A) the open channel conductance drops below that of the unmodified channel and (B) the rapid oscillation (or flickering) between an open and closed state ceases and the channel persists in the modified open state.[18,20,21] At higher concentrations, ryanodine induces a closed (non-conducting) state of the channel.

MOLECULAR MODELING

To date, molecular mechanics[23] and quantitative structure-activity relationship (QSAR) techniques have been the main tools used to model the ryanodine-receptor interaction. The principles and applications of molecular modeling and QSAR to the study of biological phenomena are described in Refs. 24–37 (with a few specific examples).

Modeling of the ryanodine-RyR interaction is hampered by lack of a crystal structure of the receptor and because the RyR is too large for homology modeling. One method to circumvent the current lack of the receptor structure is to examine the RyR from the viewpoint of the ligand by development of a quantitative structure-activity relationship (QSAR) for ryanodine analogs. In this way a complimentary physicochemical image of the binding site can be constructed. The comparative molecular field analysis (CoMFA) method[38] has been used for an initial mapping of the functional ryanodine binding site.

Over the years many intriguing structure-activity studies of ryanodine analogs have appeared.[11,12,39–49] However, development of QSAR from these data have been limited. There are multiple reasons justifying more effort in this area including, but not limited to, the following: (A) To map the structural features responsible for the multiple biological activities of ryanodine analogs including binding to the high and low affinity sites, and activation and inhibition of the channel activity of the RyR. (B) To engineer RyR isoform specific ligands that can be used in physiological and pharmacological studies of RyR function in a variety of tissue types and that can help separate the structural roles from the calcium channel functions of the RyR. (C) To engineer ryanoids that will specifically enhance or reduce RyR channel permeability and to engineer pure antagonists of ryanodine action. (D) To provide information leading to the identification of endogenous modulators acting through the ryanodine binding site. In this chapter, only QSAR studies of the ryanoids will be considered. The reader is referred to recent reviews for information about synthesis and biological activities of the many ryanodine analogs studied.[11,12]

QSAR/CoMFA

The QSAR reported to date has employed the comparative molecular field analysis technique (CoMFA).[38] This method is a member of the so-called three-dimensional quantitative structure-activity relationship techniques (3D-QSAR).[50,51] It is a proven technique with predictive power in diverse applications (see Refs. 52–57 for a few examples). The diverse and contradictory biological activities mentioned earlier have been avoided by limiting measurements to conditions where it is clear that only the high affinity binding site is involved.

MODELING THE RYANODINE BINDING SITE

Molecular Mechanics

Although quantum mechanics is preferred, molecular mechanics is generally used to determine molecular configuration because of high speed and good accuracy.[23,58–60] In simple terms, molecular mechanics treats the bonds between atomic nuclei as rods. Equations from classical mechanics are used to describe bond stretching, bending, torsion and so forth. In addition, equations are included to describe non-bonded interactions such as van der Walls and electrostatic interactions.

Conformational Analysis

Commonly, the global minimum of strain energy determined by molecular mechanics is used to determine the conformation of molecules. However, this may not be an accurate representation of a flexible molecule due to entropic contributions to the free energy of a molecule. The ryanoids used for CoMFA have been the subject of extensive conformational searching and the global minimum used in all cases reported.[14,61,62] Most of the ryanoids studied by CoMFA have limited conformational freedom,[14] to avoid the ambiguity of the conformation of the bound ligand *vs*. the conformation of the free ligand.

When some of the members possess considerable conformational space, restricting all members of the basis set to the global minimum may not be justifiable. How does one select the proper conformation under these conditions? A boot-strapping technique has been suggested. Briefly, an initial CoMFA is employed to find the conformers of additional compounds whose predicted value is closest to the measured value.[63] While the method is sensible and useful, it has the danger of being prejudicial and may propagate errors in the starting basis set. Another option is molecular dynamics. Molecular dynamics is a computational technique that combines the force field used in molecular mechanics and Newton's laws of motion to simulate molecular motion. The method aids in the prediction of the conformation of flexible ligands because it includes entropic factors.

CoMFA of the Ryanoid-Receptor Interaction

Structural elements responsible for high-affinity binding

The initial CoMFA found a strong and intricate correlation between steric bulk and high affinity binding.[14] Three isomers of ryanodine were studied in detail.[61] The results confirmed the earlier CoMFA[14] by showing that the pyrrole carbonyl is an important determinant of high affinity binding and is a major orienting or recognition factor.[61] Relocation of the pyrrole leads to alternate orientations of bound ligand that permit the pyrrole carbonyl group to occupy the same subsite in the ryanoid binding site. This conjecture was supported by analysis of ligand-induced fractional conductances.[62] See the section below titled "Orientation of ryanoids . . ." for more discussion.

The original CoMFA was found to have good predictive ability.[14] As additional, more diverse, compounds have been modeled, they have been added to the CoMFA basis set. These compounds include the ryanodine isomers 3-epiryanodine and 10-*O*-pyrrole-carbonyl-ryanodol (called 10-ryanodine for brevity). It is important to note that the conformation of flexible ryanoids was not arranged to give the best fit with the previous basis set (see Ref. 63). In every case reported here the global minimum was used for the CoMFA.

The correlations of a recent data set are shown in Figs. 7.2 and 7.3 (Welch *et al.* unpublished data). The steric and electrostatic fields make approximately equal contributions to the correlation between structure and binding. The wire frames are contours connecting points in space of equal correlation between physicochemical properties and changes in binding affinity. They, therefore, enclose volumes in which changes in structure are highly correlated with changes in biological properties. Fig. 7.2 is a diagram of the regions where increasing steric bulk are strongly correlated with changes in dissociation constant. A ball and stick representation of ryanodine is shown for reference. The right panel shows loci where steric bulk enhances binding of the ryanoids to the RyR. The major

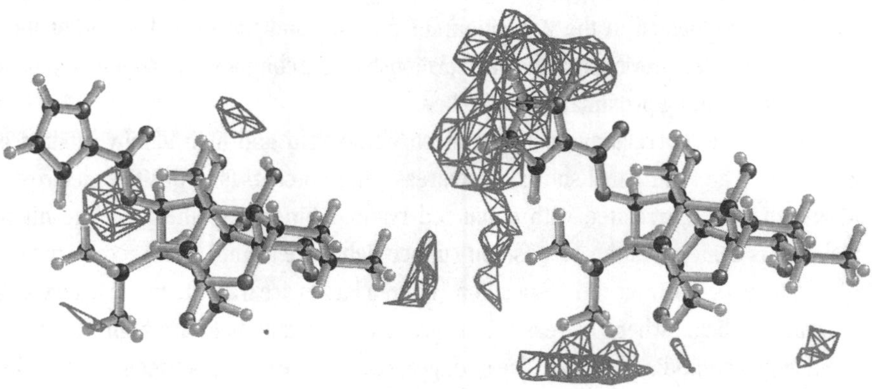

Fig. 7.2 The correlation between steric factors and binding. These figures are based on a 32 member ryanodine basis set that includes the ryanodine isomers 3-epiryanodine and 10-ryanodine (Welch, unpublished results). In each case the wire frame encloses the volume of 50% of the maximum correlation or higher. A ball-and-stick model of ryanodine is shown to orient the viewer. The left panel shows those regions where increases in steric bulk is highly correlated with decreased binding affinity. The right panel shows those areas where increased steric bulk is highly correlated with enhanced binding affinity.

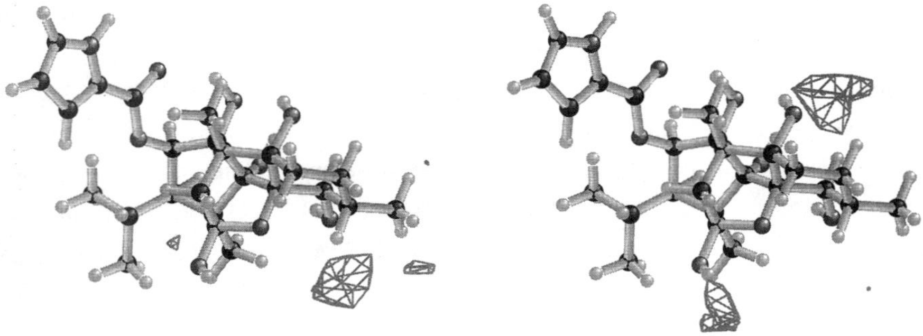

Fig. 7.3 The correlation between changes in electrostatic field and changes in dissociation constant. The wire frames are contours drawn at 25% of the maximum value of the correlation field. The left panel describes the regions where increased negative electrostatic potential is correlated with increased binding affinity (decreased dissociation constant). The right hand panel contours regions where increased positive electrostatic potential is correlated with enhanced binding to the RyR. Please see the text for more details.

areas include the entire pyrrole group (3-position) and 2-position (upper and lower left wire frames). Smaller areas exist between the 9-position and 10-position (lower right wire frame) and at the 15-position (lower middle wire frame). Conversely, the left panel illustrates those regions where increasing steric bulk is strongly correlated with loss of binding affinity. These areas are located at the 2-position, 11-position and the methyl group at the 9-position. As reported previously,[14] the interrelationships are complex with regions promoting and inhibiting binding proximal to each other.

In Fig. 7.3 the correlations between electrostatic field and high affinity binding are diagrammed. The right panel shows those areas where increases in positive electrostatic field are strongly correlated with enhanced ryanoid binding to the RyR. The highest correlation is centered at the 11-position (upper right wire frame) with secondary foci at the 1-position, 14-position and 15-position (bottom large wire frame). In the left panel the wire frames indicate where increases in negative electrostatic field are highly correlated with enhanced ryanodine binding. The strongest region is centered between the 9-position and 10-position (largest wire frame). As with the steric correlations, the electrostatic correlations are a quilt of opposing trends. Note that the CoMFA described above (that uses a much larger and much more diverse basis set) is quite similar to that reported earlier.[14] This is a reflection of the predictive power of the model.

Figures 7.2 and 7.3 summarize the correlations obtained from many ryanoid structures and binding affinities. Relating such plots to the properties of a particular compound is

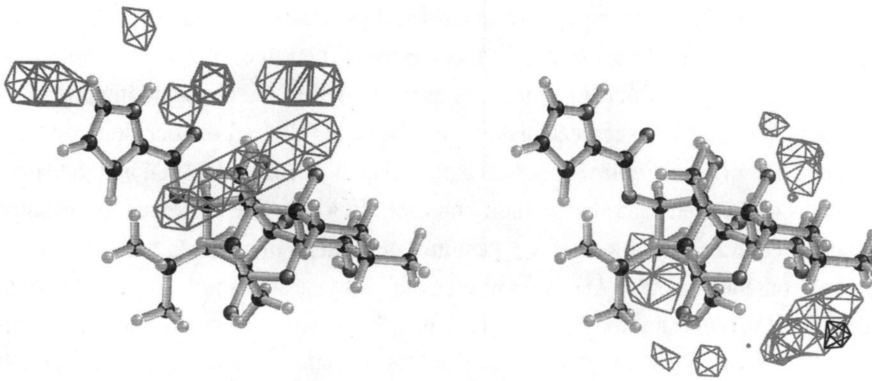

Fig. 7.4 An explanation of the binding affinity of ryanodine. The wire frames are contours at 25% of the maximum value in each case. The left panel shows those regions where steric bulk most strongly contributes to high-affinity binding of ryanodine. The wire frames in the right panel enclose those regions where electrostatic potential most strongly contributes to ryanodine binding.

often not intuitive. The CoMFA can also be used to provide an explanation of the biological properties of a particular compound (an explanatory CoMFA). One such example using ryanodine is shown in Fig. 7.4. The wire frames in the left panel enclose the atoms that contribute most of the steric factors responsible for the strong binding of ryanodine to the RyR. As would be expected from Fig. 7.2 these areas include the pyrrole, the 3-position and the hydrophobic face of ryanodine. The right panel shows the locations where the electrostatic field make the greatest contributions to high affinity binding of ryanodine. The grey wire frames denote the areas where positive electrostatic potential contributes most to high affinity binding. The black frame is the major area where negative electrostatic potential contributes to binding. These areas are mostly on the hydrophilic face of ryanodine. The notable exception is at the 15-position, the location of the sole hydroxyl group on the lipophilic face.

Interactions between the RyR and the 2-position of ryanoids

Comparison of CoMFA with data obtained by independent laboratories indicates CoMFA correctly predicts the importance of the 2(isopropyl)-position of the molecule to high affinity binding. For example, one modification of dehydroryanodine results in the formation of a double bond between the isopropyl group and the 2-position and the loss of the 2-hydroxyl group.[45] Except for the loss of the hydroxyl group, the change in configuration at the position is minor. The position of the two methyl groups is barely altered. The modification produces an increase in dissociation constant to 35 nM. (This value is obtained by converting the published data to dissociation constants using 4.6 nM as the dissociation constant of ryanodine. The results of such calculations will be called derived dissociation constants hereafter.) The original CoMFA[14] predicts a dissociation constant of 51 nM and the latest, expanded CoMFA predicts a dissociation constant of 64 nM. The increased dissociation constant is equivalent to a loss of 1.3 kcal/mol of binding energy. It is within the value expected from the loss of 6.6 $Å^2$ of solvent accessible surface area and 1-2 hydrogen bonds as a result of the removal of the 2-hydroxyl group.[64–66] These results support the hypothesis that the 2-position forms important ligand-receptor contacts. Placement of an epoxide between the 2-position and 13-position or a hydroxyl group on the 13-position also causes an increase in dissociation constant (65 and 329 nM respectively).[45]

Effect of substitutions at the 3-position

In ryanodine, a pyrrole carboxylic acid is attached to the 3-position (Fig. 7.1). As mentioned earlier, this group is found to be an important determinant for binding to the receptor.[14,61] No other substituent at this position has been reported to be as effective at promoting binding.

Ryanodyl nicotinate is a natural ryanoid[67] that illustrates the sensitivity of the dissociation constant to changes at the 3-position. Here a small increase of one carbon in the heterocyclic ring (pyrrole to pyridine) causes a 160-fold increase in dissociation constant. A similar increase in steric bulk is produced by methylating the pyrrole nitrogen.[45] The dissociation constant for this compound increased only eight-fold (derived dissociation constant equals 38 nM). The CoMFA predicts a 67 nM dissociation constant. The modification causes a loss of 1.3 kcal/mol of binding energy. At this point it is not clear how much of the reduced binding affinity to assign to the loss of a potential hydrogen bond between the pyrrole and the receptor and how much may be due to unfavorable steric interactions. The results indicate that increased molecular volume alone cannot account for the loss of binding.

Effect of substituents at the 10-position

Several workers have synthesized substituents at the 10_{eq}-position of both ryanodine and ryanodol.[11] It is notable that all such modifications have little effect on the binding of ryanodine analogs whereas these substituents have decided effects on the binding of the

Table 7.1 Dissociation constants of selected 10-substtuted ryanodine and ryanodol analogs[a]

substituent	dissociation constant (nM)	
	ryanodine series	ryanodol series
hydroxyl	7[b]	12000[b]
	6[f]	
10-*O*-(NCBZ-aminoacetyl)	1.4[c]	2300[d]
	38[e]	2000[e]
	4.8[f]	
10-*O*-(NCBZ-3-amino propanoyl)	6[f]	
	2.2[c]	1771[g]
10-*O*-(2-aminoacetyl)	0.5[c]	1400[d]
	38[e]	1300[e]
	2[f]	
10-*O*-(3-aminopropanoyl)	0.5[c]	
10-*O*-(NCBZ-4-amino butanoyl)	77[d]	490[d]
10-*O*-(2-guanidinoacetyl)	0.4[c]	
	2[f]	
10-*O*-(guanidino propanoyl)	0.3[e]	
10-*O*-pyrrolcarbonyl	12[e,h]	52[e]

[a]The dissociation constants are for skeletal muscle. Where data was orginally reported as EC_{50} they have been converted to dissociation constants.
[b]Ref. 14
[c]Ref. 39
[d]Ref. 71
[e]Ref. 61
[f]Ref. 40
[g]Welch *et al.*, unpublished results
[h]Walkiewicz *et al.*, unpublished results

ryanodol series. Table 7.I illustrates this point. Both charged and uncharged adducts to the 10-position have mixed effects on ryanodine binding; all adducts to the 10-position of ryanodol significantly enhance binding. The most dramatic example is that of 10-*O*-(NCBZ-4-aminobutanoyl)ryanodine where the adduct causes a 10-fold increase in dissociation constant in the ryanodine series but causes a 24-fold increase in the affinity in the ryanodol series. The differences may be due to the ability of the ryanodine analogs to bind to the receptor in multiple orientations. The conformational mobility of these adducts makes development of a definitive model difficult.

Interactions with the fused ring system

The original CoMFA[14] indicated that there were sites of important ligand-receptor interactions on the fused ring system. The most dramatic changes in dissociation constant were seen when the hemiketal ring in ryanodine was converted to a lactone ring. For example, the dissociation constant of anhydroryanodine ($D^{1,2}$-ryanodine–15,16,11-lactone) is 2900 nM compared to 7 nM for ryanodine (a loss of 3.7 kcal/mol). Other examples are listed in Ref. 11. Methylation of the hydroxyl group at the 4-position[45] results in a 0.8 kcal/mol loss of binding energy, consistent with the loss of a hydrogen bond. Conversion of the hydroxyl groups at the 4-position and 12-position to ketones (which causes loss of the carbon-carbon bond between these two positions) results in a substantial loss (2300 nM, 3.6 kcal/mol) in binding energy[11,44] (Ruest *et al.*, unpublished results). This compound has been named 4,12-seco-4,12-dioxyryanodine or oxoryanodine for short. The original CoMFA[14] is able to predict this dissociation constant. Another series of analogs have been synthesized where the carbon-carbon bond between the 4 and 12 positions was replaced by an oxygen bridge.[44] The 4 position and 12 position bear additional substituents ranging from hydroxyl groups (derived dissociation constant of 10 nM) to phenyl groups (derived dissociation constant of 246 nM).[44,45] Interestingly, the ketone oxygens of oxoryanodine, although smaller than most of the other substituents, had the largest effect on dissociation constant. In another example,[46] a cyclic methyl phosphonate bridging the 10 position and 12 position caused an increase in dissociation constant from 4.2 nM to 110–130 nM (derived dissociation constant). The CoMFA predicts a dissociation constant of 90 nM (Welch, unpublished results). These changes are the same magnitude as those seen for the ryanodine isomers (see section below on orientation of ryanoids). This series appears to make two contradictory points at once. First, interactions between the receptor and fused ring system are important. Second, large changes in steric bulk can be accommodated in the binding site without overwhelming loss of binding affinity. These results add further support for the hypothesis that, while the binding site interactions with the fused ring system are energetically important, the binding site is compliant and can accept multiple configurations in this region as long as sufficient polar and apolar contacts survive.[61,62]

Structural elements responsible for modulating channel properties

When ryanodine binds to the RyR, the channel persists in a long-lived subconductance state. The magnitude of the fractional conductance and the length of time the channel spends in the fractionally conducting state varies with the ryanoid structure.[48] There is an extremely strong correlation between ryanoid structure and fractional conductance but no correlation between binding affinity and fractional conductance.[62] A remarkable finding was that while substitutions at the 9-position and the 10-position had little effect on the dissociation constants of ryanodine analogs, these same substituents had profound effects on the fractional conductance of the modified RyR.[11,48,62] The strongest correlation between increasing positive electrostatic field and decreased fractional conductance is in a region centered at the 10-position (Fig. 7.5, right panel). The strongest correlation with increased negative charge and decreased fractional conductance is centered at the carbonyl oxygen near the 3-position (left panel). In general, the effect of increased steric bulk is to lower the fractional conductance (Fig. 7.6). Increased steric bulk is strongly correlated with decreased fractional conductance in two regions (right panel). One region is centered at the pyrrole. The other region is a broad correlation centered at the 10-position and includes the 9-position. The tendency of increased steric bulk to increase fractional conductance is limited and is centered at position 2 (left panel).

The CoMFA correlating changes in ryanoid structure with changes in fractional conductance are significant for two reasons. First, they support the hypothesis that the pyrrole carbonyl group can direct reorientation of the ligand in the binding site. Second, they show that fractional conductance is modulated independently of binding affinity.

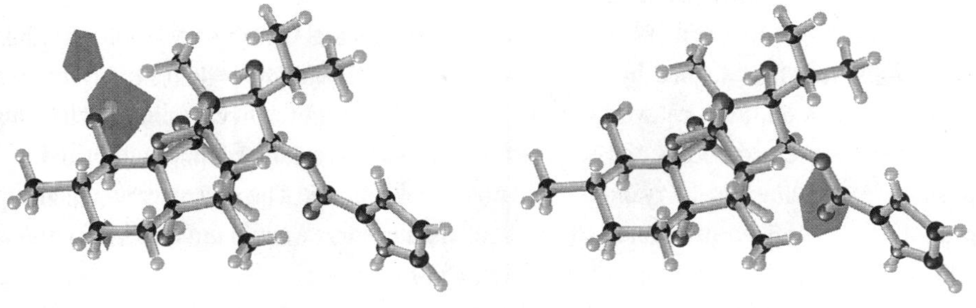

Fig. 7.5 The correlation between changes in electrostatic field and fractional conductance. The field is contoured at 50% of the maximum value of the correlation between changes in electrostatic field and fractional conductance. The left panel shows those areas where increased negative electrostatic potential is highly correlated with a decreased fractional conductance. The right panel shows those areas where increased positive electrostatic potential is highly correlated with a decreased fractional conductance.

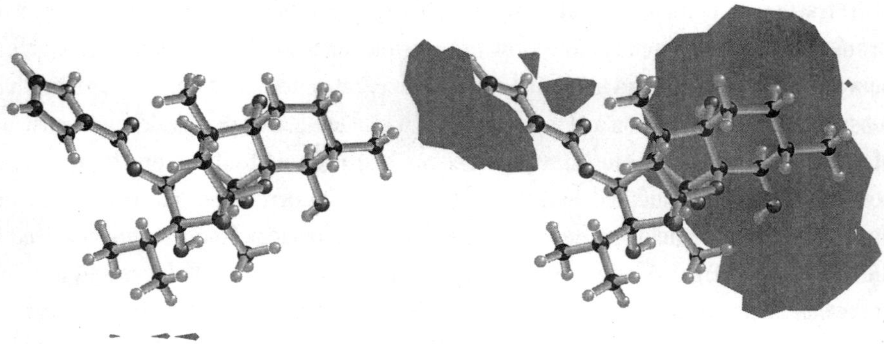

Fig. 7.6 The field is contoured at 33% of the maximum value. The left panel shows those regions where increased sterc bulk is highly correlated with an increased fractional conductance (centered at position 2 e.g. the isopropyl group). The right panel shows those regions where increased steric bulk is highly correlated with a decreased fractional conductance.

Orientation of ryanoids in the binding site

How can ryanoids bind in more than one orientation? The maps of the most important sites of interaction between ryanodine analogs and the receptor (Figs. 7.2,3,5,6) make the question of multiple orientations of the ligand important. A critical step in a CoMFA is the alignment of the members of the basis set by common structural elements. It is surmised that ligands bind to the receptor because of the similarities of the steric and electrostatic fields. Therefore, the initial alignment is commonly adjusted by field-fitting to maximize the similarities among all members of the basis set. In the original CoMFA of ryanodine binding all analogs were first aligned by atoms 1–3 (see Fig. 7.1) and then field fit.[14] Later, two isomers of ryanodine were synthesized, 3-epiryanodine and 10-ryanodine.[61] The only difference between ryanodine and 3-epiryanodine is the inversion of configuration at the 3-position. Ryanodine and 10-ryanodine are positional isomers. The pyrrolecarbonyl group is moved from the 3-position to the 10-position. Both isomers were found to bind to cardiac and skeletal muscle with a surprisingly high affinity. For example, using rabbit skeletal sarcoplasmic reticula the dissociation constants of ryanodine, 3-epiryanodine and 10-ryanodine are 7, 350, and 52 nM respectively.[61] The best agreement between experimental and predicted dissociation constants and fractional conductances was obtained when these two isomers were aligned to ryanodine using the pyrrole atoms.[61,62] These findings suggest that these two isomers bind to the RyR in orientations different from those of other members of the basis set.

It appears contradictory that the pyrrole could cause a reorientation within the binding site that would lead to many alterations in the important interactions between the RyR and both the isopropyl group at the 2-position and the fused ring system. At the outset, note that binding of ryanodine may not be completely optimum: in fact, it would be extremely fortunate if every contact between a plant alkaloid and a vertebrate receptor were complimentary. It is the net energy that leads to the observed high affinity binding. That having been said, note that neither 3-epiryanodine nor 10-ryanodine binds as tightly as does ryanodine (K_D = 350, 52, and 7 nM, respectively) and there are large differences in the dissociation constants of the three ryanoids. Clearly reorientation of the ligand causes significant losses in binding energy. The orientation of 3-epiryanodine is considerably different from10-ryanodine. This difference is reflected in the difference in the dissociation constants. A ligand will bind in the orientation that has the lowest free energy (see below). The pyrrole carbonyl group contributes about -4.2 kcal/mol of binding energy. Therefore, a ryanoid will bind in an alternate orientation if the gain of energy of interaction with the pyrrole carbonyl more than offsets the loss of interaction energy at other loci. By analyzing the common polar and hydrophobic areas, it was proposed that certain critical hydrophobic and polar contacts were required.[61] Overall ligand-receptor contacts, particularly the hydrophobic contacts, could be satisfied by a variety of groups, as long as there were no steric blocks. Assuming that the contact area between the ryanoid and the RyR is the same for all three isomers, then the major difference would be the number of hydrogen bonds made between receptor and ligand. Compared with ryanodine, the 10-isomer has 1.2 kcal/mol less binding energy whereas the 3-epi-isomer has 2.4 kcal/mol less binding energy.[61] There are ten electronegative atoms on ryanodine available to participate in one or more hydrogen bonds each. Assuming 0.5–1.5 kcal/mol for each hydrogen bond, changes in relative strength of a fraction of these potential hydrogen bonds can explain the differences in binding. Further, the steric complimentarity and hydrophobic interaction energy of 3-epiryanodine and 10-ryanodine may be greater or less than ryanodine. The point is that the magnitude of the energy changes is well within the number of potential contacts between the fused ring system of these ryanodine analogs and multiple alignments between ligand and receptor are energetically feasible.

The positional isomer 10-ryanodine makes a good example of energetic considerations. If 10-ryanodine were to bind in the same orientation as ryanodine, the interaction between the RyR and the 10-pyrrole carbonyl group would be minimal. Comparing ryanodine, 10-ryanodine, and 10-pyrrolecarbonylryanodine can quantify this point. The latter compound is ryanodine with a pyrrole carbonyl substituted at the 10-position, i.e. it bears the same substituent at both the 3-position and the 10-position. The experimental dissociation constants are 7, 52 and 12 nM respectively (Welch *et al.*, unpublished data). If all three, ligands were to bind in the same orientation, there would be a loss of 332 cal/mol of binding due to the 10-substitution: the predicted dissociation constant of 10-ryanodine is 20,571 nM. However, if 10-ryanodine were to bind in an alternate orientation that allowed the 10-pyrrole carbonyl to maximize interactions with the RyR, a gain of up to 4.2 kcal/mol of energy would be realized (see above): in this case the predicted dissociation constant of 10-ryanodine is 13 nM. In the reorientation model, this energy is more than sufficient to overcome loss of interactions elsewhere in the ligand. These lost interactions are reflected in the difference between the predicted and measured dissociation constants.

A visual inspection of the surface potential of the three ryanodine isomers (ryanodine, 3-epiryanodine and 10-ryanodine) in the common pyrrole orientation[61] was made using MOLCAD,[68,69] a method for visualizing the physicochemical properties of molecules by projecting these properties onto a Connelly solvent accessible surface. Although the pyrroles are aligned, the orientations of the rest of the molecules are much different. In spite of this, all three isomers had similar lipophilic and hydrogen bond surface potential distributions. In contrast, the electrostatic potential surface area varied greatly. While striking, the differences in electrostatic field may not be of much consequence for three reasons. First, the comparison of the explanatory CoMFA of ryanodine to the MOLCAD surface found an equal number of positive and negative electrostatic foci. Two, these isomers are neutral molecules. Therefore, the differences in surface potential are minor. Third, hydrogen bonds are likely to be the primary electrostatic interaction between ryanodine and receptor. The issue remains unresolved, however. Note that experimental data (see previous section titled "Interactions with the fused ring system") provide support for the hypothesis that the fused ring system can be reoriented in the binding site without a severe energetic penalty.

Interaction of ryanodine analogs with groups outside the binding site

Several substituents have been placed at the 9-position and the 10-equatorial positions of ryanodine.[11] These modifications produce only small perturbations in binding energy — far less than predicted from either attractive or repulsive interactions. It is unlikely that in so many different cases attractive and repulsive forces would be so well compensated. It is much more likely that these positions extend outward from the ryanodine binding pocket.[14] However, lack of steric repulsion does not eliminate ligand-RyR interactions. From measurements of fractional conductance, it is clear that the groups at the 9- and 10-positions can interact with the RyR.[48,62] Consider the experimental observation that the photoaffinity label 10-*O*-[3-(4-azidobenzamido)propionyl]ryanodine modifies the RyR.[70] The 10-pendant

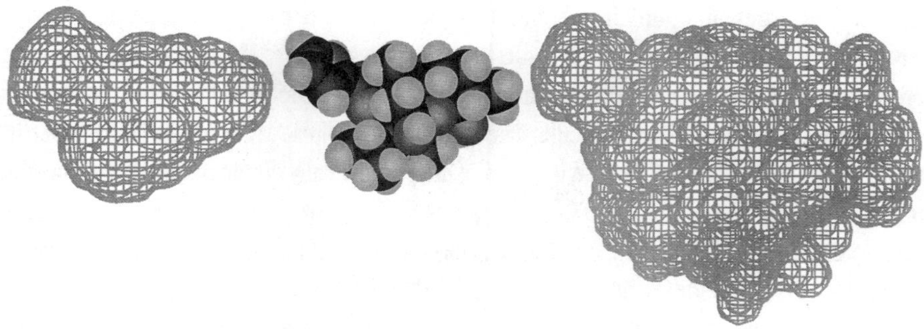

Fig. 7.7 Ryanodine and 10-*O*-[3-(4-azidobenzamido)propionyl]ryanodine were examined by molecular dynamics for 100 psec. The swept volumes of ryanodine (left) and 10-*O*-[3-(4-aidobenzamido)proponyl] ryanodine (right) are shown as wire frames. A space filling model of ryanodine (center) is shown for comparison. All molecules were aligned by positions one through four.

group is quite large and flexible. Molecular dynamics simulations (1 nsec simulation time, 37°C) indicate that the azido extends 8.9 ± 1.9 Å (average ± one standard deviation) from the 10-position. Further, as illustrated in Fig. 7.7, the mobility of this pendant group dramatically increases the swept volume of the molecule. Here, swept volume is defined as the space occupied by a molecule as a result of thermal motion at 37° C. A space-filling (CPK) representation of ryanodine is shown in the middle, the left side shows the swept volume of ryanodine, the right side shows the swept volume of 10-*O*-[3-(4-azidobenzamido)propionyl]ryanodine under identical conditions. Clearly, the 4-azidobenzamidopropionyl group can reach around and allow the terminal azido group contact with many amino acids at the periphery of the binding site. Therefore, lack of strong interactions does not preclude chemical reaction between ligand and receptor. In fact, this flexibility may provide a map of residues and peptide sequences that collaborate to form the ryanodine binding site.

Mechanism of Alterations in High Affinity Binding

The major steric and electrostatic determinants of high affinity binding of ryanodine (Fig. 7.4) were compared to the surface properties of ryanodine as estimated by MOLCAD.[68,69] Three physical properties, lipophilicity, electrostatic potential and hydrogen bonding potential were mapped to the water-accessible surface area of ryanodine. These properties are well defined in ryanodine. For example, there is an hydrophobic and hydrophilic face (or hemisphere) of the ryanodine molecule. The distribution of these MOLCAD surface properties was compared to the CoMFA-derived location of the energetically dominant forces responsible for high affinity of ryanodine (Fig. 7.4). The steric factors that contribute most to the affinity of ryanodine for the RyR are located in the lipophilic area of the molecule (and the area of negligible hydrogen bonding potential). On the other hand, there is no correlation between the steric determinants of binding and the electrostatic surface potential (this is probably significant in terms of the binding of the three ryanodine isomers discussed earlier). The converse is also true: the electrostatic factors primarily determining the strength of binding do not correlate with the lipophilic surface potential nor (surprisingly) with hydrogen bonding potential. The dominant electrostatic factors identified by CoMFA were generally concentrated in the regions of highest electrostatic surface potential (both negative and positive potential). These correlations provide some preliminary insights. An interpretation is that the primary steric contacts between receptor and ligand are within the hydrophobic area of the molecule; in contrast, the electrostatic interactions are diffuse and do not depend upon a specific steric interaction between the two moieties. Hydrogen bonds are apparently generic and do not depend upon specific pairings. Such observations are consistent with the proposal that the pyrrole group can guide the fused ring system into different orientations in the RyR binding site.

The energy analysis developed by Novotny and coworkers[64–66] can be adapted to ryanodine analog binding to gain more insight into the process. In this analysis, the free energy of ligand-receptor interaction is divided into five components: hydrophobic, electrostatic (including hydrogen bonding), conformational entropy, translational and rotational entropy, and cratic (dilutional entropy).[64–66] The dissection was developed to identify energetic interactions between a ligand and receptor where the structure of both are known. To adapt the paradigm to ryanodine analogs, only the differences in the free energy ($\Delta\Delta G$) of binding between the ryanodine analogs are considered. Since little is known about the structure of the ryanodine binding site on the RyR, protein contributions will be ignored. Because the ryanoids are all of similar mass, the last two entropic terms will remain constant and are not considered in the $\Delta\Delta G$ calculations. The change in hydrophobic energy is calculated as the change in the solvent accessible surface area (in Å^2) multiplied by 25 cal/mol. This represents the net increase in free energy arising from the increased entropy of water when the solute no longer perturbs the isotropic nature of solvent. The net change in energy due to charge-charge interactions when ligand binds to receptor is calculated as the gain or loss of ionic groups multiplied by 7 kcal/mol.[64] This is likely to be an overestimate of the energy of salt-links in an aqueous-faced binding site. Hydrogen bonding is estimated as a net gain of 0.5 kcal/mol when neither the acceptor nor donor atoms carry a formal charge and as 1.5 kcal/mol where the participating atom on the ryanodine analog carries a formal charge. The conformational entropy change is estimated by -NRln(3) where N is the number of rotatable bonds, R is the gas constant and 3 represents the number of rotamers per rotatable bond. The change in conformational entropy is estimated by the difference in the number of rotatable bonds between ryanoids. Using molecular dynamics, the loss or gain in the number of rotamers in a rotatable bond in the ryanodine analogs can be determined. Therefore, the change in conformational entropy can be measured. The relationship between free energy and entropy ($\Delta G = \Delta H - T\Delta S$) indicates that the loss of motional freedom of rotatable bonds will oppose binding. Ryanodine has little motional freedom;[14] therefore, changes in conformational entropy are expected to have little contribution to the binding free energy. Other ryanodine analogs have considerable motional freedom and this term is potentially significant if there are extensive receptor-ligand interactions with these regions of the ryanoid. Only a few applications of this analysis will be described here.

First, for purposes of scaling, consider the relationship between the experimentally determined binding free energy (ΔG) to that estimated from using the paradigm by Novotny.[64–66] A dissociation constant of 7 nM yields a free energy of –11.6 kcal/mol. The solvent accessible surface area of ryanodine is 359 Å^2. The maximum hydrophobic energy is estimated to be –8.975 kcal/mol (the receptor may contribute an additional –4.5 kcal/mol). If each electronegative atom on ryanodine were to participate in an –0.5 kcal/mol hydrogen bond with the receptor, an additional –5 kcal/mol is available for binding. Because there is little conformational freedom in ryanodine,[14] changes in conformational entropy are likely to be minor. The total estimate of 14 kcal/mol is considerably larger than experimental value and is, therefore, consistent with a model in which at least part of the ryanodine molecule extends out of the binding site.[61]

Substituents at the 3-position

The pyrrole carbonyl at the 3-position has enormous effect on binding affinity. The $\Delta\Delta G$ of binding between 3-deoxyryanodol (cinnzeylanol) and ryanodol is 0.2 kcal/mol. (In 3-deoxyryanodol, the 3-hydroxyl group of ryanodol has been replaced by a hydrogen. Ryanodol has the larger dissociation constant). The increased solvent accessible surface area (3 Å^2) predicts a 0.075 kcal/mol increase in binding energy and the hydroxyl group can potentially serve as an hydrogen bond donor and acceptor (–1 to –3 kcal/mol). Therefore, either the 3-hydroxyl group has weak interactions with the RyR or the favorable interactions are opposed by a conformational entropy change (potentially as much as +0.68 kcal/mol). Addition of the pyrrole carbonyl ester to 3-deoxyryanodol (to form ryanodine) is accompanied by an increase in the free energy of binding. The exact value varies with the particular RyR isoform. The $\Delta\Delta G = -4.0$ kcal/mol for mammalian cardiac muscle,[71] $\Delta\Delta G = -3.8$ for avian skeletal muscle and $\Delta\Delta G = -4.6$ for rabbit skeletal muscle[14,61] (average $\Delta\Delta G = -4.2$ kcal/mol). The increase in solvent accessible surface area accounts for –1.6 kcal/mol. The oxygen bridge does not appear to be solvent accessible and may be unable to participate in hydrogen bonding. If the pyrrole nitrogen can serve as a hydrogen bond donor and the carbonyl oxygen can serve as an acceptor for two hydrogen bonds, the total DDG of binding is estimated to be –3.1 kcal/mol or about a kilocalorie short of the experimental value. This additional energy may arise from the hydrophobic energy contributed by the receptor (0.8 kcal/mol), an interaction of the pyrrole with a charged amino acid side chain or may be related to the large dipole of pyrrole.

Ryanodyl nicotinate is a homolog of ryanodine where pyridine has replaced the pyrrole radical at the three position. This compound has 1.0 kcal/mol more binding energy than ryanodol. The 77 $Å^2$ increase in solvent accessible surface area predicts a 1.9 kcal/mol increase in binding energy. The relatively weak binding of ryanodyl nicotinate is probably due to steric conflicts at the binding site.[14]

Mechanism of Modulation of Channel Function: Occlusion of the Channel or Allosteric Modulation?

It has been generally accepted that the binding of ryanodine to a single high affinity site interrupts the normal conformational interconversions that gate the channel and prevents the channel from attaining the normal fully open conformation. In the presence of ryanodine, the channel has an open probability (for the partially conducting state) near unity. Occasional closings are observed but openings to the fully (unmodified) conducting state are not observed.[20,21] Two mechanisms are possible. In one case, the ryanodine analog binds within the conduction pathway thus directly blocking the permeant ion (accounting for the fractional conductance) and preventing the channel from fully conducting. This would be like a person blocking a doorway by standing in it (Fig. 7.8). The presumption would be that the channel can still close with the ryanoid in the conduction pathway or that the ryanoid dissociates momentarily. An alternate mechanism is that of an allosteric modulator. In this case the

Fig. 7.8 A cartoon representing two possible models of ryanodine modulation of RyR channel conductance as described in the text. Passage through the left door is blocked directly by the lgand. Conductance is reduced because ions must push pass the bulky agent. At the same time the presence of the ligand blocks the door open as the agent must move out of the doorway before it can close. In contrast, passage through the right door is blocked indirectly by the ligand which acts as a door stop preventing the door from fully opening. Because the door stop is wedged in the door, the door cannot easily close.

ryanoid would be like a door stop. The door stop does not directly block the path, but acts to keep the door from fully opening. Like a door wedge, the ligand does not block channel closure; but, like a wedge, friction may make closure unlikely. In this allosteric model, the various ryanoid structures stop the door open in various positions. (See Ref. 10 for a similar idea.) The model in Fig. 7.8 ignores interactions of the permeant ion with the channel: the cation may not be a passive component of the conduction process.[72] Structure-function studies where both cation and ryanoid are varied will be required to delineate the interaction between these two participants of channel function.

The allosteric mechanism has been favored[62] because structure-activity relationships are not consistent with the fractional conductance arising from direct blockage of the cation channel by the ryanoid. Briefly, the decrease in fractional conductance was too insensitive to both increased steric bulk and increased positive charge.

In an allosteric model, ryanodine binds preferentially to a particular conformer of the RyR or a portion of the ryanodine binding energy is directed toward induction of a change in RyR conformation. There is no direct evidence for ryanodine serving as an allosteric modulator; however, several attributes of the RyR indicate it is a conformationally flexible molecule and exhibits allosteric behavior.[5,8,10,11,16,21,22,73]

Cartoon of the Binding Site

Figure 7.9 is a cartoon of the binding site to illustrate what is currently believed to be the essential ryanoid-receptor interactions. Ryanodine is shown in the site for orientation. The binding site may be a hole or may be a narrow crevice (such as might be formed between two domains). The 3-position with the pyrrole, the 2-position with the isopropyl group and much of the fused ring structure are shown buried within the binding site. These are so located to reflect the sensitivity of binding to changes in these positions. The 9-position and 10-position are shown as extending away from the binding site to reflect the insensitivity of the dissociation constant to modifications at these positions. Substituents at this position are free to interact with remote parts of the molecule but such interactions do not involve tight steric complimentarity. Examination of the current CoMFA basis set indicates a common polar area encompassing the polar edge of the pyrrole and the carbonyl oxygen of the ester link. Two common apolar areas include the hydrophobic edge of the pyrrole and the 2-position. These are considered to be the essential interactions for tight binding of the ligand to the ryanodine binding site: molecules that present these features without excessive steric hindrances will be tightly bound.

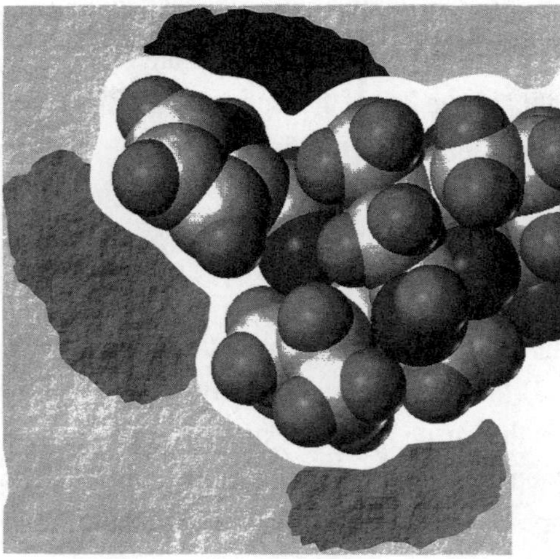

Fig. 7.9 A cartoon of the ryanodine binding site. A space filling model of ryanodine is shown occupying the site. The black area represents the essential polar ligand-receptor interaction; the dark grey areas represent the essential hydrophobic ligand-receptor interactions.

FUTURE TASKS

Much work remains to understand the RyR. Some of the future tasks of molecular modeling of the RyR include: (A) Construction of a CoMFA for low affinity binding. (B) Analysis of the correlation between structure and ryanoid-induced channel closure. (C) Analysis of ryanoid interactions with truncated (C-terminal 130 kDa) RyR.[74] (D) Utilization of other CoMFA descriptor fields.[26,75] (E) Inclusion of non-ryanoids in the CoMFA. (F) Development of isoform specific ryanoids. (G) Elucidation of the mechanism of modulation of RyR channel activity. (H) Development of models of conformationally mobile analogs.

SUMMARY

Application of molecular modeling has provided new insights into the relationship between ryanodine and the RyR. First, CoMFA is able to predict dissociation constants of, and modulation of channel function by, novel ryanoids. This critical step allows specific engineering of ligand properties. Comparison of surface properties and CoMFA identified three essential ligand-RyR interactions required for high-affinity binding and led to a preliminary model of the ryanodine binding site. A semiquantitative energy analysis is consistent with the proposed orientation of ryanoids in the RyR binding site. Never-the-less, a quantitative description of the RyR of the kind prescribed by Lord Kelvin remains in the future. The focus of current work is on meeting that goal.

ACKNOWLEDGEMENTS

The author thanks the American Heart Association (93012790) and the National Science Foundation (MCB-9317648) for their financial support. Special thanks to Prof Luc Ruest for producing the ryanoids that made our work possible, Prof John Sutko for a critical reading of a draft of this manuscript, Ms. Ruth Gault for drawing Fig. 7.8 and to the current members of our laboratory (Jennifer Herrick, Peter Thorkildson, Chris Lindekin, Tom Walkiewicz) for all their good work.

REFERENCES

1. Sorrentino V, Volpe P. Ryanodine receptors: how many, where and why? *Trends Pharmacol Sci* (TIPS) 1993; **14**:98–103.
2. Sutko JL, Airey JA. Ryanodine receptor Ca^{2+} release channels: Does diversity in form equal diversity in function? *Physiol Rev* 1996; **76**:1027–1071.
3. Williams, A. Ion conduction and discriminations in the SR ryanodine receptor/Ca^{2+}-release channel. *J Muscle Res Cell Motil* 1992; **13**:7–26.
4. McPherson PS, Campbell KP. The ryanodine receptor/Ca^{2+} release channel. *J Biol Chem* 1993; **268**:13765–13768.
5. Ogawa Y. Role of ryanodine receptors. *Crit Rev Biochem Mol Biol* 1994; **29**:229–274.
6. Coronado R, Morrissette J, Sukhareva M, Vaughan DM. Structure and function of ryanodine receptors. *Am J Physiol* 1994; **266**:C1485–C1504.
7. Meissner, G. Ryanodine receptor/Ca^{2+} release channels and their regulation by endogenous effectors. *Ann Rev Physiol* 1994; **56**:485–508.
8. Dulhunty AF, Junankar PR, Eager KR, *et al.* Ion channels in the sarcoplasmic reticulum of striated muscle. *Acta Physiol Scand* 1996; **156**:375–385.
9. Ishide N. Intracellular calcium modulators for cardiac muscle in pathological conditions. *Jpn Heart J* 1996; **37**:1–17.
10. Kasai M, Ide T. Regulation of calcium release channel in sarcoplasmic reticulum. *Ion Channels* 1996; **4**:303–301.
11. Sutko JL, Airey JA, Welch W, Ruest L. The pharmacology of ryanodine and related compounds. *Pharmacol Rev* 1997; **49**:53–98.

12. Jefferies, PR, Casida JE. Ryanoid chemistry and action. In: Hedin PA, Menn JJ, Hollingworth RM, eds. *Natural and Engineered Pest Management Agents*. ACS Symposium Series No 551. Washington, DC: American Chemical Society, **1994**:130–144.
13. Zucchi R, Ronca-Testoni S. The sarcoplasmic reticulum Ca^{2+} channel/ryanodine receptor: modulation by endogenous effectors, drugs and disease states. *Pharmacol Rev* 1997; **49**:1–51.
14. Welch W, Ahmad S, Airey JA, *et al.* Structural determinants of high-affinity binding of ryanoids to the vertebrate skeletal muscle ryanodine receptor: A comparative molecular field analysis. *Biochemistry* 1994; **33**:6074–6085.
15. Lai FA, Erickson HP, Rousseau E, *et al.* Purification and reconstitution of the calcium- release channel from skeletal muscle. *Nature* (Lond.) 1988; **331**:315–319.
16. Lai FA, Misra M, Xu L, *et al.* The ryanodine receptor-Ca^{2+} release channel complex of skeletal muscle sarcoplasmic reticulum: evidence for a cooperatively coupled, negatively charged homotetramer. *J Biol Chem* 1989; **264**:16776–16785.
17. Carroll S, Skarmeta JG, Yu X *et al.* Interdependence of ryanodine binding, oligomeric receptor interactions, and Ca^{2+} release regulation in junctional sarcoplasmic reticulum. *Arch Biochem Biophys* 1991; **290**:239–247.
18. Pessah IN, Zimanyi I. Characterization of multiple [^{3}H]ryanodine binding sites on the Ca^{2+} release channel of sarcoplasmic reticulum from skeletal and cardiac muscle: Evidence for a sequential mechanism in ryanodine action. *Mol Pharmacol* 1991; **39**:679–689.
19. Wang, JP, Needleman DH, Hamilton SL. Relationship of low affinity [^{3}H] ryanodine binding sites to high affinity sites on the skeletal muscle Ca^{2+} release channel. *J Biol Chem* 1993; **268**:20974–20982.
20. Rousseau E, Smith JS, Meissner G. Ryanodine modifies conductance and gating behavior of single Ca^{2+} release channel. *Am J Physiol* 1987; **253**:C364–C368.
21. Smith JS, Imagawa T, Ma J, *et al.* Purified ryanodine receptor from rabbit skeletal muscle is the calcium-release channel of sarcoplasmic reticulum. *J Gen Physiol* 1988; **92**:1–26.
22. Netticadan, T, Xu A, Narayanan N. Divergent effects of ruthenium red and ryanodine on Ca^{2+}/calmodulin-dependent phosphorylation of the Ca^{2+} release channel (ryanodine receptor) in cardiac sarcoplasmic reticulum. *Arch Biochem Biophys* 1996; **333**:368–376.
23. Burkert U, Allinger NL. *Molecular Mechanics.* ACS Monograph 177. Washington DC:American Chemical Society, 1982:339pp.
24. van de Waterbeemd H, ed. *Chemometric Methods in Molecular Design. Methods and Principles in Medicinal Chemistry,* Vol. 2. New York:Weinheim VCH, 1995:359 pp.
25. van de Waterbeemd H, ed. *Advanced Computer-assisted Techniques in Drug Discovery. Methods and Principles in Medicinal Chemistry,* Vol. 3; New York:Weinheim VCH, 1995:343 pp.
26. Hansch C, Leo A. *Exploring QSAR. Volume 1: Fundamentals and Applications in Chemistry and Biology.* Washington, DC: American Chemical Society,1995:557pp.
27. Leo A, Hansch C, Hoekman D. *Exploring QSAR. Vol. 2: Hydrophobic, Electronic, and Steric Constants.* Washington DC: American Chemical Society, 1995:1000pp.
28. Cramer CJ, Truhlar DG, (Eds) *Structure and Reactivity in Aqueous Solution: Characterization of Chemical and Biological Systems.* ACS Symposium Series No. 568, Washington DC. American Chemical Society, 1994:438pp.

29. Stone AJ. *The Theory of Intermolecular Forces.* Oxford:Clarendon Press, 1996:264pp.
30. Comba P, Hambley TW. *Molecular Modeling.* New York:Weinheim VCH, 1995:197 pp.
31. Smith DA, ed. *Modeling the Hydrogen Bond.* ACS Symposium Series No. 569. Washington DC: American Chemical Society, 1994:300pp.
32. Kumosinski TF, Liebman MN, eds. *Molecular Modeling: from Virtual Tools to Real Problems.* ACS Symposium Series No. 576, Washington DC: American Chemical Society, 1994:516pp.
33. Richards WG, ed. *Computer-aided Molecular Design.* London:IBC Technical Services Ltd., 1989: 266pp.
34. Gilson MK, Given JA, Bush BL, McCammon JA. The statistical-thermodynamic basis for computation of binding affinities: A critical review. *Biophys J* 1997; **72**:1047–1069.
35. Breed J, Sankararamakrishnan R, Kerr ID, Sansom MSP. Molecular dynamics simulations of water within models of ion channels. *Biophys J* 1996; **70**:1643–1661.
36. Smart OS, Breed J, Smith GR, Sansom MSP. A novel method for structure-based prediction of ion channel conductance properties. *Biophys J* 1997; **72**:1109–1126.
37. Zeidel ML Low permeabilities of apical membranes of barrier epithelia: what makes watertight membranes watertight? *Am J Physiol* 1996; **271(2 pt 2)**:F243–245.
38. Cramer III RD, Patterson DE, Bunce JD. Comparative molecular field analysis (CoMFA). 1. Effect of shape on binding of steroids to carrier proteins. *J Am Chem Soc* 1988; **110**:5959–5967.
39. Bidasee KR, Besch Jr HR, Gerzon K, Humerickhouse, RA. Activation and deactivation of sarcoplasmic reticulum calcium release channels: molecular dissection of mechanisms via novel semi-synthetic ryanoids. *Mol Cell Biochem* 1995; **149/150**:145–160.
40. Gerzon K, Humerickhouse RA, Besch Jr HR, *et al.* Amino- and guanidinoacylryanodines: basic ryanodine esters with enhanced affinity for the sarcoplasmic reticulum Ca^{2+}-release channel. *J Med Chem* 1993; **36**:1319–1323.
41. Humerickhouse RA, Besch Jr HR, Gerzon K, *et al.* Differential activating and deactivating effects of natural ryanodine congeners on the calcium release channel of sarcoplasmic reticulum: evidence for separation of effects at functionally distinct sites. *Mol Pharmacol* 1993; **44**:412–421.
42. Humerickhouse RA, Bidasee KR, Gerzon K, *et al.* High affinity C_{10}-O_{eq} ester derivatives of ryanodine: activator-selective agonists of the sarcoplasmic reticulum calcium release channel. *J Biol Chem* 1994; **269**:30243–30253.
43. Jefferies PR, Toia RF., Brannigan B, *et al.* Ryania insecticide: analysis and biological activity of 10 natural ryanoids. *J. Agric. Food Chem.* 1992; **40**:142–146.
44. Jefferies, PR, Lemberg E, Lam W-W, Casida, JE. Bioactive ryanoids from nucleophilic additions to 4,12-seco-4,12-dioxoryanodine. *J Med Chem* 1993; **36**:1128–1135.
45. Jefferies PR, Blumenkopf TA, Gengo PJ, *et al.* Ryanodine action at calcium release channels. 1—importance of hydroxyl substituents. *J Med Chem* 1996; 39:2331–2338.
46. Jefferies PR, Gengo PJ, Watson MJ, Casida JE. Ryanodine action at calcium release channels. 2—relation to substituents of the cyclohexane ring. *J Med Chem* 1996; **39**:2339–2346.
47. Kahl SD, McPherson PS, Lewis T, *et al.* Radioimmunoassay for the calcium release channel agonist ryanodine. *Anal Biochem* 1994; **218**:55–62.

48. Tinker A, Sutko JL, Ruest L, *et al.* Electrophysiological effects of ryanodine derivatives on the sheep cardiac sarcoplasmic reticulum calcium-release channel. *Biophys J* 1996; **70**: 2110–2119.
49. Waterhouse AL, Pessah IN, Francini AO, Casida JE. Structural aspects of ryanodine action and selectivity. *J Med Chem* 1987; **30**:710–716.
50. Marshall GR, Cramer III RD. Three-dimensional structure-activity relationships. *Trends Pharmacol Sci* (TIPS) 1988; **9**:285–289.
51. Green SM, Marshall GR. 3D-QSAR: a current perspective. *Trends Pharmacol Sci* (TIPS) 1995; **16**:285–291.
52. Waller CL, Juma BW, Gray Jr LE, Kelce WR. Three-dimensional quantitative structure-activity relationships for androgen receptor ligands. *Toxicol Applied Pharmacol* 1996; **137**:219–227.
53. Carotti A, Altomare C, Cellamare S *et al.* LFER and CoMFA studies on optical resolution of a-alkyl a-aryloxy acetic acid methyl esters on DACH-DNB chiral stationary phase. *J Computer-Aided Molecular Des* 1995; **9**:131–138.
54. Thull U, Kneubuhler S, Gaillard P, *et al.* Inhibition of monoamine oxidase by isoquinoline derivatives: qualitative and 3D-quantitative structure-activity relationships. *Biochem Pharmacol* 1995; **50**:869–877.
55. Brandt W, Lehmann T, Willkomm C *et al.* CoMFA investigations on the series of artificial peptide inhibitors of the serine protease thermitase. Synthesis of an inhibitor of predicted greater potency. *Int J Pept Protein Res* 1995; **46**:73–78.
56. Briens F, Bureau R, Rault S, Robba M. Applicability of CoMFA in ecotoxicology: a critical study on chlorophenols. *Ecotoxicol Environ Saf* 1995; **31**:37–48.
57. Kim KH, Martin YC. Direct prediction of dissociation constants (pK_A's) of clonidine-like imidazolines, 2-substituted imidazoles and 1-methyl-2-substituted-imidazoles from 3D structures using a comparative molecule field analysis (CoMFA) approach. *J Med Chem* 1991; **34**:2056–2060.
58. Boyd DB, Lipkowitz KB. Molecular mechanics: the method and its underlying philosophy. *J Chem Ed* 1982; **59**:269–274.
59. Cox PJ. Molecular mechanics: illustrations of its application. *J Chem Ed* 1982; **59**:275–277.
60. Clark M, Cramer III RD, Van Opdenbosch N. Validation of the general purpose Tripos 5.2 force field. *J Computational Chem* 1989; **10**:982–1012.
61. Welch W, Sutko JL, Mitchell KE, *et al.* The pyrrole locus is the major orienting factor in ryanodine binding. *Biochemistry* 1996; **35**:7165–7173.
62. Welch W, Williams AJ, Tinker A, *et al.* Structural components of ryanodine responsible for modulation of sarcoplasmic reticulum calcium channel function. *Biochemistry* 1997; **36**:2939–2950.
63. Nicklaus MC, Milne GW, Burke Jr TR. QSAR of conformationally flexible molecules: comparative molecular field analysis of protein-tyrosine kinase inhibitors. *J Computer-Aided Molecular Design* 1992; **6**:487–504.
64. Novotny J, Bruccoleri RE, Saul FA. On the attribution of binding energy in antigen-antibody complexes McPC603, D1.3 and HyHEL-5. *Biochemistry* 1989; **28**:4735–4749.

65. Krystek S, Stouch T, Novotny J. Affinity and specificity of serine endopeptidase-protein inhibitor interactions: empirical free energy calculations based on X-ray crystallographic structures. *J Mol Biol* 1993; **234**:661–679.
66. Tulip WR, Harley VR, Webster RG, Novotny J. N9 neuraminidase complexes with antibodies NC41 and NC10: empirical free energy calculations capture specificity trends observed with mutant binding data. *Biochemistry* 1994; **33**:7986–7997.
67. Jefferies PR, Toia RF, Casida JE. Ryanodyl 3-(pyridine-3-carboxylate): a novel ryanoid from *Ryania* insecticide. *J Nat Prod* (Lloydia) 1991; **54**:1147–1149.
68. Heiden W, Goetze T, Brickmann J. Fast generation of molecular surfaces from 3D data fields with an enhanced "marching cube" algorithm. *J Comp Chem* 1993; **14**:246–250.
69. Heiden W, Moeckel G, Brickmann J. A new approach to the display of local lipophilicity/hydrophilicity mapped on molecular surfaces. *J Comp-Aided Mol Design* 1993; **7**:503–514.
70. Witcher DR, McPherson PS, Kahl SD, *et al.* Photoaffinity Labeling of the Ryanodine Receptor/Ca^{2+} release channel with an azido derivative of ryanodine. *J Biol Chem* 1994; **269**: 13076–13079.
71. Mitchell KE, Velasco J, Ruest L, *et al.* The structural determinants for ryanodine binding are different than the molecular features which correlate to channel activation. *Biophys J* 1996; **70**:A388.
72. Lindsay ARG, Tinker A, Williams AJ. How does ryanodine modify ion handling in the sheep cardiac sarcoplasmic reticulum Ca^{2+}-release channel? *J Gen Physiol* 1994; **104**:425–447.
73. Zorzato F, Menegazzi P, Treves S, Ronjat M. Role of malignant hyperthermia domain in the regulation of Ca^{2+} release channel (ryanodine receptor) of skeletal muscle sarcoplasmic reticulum. *J Biol Chem* 1996; **271**:22759–22763.
74. Bhat MB, Zhao J, Takeshima H, Ma J. The carboxyl-terminal portion of skeletal muscle ryanodine receptor forms a regulated calcium release channel. *Biophys J* 1997; **72**:A333.
75. Kellogg GE, Semus SF, Abraham DJ. HINT: a new method of empirical hydrophobic field calculation for CoMFA. *J Comput-Aided Mol Des* 1991; **5**:545–552.

Chapter 8

Phosphorylation of Ryanodine Receptors

Judith A. Airey and Rong He

INTRODUCTION

For numerous cellular processes phosphorylation has been shown to regulate protein function. This is particularly true for cell signalling; a field which has exploded with the identification of numerous pathways, kinases and substrates. It is natural, therefore, to ask whether ryanodine receptors (RyR), intracellular calcium-release channels, which participate in signalling in many cell types are controlled by phosphorylation, and if so, what is the physiological significance of this phosphorylation? A number of physiological conditions have been identified in both cardiac and skeletal muscle where SR calcium-release is altered. For instance phosphorylation could play a regulatory role in RyR channel function during β-adrenergic stimulation of the heart, and in skeletal muscle during tetanus and fatigue.[1–5]

Phosphorylation of the RyR calcium-release channels has been studied most extensively in cardiac and skeletal muscle, however, this work has not yet yielded a cohesive picture. There is not a general agreement as to the stoichiometries and functional consequences of phosphorylation of the RyR isoforms and in many instances the data are contradictory.

This chapter contains a brief survey of the literature on phosphorylation of RyR isoforms and a discussion of some of the problems associated with the study of phosphorylation of these proteins.

Kinases Relevant to SR Phosphorylation

Analysis of RyR phosphorylation has involved both endogenous and exogenous kinases. Exogenous kinases are those added to the ryanodine receptor preparations. During purification of RyR any kinases not tightly associated with the SR membrane will be removed. These kinases are typically found in the cytoplasm; and could be translocated to the SR, associated with other membranes in the vicinity of the SR, or localized close to the SR. Endogenous kinases are those activities that remain associated with SR membranes, and are therefore positioned in close proximity to SR proteins. This category may be somewhat artificial since the ionic strength or pH of the buffers used in the preparation of the membranes may affect which, and at what level, kinases remain associated with the SR. Several kinase activities have been reported to be associated with SR membranes. The calcium/calmodulin-dependent kinase (CaMK) has been extensively studied in skeletal muscle and found to be a 60 kDa protein which can autophosphorylate.[6–10] As well as being membrane bound, CaMK is found in a soluble form[11] and has been isolated from both cardiac[12] and skeletal[13] muscle. In addition, phosphorylation of the skeletal muscle RyR by an endogenous kinase(s) associated with the SR has been demonstrated in the absence of calcium and/or calmodulin,[14] indicating the existence of additional endogenous kinases in the SR.

A further finding, that may affect regulation by phosphorylation, is the suggestion that ATP synthesis is compartmentalized in skeletal muscle triads.[15] Thus, in addition to the kinase localization, the availability of ATP may influence the phosphorylation of SR proteins. It should be noted that in some of the experiments described below, exogenously added kinases were purified from tissues and species other than those being studied. It is not known whether 'foreign' kinases completely mimic the effects of a kinase native to a particular tissue and species. Additionally it is unknown if, during purification of kinases, accessory proteins that may regulate a kinase are removed.

Initial Studies of RyR Phosphorylation

In 1984 it was demonstrated by SDS polyacrylamide gel electrophoresis and autoradiography that high molecular mass proteins were phosphorylated in both cardiac and skeletal junctional SR.[16] Subsequent sequencing of the proteins now known as the are RyR[17–21] showed that they are large proteins and therefore, the possibility existed that the RyR were the high molecular mass proteins that were phosphorylated. In mammals three isoforms of the RyR have been identified.[21,22] The predominant skeletal muscle isoform (type 1) and the cardiac isoform (type 2) have been extensively studied. Less is known about the brain isoform (type 3), consequently this isoform is discussed only briefly. All three isoforms contain numerous potential phosphorylation sites.[17–21]

CARDIAC MUSCLE RyR

There is a general consensus that the RyR in cardiac muscle is phosphorylated to a significant extent, and that phosphorylation regulates its calcium channel activity. Early evidence from Takasago *et al.* showed that the cardiac RyR was phosphorylated in vitro by exogenous PKA.[23] After phosphorylation of SR membranes, RyR were immunoprecipitated. Approximately 90% of the [^{3}H]ryanodine that had been prebound to the membranes was precipitated, verifying that the phosphoprotein identified was RyR. An effect of phosphorylation on RyR channel activity was suggested by the observation that phosphorylation increased [^{3}H]ryanodine binding by 30%. A stoichiometry of 1.59 mol phosphate/mol RyR tetramer, or 0.4 mol phosphate/mol monomer of receptor (see Table 8.1) was estimated. The issue of the effect of phosphorylation on [^{3}H]ryanodine binding, and hence on the stoichiometries of phosphorylation is further discussed later in this chapter.

Since the work of Takasago *et al.*[23], there have been many studies of phosphorylation of the cardiac RyR. Witcher *et al.*[24] identified Ser 2809 as the site phosphorylated by both exogenous and endogenous CaMKs. This serine is within consensus sequences for phosphorylation by both CaMK and PKA. Although Ser 2809 is not close to the putative membrane spanning region of the receptor in the primary sequence of RyR, it is in a region that contains several sequences suggested to be involved in regulation of the RyR channel.[19,20] However, other studies using peptide mapping[25] showed that the major phosphopeptide in a tryptic digest of cardiac RyR phosphorylated by exogenous CaMK is not the same peptide that is obtained after phosphorylation by exogenous PKA, PKC or PKG. The major phosphopeptide was the same for PKA, PKC and PKG suggesting that these kinases share a site of phosphorylation that differs from that acted on by exogenous CaMK in vitro. Thus these results suggest that serine 2809 may not be a site of phosphorylation for PKA, PKC or PKG in the cardiac RyR.

Table 8.1 Ryanodine Receptor Phosphorylation Stoichiometries Obtained in Vitro and in Vivo

	SOURCE	PHOS SUBSTRATE	$CaMK_{end}$	$CaMK_{exog}$	PKA	PKG	PKC	Ref
In vitro								
Cardiac								
	dog	SR[a]	1.1–1.3	4.3-5.2	0.9–1.1	–	–	24
	pig	SR[a]	2.0	–	1.2	–	–	27
	dog	SR[a]	3.4–6.9		1.0–2.0	0.9–1.8	0.8–1.6	25
	dog	SR[a]	–	–	0.7[a]	–	–	23
	dog	RyR[b]	–	–	0.3[b]	–	–	
	cow	SR[a]	0.4	2.0	1.0	0.4	–	26
	cow	RyR[b]	0.2	2.1	0.8	0.2	–	26
	dog	SR[a]	–	1.2	1.6	–	–	31
Skeletal								
	pig	SR[a]	0.2	–	0.02	–	–	27
	rabbit	RyR[b]	–	0.8	1.8	0.8	–	44
	rabbit	RyR[b]	–	–	3.5	2.4	–	44
	rabbit	SR[a]	–	0.9	1.9	–	–	45
	rabbit	SR[a]	–	0.9	1.9	–	0.9	46
In vivo								
Cardiac								
	rat	vent myocytes[ac]	–	–	0.45[d]	0.15[e]	–	42
Skeletal								
	rabbit	fast twitch[b]	–	(1.6,2.8,3.6)[f]	–	–	–	44

Abbreviations: phos substrate = preparation used in phosphorylation reaction; $CaMK_{end}$ = CaMK activity endogenous to SR membrane; $CaMK_{exog}$ = CaMK added to phosphorylation reaction; PKA = cAMP-dependent protein kinase; PKG = cGMP-dependent protein kinase; PKC = phospholipid, Ca^{2+}-dependent protein kinase; SR = sarcoplasmic reticulum membranes; RyR = purified ryanodine receptor.

All stoichiometries expressed as mol Pi/mol RyR tetramer

[a]Stoichiometry based on [^{3}H]ryanodine binding.

[b]Stoichiometry based on mass of RyR monomer added to reaction.

[c]Ventricular myocytes permeabilized using saponin.

[d]In presence of added cAMP.

[e]In absence of added cAMP.

[f]Values obtained for different muscle samples for Ser 2843. No exogenous added stimuli.

The stoichiometries obtained for phosphorylation of the cardiac RyR by different laboratories are not consistent (see Table 8.1). For instance, for endogenous CaMK, stoichiometries range from 0.5[26] to 4.0[25] mol phosphate/mol RyR tetramer. The reasons for this variation are not clear, and are considered further below. It has been reported that the phosphorylation of PKA and endogenous CaMK were additive[27] and that exogenous CaMK and PKA were not additive.[26] These last observations are agreement with the mapping studies[25] suggesting that exogenous CaMK and PKA do not phosphorylate the same site, but do not agree with the data that both endogenous and exogenous CaMK phosphorylate Ser 2809.[24]

Effects of Phosphorylation on Cardiac RyR Function

A caveat related to some of the studies discussed in this section is that in many studies it was not determined in the same study whether phosphorylation of RyR, or of a protein associated with RyR, effectuated the change in receptor function (see Table 8.2).

Bilayer Studies

Calmodulin can have an inhibitory effect on calcium release.[28–30] In studies, using canine junctional SR membranes incorporated into bilayers, Witcher *et al.*[24] showed that addition of exogenous CaMK, to the cis (cytoplasmic) side of the channel, reversed the inhibition of RyR by calmodulin. The decreased open probability (P_o) caused by calmodulin was reversed by the addition of CaMK to the cis chamber. Subsequent experiments by Hain and coworkers[31] confirmed these findings and further, found that addition of PKA could not reactivate the channel after calmodulin inhibition. This difference in the effects of exogenous CaMK and PKA is additional evidence that these kinases may phosphorylate different sites in the cardiac RyR.

Data of Hain *et al.*[31] using canine cardiac SR incorporated into bilayers, indicate that phosphorylation can reverse the inhibition of the RyR channel by Mg^{2+}. They showed that in the presence of mM Mg^{2+} most RyR are blocked. Addition of exogenous CaMK, alleviated the Mg^{2+} inhibition. Application of potato acid phosphatase or protein phosphatase 1 restored the inhibition. Channels insensitive to Mg^{2+} inhibition prior to the addition of exogenous CaMK, were inactivated by the application of protein phosphatase 1, indicating that some of the channels were already phosphorylated. Hain *et al.*[31] also suggested that exogenous PKA was more effective than exogenous CaMK in reversing Mg^{2+} inhibition as application of

Table 8.2 Functional Consequences of in Vitro Ryanodine Receptor Phosphorylation

	SOURCE	PHOS SUBSTRATE	KINASE/ PPase	[^{3}H]RY BINDING	BILAYER CHANNEL	SR Ca^{2+} PERM	RYR PHOS	FORCE	REF
Cardiac									
	dog	SR/bilayer	$CaMK_{exog}$	–	↑P_o [a]	–	yes	–	24
	dog	SR	PKA	↑B_{max}, ↑Ca-S	–	–	yes	–	23,35
	dog	SR	PKG	↑B_{max}	–	–	yes	–	25
	dog	SR	PKC	↑B_{max}	–	–	yes	–	25
	dog	SR	$CaMK_{end}$	↓B_{max}, ↓Ca-S	–	–	yes	–	25
	dog	SR/bilayer	PKA	–	↑P_o [b]	–	yes	–	31
	dog	RyR/bilayer	$CaMK_{exog}$	–	↑P_o [ab]	–	yes	–	31
	dog	SR/bilayer	$CaMK_{end}$	–	↓P_o	–	no	–	31
	dog	SR/bilayer	PKA	–	↑P_o[c],↑adapt[d]	–	no	–	34
	pig/rabbit	SR, SR/bilayer	$CaMK_{exog}$	↓B_{max}	↓P_o	–	no	–	32
	pig/rabbit	SR, SR/bilayer	acid PPase	↓K_d	↑P_o	–	no	–	32
	rat	trabeculae	ND		–	–	no	↑	39
Skeletal									
	dog	SR/bilayer	$CaMK_{exog}$	–	no effect	–	–	–	24
	frog	SR/patch	$CaMK_{end}$	–	↓P_o	–	no	–	48
	rabbit	SR	$CaMK_{end}$	–	–	no effect	yes	–	9
	rabbit	SR/bilayer	PKA	–	↑P_o [b]	–	yes	–	45
	rabbit	SR/bilayer	$CaMK_{exog}$	–	↑P_o [bc]	–	yes	–	45
	rabbit	SR/bilayer	$CaMK_{end}$	–	↓P_o [d]	–	no	–	45
	rabbit	SR	PKA	–	–	↑	yes	–	46
	rabbit	SR	$CaMK_{exog}$	–	–	↑	yes	–	46
	rabbit	SR	PKC	–	–	↑	yes	–	46
	rabbit	SR/bilayer	ND	–	↑P_o,↑Ca-S	–	yes	–	50
Brain									
	rabbit	ER	PKA	↑B_{max}	–	–	no	–	72

Abbreviations not used in Table 1: [^{3}H] ry binding = [^{3}H] ryanodine binding; bilayer channel = ryanodine receptor channels recorded in planar lipid bilayers; SR Ca^{2+} perm = permeability of SR membranes to Ca^{2+}; RyR phos = determined that ryanodine receptor was phosphorylated; SR/bilayer = SR membranes fused with bilayers; ND = not determined; P_0 = probablity of channel opening; adapt = adaptation response; Ca-S = sensitivity to Ca^{2+}; PPase = phosphatase; trabeculae = saponin permeabilised ventricular trabeculae; ↑ = increased; ↓ = decreased

[a]Increased channel activity in the presence of calmodulin

[b]Increased channel activity in the presence of 1 mM Mg^{2+}

[c]Increased channel activity in response to 10 μM Ca^{2+}

[d]Increased rate of adaption to Ca^{2+} activation

exogenous CaMK resulted in lower channel activity. Contrary to the effects of exogenous CaMK, it was found that endogenous CaMK inhibits RyR channel activity (in the absence of Mg^{2+}) and that this inhibition could be reversed by addition of phosphatases. Therefore, in this experimental system exogenous and endogenous CaMK have opposite effects, and depending on which kinase is used to phosphorylate protein in the bilayer, treatment with phosphatases can also have opposing effects on channel function. Since the Mg^{2+} concentrations that inhibit the opening of RyR in bilayers are those that exist in intact cardiac muscle, Hain *et al.*[26] suggest that phosphorylation by exogenous CaMK or PKA is required for activity of cardiac RyR channels in vivo.

The role of phosphorylation in regulating RyR channel activity has also been addressed by Lokuta *et al.*[32] and their results do not agree with Hain *et al.*[31] In rabbit and pig cardiac SR exogenous CaMK decreased the P_o of RyR. This effect was reversed by the addition of acid phosphatase; with the channels exhibiting a greater P_o after phosphatase treatment than observed prior to the exogenous CaMK treatment. Based on these data, the authors suggested that there are two phosphorylation sites (one of which is a CaMK site) that control the activity of RyR, and that acid phosphatase is able to dephosphorylate both sites. An alternative explanation for these results is that some of the channels in their SR preparations were already phosphorylated and hence addition of phosphatase dephosphorylated all channels, not just those phosphorylated in vitro. The results of Hain and coworkers[31] and Lokuta and coworkers[32] are ostensibly opposite. However, it should be noted that the buffer used by Lokuta *et al.*[32] does not appear to contain Mg^{2+}.

In other bilayer studies the effects of phosphorylation on adaptation of the cardiac RyR were investigated. Adaptation is a phenomenon in which successive increases in $[Ca^{2+}]$ result in a transient activation of RyR (see Chapters 13 & 14 for detailed discussion). Excitation-contraction coupling has a time course of milliseconds, but initial studies on adaptation indicated that this process had a time course of seconds and therefore was too slow to be operative during the normal cardiac cycle.[33] However, Valdivia *et al.*[34] found that addition of PKA to the cis chamber of bilayers containing cardiac RyR (from fusion of SR membranes) increased the rate of the adaptive response placing it on a more physiological time scale. This result suggests that phosphorylation may be required for fast adaptive responses of the cardiac RyR. It should be noted that the existence of adaptation is controversial (for e.g. see Ref. 35 and Chapter 14). Moreover, Mg^{2+} was found to produce a similar increase in the rate of adaption. The similarity of the effects of Mg^{2+} and PKA treatment in this case,[28] is unlike the opposing effects on cardiac RyR function observed by Hain *et al.*[31]

[^{3}H]ryanodine Binding Studies

Since ryanodine is thought to bind with high affinity to a conformation of RyR associated with an open state of the channel, it is suggested that changes in measured [^{3}H]ryanodine binding reflect changes in the channel activity and hence indicate the functional state of the receptor (see Table 8.2).[36,37] Treatment of canine ventricular SR with PKC and PKG has been reported to increase [^{3}H]ryanodine binding,[20] whereas exposure to PKA has been found to either increase[23,25] or decrease[34] binding. Exogenous CaMK was reported to decrease binding,[32] as was endogenous CaMK.[23,25] The observed changes in [^{3}H]ryanodine binding do not agree with some of the results obtained in bilayer experiments where an increased P_o would be expected to correlate with an increased level of [^{3}H]ryanodine binding, and, conversely, a decreased P_o would be expected to be associated with decreased binding.

Intact and Semi-Intact Cardiac Muscle Studies

β-adrenergic stimulation of cardiac cells increases the rate of force development, developed force and the rate of relaxation. It is thought that the increase in force is due to an increase in release of calcium and that the increased rate of relaxation is mainly due to an increased rate of pumping of the SR Ca^{2+} ATPase. β-adrenergic stimulation causes a rise in cAMP, and hence an increase in the activity of PKA. Using saponin permeabilized myocytes and intact myocytes metabolically labelled with ^{32}Pi, Yoshida *et al.*[38] showed that isoproterenol treatment resulted in an increased phosphorylation of RyR (confirmed by immunoprecipitation). This effect was blocked by propranolol, a β-adrenergic antagonist. Addition of cAMP to the permeabilised myocytes also increased phosphorylation of RyR. Interestingly, the increase after addition of cAMP was greater than that seen with isoproterenol in the permeabilised myocytes. Using saponin skinned ventricular trabeculae from rats, Patel *et al.*[39] showed that for a given "cytoplasmic" calcium concentration, exposure to isoproterenol resulted in an increased isometric tension, relative to unexposed tissue. After ruling out a change in the sensitivity of contractile proteins to Ca^{2+}, it was concluded that there was an increased release of SR calcium. Both treated and untreated muscle released the same amount of Ca^{2+} in response to caffeine indicating that isoproterenol treatment had not affected the size of the releasable Ca^{2+} stores. Treatment with alkaline phosphatase returned the calcium release in isoproterenol treated muscles to control levels. These data suggest that phosphorylation by PKA results in an increased release of SR

Ca^{2+} that can lead to an increased force of contraction during stimulation of the heart by β agonists.

Phosphorylation of the Cardiac RyR in Nervous Tissue

The cardiac isoform of RyR (type 2) is expressed in nervous tissue.[40,41] In canine brain this isoform has been reported to be phosphorylated by both endogenous and exogenous CaMKs, and tryptic mapping confirmed that the major phosphorylation site is Ser 2809, the same site phosphorylated in cardiac tissue.[40] PKA and PKG have also been shown to phosphorylate the cardiac isoform in neuronal tissue.[42] Since RyR may be involved in different types of signalling processes in neuronal and cardiac tissue, it will be interesting to see if the regulation of the cardiac RyR isoform is similar in both tissues.

SKELETAL MUSCLE RyR

As just described for the cardiac RyR isoform, published results for phosphorylation of the skeletal RyR in vitro are contradictory. However, unlike the cardiac RyR where it is generally agreed that phosphorylation is a physiologically relevant event, for the skeletal muscle RyR isoform this is unclear. In earlier studies, where the cardiac RyR was found to be phosphorylated to a significant level by exogenous kinases, the skeletal muscle RyR was only phosphorylated to very low levels.[23,24] This was true for both fast and slow twitch muscles.[24] Moreover, it was suggested that the high molecular weight band that is phosphorylated in skeletal muscle SR[16] is not RyR but rather a protein with a similar molecular mobility on SDS polyacrylamide gels.[24,43] It should be noted in both skeletal and cardiac tissue, there is a high molecular mass peptide with a slightly fster mobility than RyR which is phosphorylated but not recognized by anti-RyR antibodies.[27]

In contrast, more recent studies have found the skeletal RyR is phosphorylated by PKA,[27,44–46] exogenous CaMK,[44–46] endogenous CaMK,[27,45] PKC[46] and PKG.[26,44] In some studies the skeletal RyR was phosphorylated to significant levels by only some of these kinases. For example, Takasago *et al.*[25] reported that the skeletal muscle RyR was phosphorylated to a significant extent by PKG, but not by PKA, exogenous CaMK and PKC. In contrast, Suko *et al.*[44] found that all four kinases phosphorylated the skeletal RyR to significant levels (see Table 8.1).

The stoichiometries calculated for skeletal muscle RyR phosphorylation by the kinases vary widely between laboratories (see Table 1). For example, the values for PKA range from 0.02 to 3.52 mol phosphate/mol of RyR tetramer.[27,44] Generally, the stoichiometries calculated for the skeletal muscle RyR are lower than those determined for the cardiac isoform. There could be two principal reasons for lower stoichiometries. Either the skeletal RyR cannot be phosphorylated to high stoichiometries under the conditions used, or the receptor is already significantly phosphorylated. Potentially the question of whether RyR are already phosphorylated could be answered by phosphatase studies. However, as considered below, the experimental conditions used, and consequently the activity state of RyR, may influence the answer obtained.

Using purified skeletal muscle RyR, Suko *et al.*[44] showed that PKA, PKG and exogenous CaMK phosphorylate Ser 2843. This serine corresponds in sequence alignments to Ser 2809 in the cardiac RyR. Exogenous CaMK was suggested to phosphorylate additional amino acids, including an undefined threonine and Ser 4089. Ser 4809 is not located within a known phosphorylation consensus sequence.[44] From phosphoamino acid analysis, it was estimated that the ratio of phosphorylation of exogenous CaMK was 4:1 serine:threonine. In another study, Ser 2843 in the skeletal RyR was shown to be phosphorylated by endogenous kinase(s) in the absence of added Ca^{2+} and calmodulin.[47]

Functional Effects of Phosphorylation on the Skeletal RyR

Patch clamp. In a study of frog skeletal muscle, Wang and Best[48] used the patch clamp technique to record RyR activity on extruded blebs (sarcoballs)[49] of the SR. In this method the cytoplasmic face of the SR is exposed to the solution in the pipette. With 0.12 mM Ca^{2+} and 1 mM ATP in the pipette, there were few channel openings. However, upon treatment with acid phosphatase and 0 mM ATP, channel activity appeared within a few minutes. RyR channel activity in phosphatase-treated patches was decreased upon re-exposure to ATP but not after the addition of a nonhydrolyzable analogue of ATP (AMP-PNP). This suggests that hydrolysis of ATP is required for the effect on channel activation, since AMP-PNP like ATP will bind to and activate the channel. The role of phosphorylation in this effect was established further by the use of ATP-γS (thiophosphate groups are resistant to phosphatases). Application of phosphatase to ATP-γS treated patches did not result in channel reactivation. To identify the kinase involved, a specific peptide inhibitor of CaMK, that mimics the autoinhibitory domain of the kinase was used. This inhibitor was found to prevent the ATP-dependent inactivation of the channel. These results indicate that an endogenous CaMK associated with the SR regulates RyR channel activity and suggest that RyR and the kinase are in close physical proximity within the membrane. Whether RyR is a direct target of the endogenous kinase or is regulated by another protein that is the target of the kinase remains to be determined.

Bilayer Studies

Functional studies of the role of phosphorylation have been carried out with both purified skeletal receptors and SR membranes reconstituted into bilayers. Herrmann-Frank and Varsányi[50] purified RyR from skeletal muscle SR membranes that had been phosphorylated by endogenous kinases. The receptors were purified by density gradient centrifugation and both kinase-treated, and untreated receptors were incorporated into bilayers. Their results show that phosphorylation of the receptor correlated with an enhanced P_o and an increased sensitivity to activation by cytoplasmic (cis) Ca^{2+}. For example, in 50 μM Ca^{2+} mean open times were measured to be 1.1±0.6 ms for non-phosphorylated RyR and 3.2±1.0 ms after phosphorylation. Addition of protein phosphatase 2A resulted in the receptor channels becoming less active if they had been previously phosphorylated, but did not have an effect on the activity of channels that had not been phosphorylated in vitro. Channel activity that was deactivated by treatment with phosphatase was reactivated by addition of ATP.

In experiments that parallel those described above for the cardiac RyR, Hain *et al.*[45] obtained similar results for the skeletal RyR as for the cardiac RyR; exogenous CaMK or PKA alleviated Mg^{2+} inhibition and endogenous CaMK caused an inhibition of channel activity (in the absence of Mg^{2+}). Interestingly, the effects of endogenous CaMK were reversed by exogenous CaMK (in the absence of Mg^{2+}). These data suggest that the regulation of the skeletal RyR channel function by phosphorylation is complex and is inconsistent with the number of phosphorylation sites identified in RyR to date.

Calcium Fluxes across SR Membranes

The effect of phosphorylation on Ca^{2+} loading of isolated SR terminal cisternae was studied by Mayrleitner *et al.*[46] Exposure of membranes to exogenous PKA, PKC and CaMK reduced the rate of loading of vesicles in the presence of 1 mM Mg^{2+}. This was suggested to be due to an increased opening of RyR channels with a concomitantly greater leak of Ca^{2+} leading to a decreased Ca^{2+} content (or net uptake) of the terminal cisternae vesicles.

These results, in combination with bilayer data, suggest that both the skeletal and the cardiac RyR need to be phosphorylated to function in the presence of physiological concentrations of Mg^{2+}. Importantly, the flux data which represent measurements on a population of vesicles confirm the bilayer data which are based on results obtained from a relatively small number of channels, that are selected by their ability to fuse with a bilayer.

Chu *et al.*[9] studied phosphorylation of proteins in longitudinal and junctional SR from skeletal muscle. In the presence of ATP and Mg^{2+}, several proteins but not RyR were phosphorylated by endogenous kinases. The addition of Ca^{2+} and calmodulin resulted in phosphorylation of numerous proteins, including RyR. Dephosphorylation was slow in the absence of added phosphatase suggesting that exogenous (or cytoplasmic) phosphatases are required for this event. Phosphorylation of isolated junctional SR in the presence of Ca^{2+} and calmodulin was found to have no effect on calcium-induced calcium release, as determined by rapid filtration. Thus, data for Ca^{2+} fluxes across SR membranes, suggest that exogenous PKA, PKC and CaMK increase RyR channel activity[46] while phosphorylation by endogenous CaMK is without effect on SR Ca^{2+} permeability and presumably the activity of RyR channels. The latter finding is in conflict with the ability of phosphorylation by endogenous CaMK to inhibit skeletal ryanodine receptor channel activity in extruded blebs[48] and bilayers.[45]

Skeletal RyR Phosphorylation in Vivo

There is very little in vivo data for the skeletal RyR. The data of Varsányi and Meyer[47] indicate variable phosphorylation of the skeletal RyR in different preparations. These authors suggest that the amino acid, which can be phosphorylated, may under some conditions not be accessible to kinases. The results of Hain *et al.*[45] also indirectly suggest that skeletal RyR are phosphorylated in vivo. It is clear that further in vivo studies of skeletal receptor regulation are required to determine if phosphorylation regulates the channel activity of the skeletal RyR.

PHOSPHORYLATION OF NON RyR SR POTEINS THAT AFFECT RyR FUNCTION

Two nonRyR proteins that are phosphorylated in skeletal muscle SR are the 150/160 kDa protein,[14,47,51,52] identified as sarcalumenin[53,54] and the 160/170 kDa protein, the

Table 8.3 Non Ryanodine Receptor Phosphorylation Events Associated with Altered Skeletal Muscle Ryanodine Receptor

SOURCE	PHOS SUBSTRATE	KINASE	PROTEIN PHOS	RyR PHOS	RyR FUNCTION	REF
rabbit	SR	$CaMK_{end}$	60 kDa[a]	N.D.	↓Ca^{2+} release	73
rabbit	SR	Casein Kinase II	150/160 kDa[b]	no	↓[^{3}H]ryanodine binding	51,54
rabbit	SR	Casein Kinase II[c]	160/170 kDa[cd]	no	↓[^{3}H]ryanodine binding	51,54

Abbreviations not in Tables 1 and 2: protein phos = protein that was phosphorylated;
[a]This protein has been identified as phosphoglucomutase
[b]This protein has been identified as sarcalumenin
[c]Note Damiani *et al.*[53] report that this protein is phosphorylated by cAMP
[d]This protein has been identified as the histidine-rich Ca^{2+} binding protein (see also Ref. 53)

histidine-rich Ca^{2+} binding protein[9,14,51–56] (see Table 8.3). It has been reported by Orr and Shoshan-Barmatz[14] that phosphorylation of these proteins leads to inhibition of high affinity [^{3}H]ryanodine binding, decreasing the K_D by 1.8 fold and the B_{max} by 8%. These authors suggest that the endogenous kinase responsible for phosphorylation of these proteins is casein kinase II, and that this kinase is localized in the lumen of the SR.[54] This localization is not in agreement with other data that suggested that casein kinase II is at low levels in the SR and that luminal SR proteins that are substrates for casein kinase II are phosphorylated during processing and transport prior to their localization in the SR.[57] The data of Damiani *et al.*[56] indicate that the 160/170 kDa protein is phosphorylated by endogenous CaMK. Orr and Shoshan-Barmatz[14] reported that at pH 7.2, phosphorylation of these two proteins had no effect on the SR Ca^{2+} permeability (with effects only being seen when phosphorylation was carried out at pH 8.2). This lack of effect is surprising since it would be expected that the decrease in [^{3}H]ryanodine binding (measured in these experiments at pH 7.4) would be consistent with a change in SR calcium permeability, since a decrease in [^{3}H]ryanodine binding is thought to indicate a diminished opening of RyR.

RyR 3

There is little information about the phosphorylation of the third mammalian isoform of RyR, type 3 (often called the brain isoform, although the cardiac isoform (type 2) may be the predominant isoform in the brain). This lack of data is due to the low levels of this receptor isoform in mammalian tissues which makes studies of this isoform difficult. The β isoform in non-mammalian vertebrate tissues is a putative homologue of the type 3 RyR.[58–60] The β RyR isoform is expressed in fast twitch skeletal muscle at levels almost equivalent to those of the α RyR, the putative homologue of the skeletal RyR (type 1).[58,61] In chicken skeletal muscle, in vitro, both the α and β isoforms were found to be phosphorylated, with the β RyR being phosphorylated to a greater level than the a receptor by both exogenous PKA and CaMK.[43] Interestingly the reverse situation was observed in frog skeletal muscle, where phosphorylation of the α RyR was greater than the β RyR.[43] As the two α RyR and the two β RyR are homologues at the sequence level, the phosphorylation data suggest that these isoform pairs may not be complete functional homologues.

DOES PHOSPHORYLATION BY PKA REGULATE CALMODULIN BINDING?

Guerrini *et al.*[62] delineated calmodulin binding domains in all three mammalian RyR isoforms. One domain, present in all three isoforms, contains a serine that is phosphorylated in vitro by PKA. This is Ser 3020 of the cardiac RyR, Ser 3055 of the skeletal RyR, and Ser 2919 of RyR3. In phosphopeptide mapping studies[40,47] none of these serines have been identified as amino acids that are phosphorylated. Using peptides derived from the ryanodine receptors the authors[62] showed that phosphorylation by PKA inhibited calmodulin binding, and conversely calmodulin binding inhibited phosphorylation of all three isoforms. These data suggest that phosphorylation could regulate the binding of calmodulin to RyR.

ARE THE EFFECTS OF PHOSPHATASES AND KINASES DEPENDENT ON THE FUNCTIONAL STATE OF RyR?

As discussed above, it was concluded by some workers that RyR are phosphorylated in vivo.[27,31,45] This conclusion is based on the effects produced by phosphatase treatment.

Experiments addressing the role of phosphorylation in alleviation of Mg^{2+} inactivation[31,45] and those showing that addition of acid phosphatase to cardiac SR microsomes increased [^{3}H]ryanodine binding in a calcium dependent manner,[27] indicate that exogenous phosphatases can dephosphorylate RyR (or other SR proteins) and affect the function of the RyR channel. However, other workers have found that treatment with phosphatases did not change the levels of phosphorylation that could be achieved in vitro.[24,27] It has been suggested that acid phosphatase requires the channel to be in an open state, to exert an effect,[44] and such a requirement could at least in part account for the differences in the observed effects of phosphatases. If this is the case, it is possible that the conditions used for dephosphorylation in some experiments may not have favoured the appropriate, open conformation of RyR.

Netticadan *et al.*[63] studied the effects of ruthenium red, an RyR inhibitor, and ryanodine (0.25-1 μM) which opens RyR, on phosphorylation of SR membranes from rabbit ventricular myocardium. The addition of ruthenium red resulted in a decrease in RyR phosphorylation by both endogenous and exogenous CaMK, while addition of ryanodine resulted an increased phosphorylation of RyR by endogenous CaMK (the effects of exogenous CaMK were not described). These results suggest that when RyR is closed, the channel is refractory to phosphorylation by both CaMKs and that opening of the channel allows the channel to be phosphorylated by endogenous CaMK. Therefore, the activation (and conformational) state of the channel may affect the efficacy with which both kinases and phosphatases can phosphorylate or dephosphorylate RyR.

STOICHIOMETRIES OF PHOSPHORYLATION

As described above, and shown in Table 8.1, the stoichiometries of phosphorylation of RyR vary. What might these different numbers mean to RyR function? In cases of substoichiometric phosphorylation of RyR tetramers does just one monomer of RyR tetramer need to be phosphorylated to effect a change? A model for this could be high affinity [^{3}H]ryanodine binding, where there one ryanodine molecule is thought to bind per tetramer. Since high affinity [^{3}H]ryanodine binding can affect channel function, one could argue by analogy that phosphorylation of only one receptor monomer could affect channel function. If only one monomer needs to be phosphorylated does this site have to be located close to the membrane spanning region of the molecule? In the primary sequence of the receptor, unlike the suggested site for the high affinity [^{3}H]ryanodine binding site,[17–20] the putative major phosphorylation site at Ser 2809 (cardiac) or Ser 2843 (skeletal) is not close to the membrane spanning region. In contrast, Ser 4089 in the skeletal RyR suggested by Suko *et al.*[44] is much closer to the membrane spanning region. The phosphorylation of only one monomer of the RyR tetramer could result an on/off switch-like regulation of receptor channel activity. In contrast phosphorylation of multiple monomers, when more than one monomer is phosphorylated at the same amino acid would produce a more graded regulation. These two scenarios would result in very different types of control of RyR channel activity. To understand fully the functioning of RyR it will be important to define the stoichiometries of phosphorylation of these proteins that are achieved in vivo.

In many studies the calculation of stoichiometries (mol phosphate/mol RyR tetramer) was based on measurement of high affinity [^{3}H]ryanodine binding and on the existence of one mol of high affinity binding site per mol RyR tetramer. However, this approach may be complicated by data described above suggesting that the phosphorylation state of the receptor can affect [^{3}H]ryanodine binding. Some of the observed changes in [^{3}H]ryanodine binding are significant, such as a 30% increase for PKA,[18] or a 53% decrease in [^{3}H]ryanodine binding in the presence of Ca^{2+}, calmodulin and ATP (ie endogenous CaMK)[25] for the cardiac RyR. Thus phosphorylation of RyR may affect the level of [^{3}H]ryanodine binding and hence the calculated stoichiometries. A further complication with regards to calculation of stoichiometries using this method is whether the same population of RyR that bind [^{3}H]ryanodine are phosphorylated. Within a preparation a proportion of receptors are going to be in a different conformation. Which conformations are substrates for phosphorylation?

POTENTIAL CAUSES OF THE VARIABILITY IN THE EXTENT OF ESTIMATES OF RyR PHOSPHORYLATION

The varying results obtained by different laboratories, and even the differences sometimes found between preparations in the same laboratories, are puzzling. As noted by Varsányi and Meyer[47] preparations of rabbit SR membranes, in which RyR was shown to be only partially phosphorylated in vivo, significant additional phosphorylation was not achieved in vitro. One explanation for this result involves the state of the animal from which the tissue is extracted. For instance, if the animal was excited the release of adrenalin may have affected the state of the muscle and hence the phosphorylation state of RyR. It is known that some anaesthetics affect RyR, most notably the halothane anaesthetics and their role in malignant hyperthermia (see Chapter 12). Does the type and dose of an anaesthetic given to an animal affect the functional state of RyR or the phosphorylation state? During the isolation of SR membranes, does the way the muscle is handled and which protease inhibitors are present, have an effect. What is the effect of buffers, and the concentrations of monovalent ions present.[64] The majority of studies of RyR have been carried out in vitro using heavy SR membranes. The use of these membrane preparations has the advantage that RyR is at least in a partially correct context within the membrane, with native SR lipids etc. However, it is not clear what the effects of removal of RyR from the context of the triad junction may be on the receptor or ancillary proteins. The extent of purification of RyR may affect the level of RyR phosphorylation that can be achieved. As RyR is purified to a greater extent, both ancillary activators and inhibitors of RyR channels may be lost.

An additional consideration is that there may be conformational changes that may alter the extent to which RyR can serve as a substrate for kinases or phosphatases. As discussed above the functional state of RyR, such as whether the channel is open or closed, may influence the ability of RyR to be either phosphorylated or dephosphorylated.[63] In preliminary studies, using PKA, we have found conditions that alter RyR conformation substantially increases the level of phosphate incorporation in all RyR isoforms tested (He and Airey, unpublished observations). Therefore, it is not clear how differences in isolation protocols may affect the functional state of RyR so that some of the differences in data may be due to variables in experimental protocols.

The data suggesting that the α and β isoforms in frog and avian skeletal muscle may not be complete functional homologues[43] raises the question of whether there are species differences between the mammalian RyR and accessory proteins. Does the importance of phosphorylation by the kinases differ between species? There are some known functional differences between mammalian striated muscles. For instance, the contractile activity of rat hearts depends to a greater extent on release of Ca^{2+} from SR stores, than rabbit hearts.[65,66] Therefore, if phosphorylation alters the release characteristics of RyR, then the mechanical performance of rat heart may be more affected than that of rabbit heart by the phosphorylation event. Other reasons for variability in estimates of phosphorylation for striated muscle could be the composition of the muscle in different preparations, for example differences in SR calcium release between fast and slow twitch muscle have been described.[1] Other factors that can be controlled experimentally, that may have functional consequences, include fatigue or tetanus.[2,5]

IS THE PHOSPHORYLATION OF RyR OR THE PHOSPHORYLATION OF ACCESSORY PROTEINS PHYSIOLOGICALLY IMPORTANT?

Whether a RyR itself is phosphorylated or another protein that interacts with RyR is phosphorylated was not determined in many of the studies described here. The number of proteins that interact with RyR, and their role in RyR function is unclear. For instance, it has been shown that the RyR is modulated by FK-binding protein (FKBP).[67,68] In addition, calcineurin has been shown to be associated with RyR in skeletal muscle.[69] Does this phosphatase interact with the skeletal RyR or does it act upon another protein associated with RyR? It is possible that several modulators of RyR function may act in concert to modulate RyR. This possibly may complicate in vitro studies of RyR phosphorylation, particularly when purified receptor preparations are used.

FUTURE DIRECTIONS

Additional studies are required to understand the functional role of phosphorylation of RyR. It needs to be established whether phosphorylation of RyR is required for Ca^{2+} release under basal conditions or exerts an effect under situations such as muscle fatigue or adrenergic stimulation of the heart. Krebs and Beavo[70] proposed four criteria that are requisite for physiological significance of a phosphorylation event:

1. Demonstration in vitro that the protein can be phosphorylated stoichiometrically at a significant rate in a reaction(s) catalyzed by an appropriate protein kinase(s) and dephosphorylated by a phosphoprotein phosphatase(s).
2. Demonstration that functional properties of the protein undergo meaningful changes that correlate with the degree of phosphorylation.
3. Demonstration that the protein can be phosphorylated and dephosphorylated in vivo or in an intact cell system with accompanying functional changes.
4. Correlation of cellular levels of protein kinase and/or phosphoprotein phosphatase effectors and the extent of phosphorylation of the enzyme.

At present, none of these criteria have been answered in a definitive manner for RyR.

Whatever the reasons for the wide range of results obtained in the study of phosphorylation of RyR in vitro, it is clear that these results cannot be extrapolated to what occurs in vivo. Therefore, to understand the role of phosphorylation in regulation of RyR the protein needs to be studied in vivo. Whether RyR are significantly phosphorylated under stimulated and nonstimulated conditions can only be determined by studying the receptors in their native condition. Until there is a body of information of studies of the RyR in vivo, which of the results obtained in vitro are congruent with living tissue it will not be known. As stated in 1980 by Manning *et al.*[71] "ultimately, it is by conjoint analyses in vitro and in vivo that the relevance of a particular protein phosphorylation-dephosphorylation to cellular function and the factors governing its occurrence can be elucidated".

ACKNOWLEDGEMENTS

The authors laboratory has been funded by National Science Foundation Grants MCB 9506257 and IBN 9306850 and National Heart, Lung and Blood Grant HL-27470. We thank Dr John Sutko for support and critical reading of the manuscript.

REFERENCES

1. Lee YS, Ondrias K, Duhl AJ *et al.* Comparison of calcium release from sarcoplasmic reticulum of slow and fast twitch muscles. *J Membr Biol* 1991; **122**:155–163.
2. Kim A, DiFranco M, Vergara JL. Tetanic stimulation: a physiological tool to critically investigate excitation-contraction coupling in skeletal muscle. *Biophys J* 1996; **70**:A234.
3. Baker AJ, Longuemare MC, Brandes R *et al.* Intracellular tetanic calcium signals are reduced in fatigue of whole skeletal muscle. *Am J Physiol* 1993; **264**:C577–C582.
4. Györke S. Effects of repeated tetanic stimulation on excitation-contraction coupling in cut muscle fibres of the frog. *J Physiol* (Lond) 1993; **464**:699–710.
5. Westerblad H, Lee JA, Lannergren J *et al.* Cellular mechanisms of fatigue in skeletal muscle. *Am J Physiol* 1991; **261**:C195–C209.
6. Campbell KP, MacLennan DH. A calmodulin-dependent protein kinase system from skeletal muscle sarcoplasmic reticulum. Phosphorylation of a 60,000-dalton protein. *J Biol Chem* 1982; **257**:1238–1246.
7. Tuana BS, MacLennan DH. Calmidazolium and compound 48/80 inhibit calmodulin-dependent protein phosphorylation and ATP-dependent Ca^{2+} uptake but not Ca^{2+}-ATPase activity in skeletal muscle sarcoplasmic reticulum. *J Biol Chem* 1984; **259**:6979–6983.
8. Tuana BS, MacLennan DH. Isolation of the calmodulin-dependent protein kinase system from rabbit skeletal muscle sarcoplasmic reticulum. *FEBS Lett* 1988; **235**:219–223.
9. Chu A, Sumbilla C, Inesi G *et al.* Specific association of calmodulin-dependent protein kinase and related substrates with the junctional sarcoplasmic reticulum of skeletal muscle. *Biochemistry* 1990; **29**:5899–59.
10. Leddy JJ, Murphy BJ, Qu-Yi *et al.* A 60 kDa polypeptide of skeletal-muscle sarcoplasmic reticulum is a calmodulin-dependent protein kinase that associates with and phosphorylates several membrane proteins. *Biochem J* 1993; 849–856.
11. Schulman H, Lou LL. Multifunctional Ca^{2+}/calmodulin-dependent protein kinase: domain structure and regulation. *Trends Biochem Sci* 1989; **14**:62–66.
12. Iwasa T, Inoue N, Fukunaga K *et al.* Purification and characterization of a multifunctional calmodulin-dependent protein kinase from canine myocardial cytosol. *Arch Biochem Biophys* 1986; **248**:21–29.
13. Woodgett JR, Davison MT, Cohen P. The calmodulin-dependent glycogen synthase kinase from rabbit skeletal muscle. Purification, subunit structure and substrate specificity. *Eur J Biochem* 1983; **136**:481–487.

14. Orr I, Shoshan-Barmatz V. Modulation of the skeletal muscle ryanodine receptor by endogenous phosphorylation of 160/150-Kda proteins of the sarcoplasmic reticulum. *Biochim Biophys Acta* 1996; **1283**:80–88.
15. Han J-W, Thieleczek R., Varsányi M., Heilmeyer IMG Jr. Compartmentalized ATP synthesis in skeletal muscle triads. *Biochemistry* 1992; **31**:377–384.
16. Seiler S, Wegener AD, Whang DD *et al.* High molecular weight proteins in cardiac and skeletal muscle junctional sarcoplasmic reticulum vesicles bind calmodulin, are phosphorylated, and are degraded by Ca^{2+}-activated protease. *J Biol Chem* 1984; **259**:8550–8557.
17. Takeshima H, Nishimura S, Matsumoto T *et al.* Primary structure and expression from complementary DNA of skeletal muscle ryanodine receptor. *Nature* 1989; **339**:439–445.
18. Zorzato F, Fujii J, Otsu K *et al.* Molecular cloning of cDNA encoding human and rabbit forms of the Ca^{2+} release channel (ryanodine receptor) of skeletal muscle sarcoplasmic reticulum. *J Biol Chem* 1990; **265**:2244–2256.
19. Nakai J, Imagawa T, Hakamata Y *et al.* Primary structure and functional expression from cDNA of the cardiac ryanodine receptor/calcium release channel. *FEBS Lett* 1990; **271**:169–177.
20. Otsu K, Willard HF, Khanna VK *et al.* Molecular cloning of cDNA encoding the Ca^{2+} release channel (ryanodine receptor) of rabbit cardiac muscle sarcoplasmic reticulum. *J Biol Chem* 1990; **265**:13472–13483.
21. Hakamata Y, Nakai J, Takeshima H *et al.* Primary structure and distribution of a novel ryanodine/calcium release channel from rabbit brain. *FEBS Lett* 1992; **312**:229–235.
22. Giannini G, Clementi E, Ceci R *et al.* Expression of a RyR-Ca^{2+} channel that is regulated by TGF-β. *Science* 1992; **257**:91–94.
23. Takasago T, Imagawa T, Shigekawa M. Phosphorylation of the cardiac ryanodine receptor by cAMP-dependent protein kinase. *J Biochem* (Tokyo) 1989; **106**:872–877.
24. Witcher DR, Kovacs RJ, Schulman H *et al.* Unique phosphorylation site on the cardiac ryanodine receptor regulates Ca^{2+} channel activity. *J Biol Chem* 1991; **266**:11144–11152.
25. Takasago T, Imagawa T, Jurukawa K *et al.* Regulation of the cardiac ryanodine receptor by protein kinase-dependent phosphorylation. *J Biochem* (Tokyo) 1991; **109**:163–70.
26. Hohenegger M, Suko J. Phosphorylation of the purified cardiac ryanodine receptor by exogenous and endogenous protein kinases. *Biochem J* 1993; **296**:303–308.
27. Strand MA, Louis CF, Mickelson JR. Phosphorylation of the porcine skeletal and cardiac muscle sarcoplasmic reticulum ryanodine receptor. *Biochim Biophys Acta* 1993; **1175**:319–326.
28. Meissner G. Evidence of a role for calmodulin in the regulation of calcium release from skeletal muscle sarcoplasmic reticulum. *Biochim Biophys Acta* 1986; **25**:244–251.
29. Tripathy A, Xu L, Mann G *et al.* Calmodulin activation and inhibition of skeletal muscle Ca^{2+} release channel (ryanodine receptor). *Biophys J* 1995; **69**:106–119.
30. Ikemoto T, Iino M, Endo M. Enhancing effect of calmodulin on Ca^{2+}-induced Ca^{2+} release in the sarcoplasmic reticulum of rabbit skeletal muscle fibres. *J Physiol* 1995; **487**:573–582.
31. Hain J, Onoue H, Mayrleitner M *et al.* Phosphorylation modulates the function of the calcium release channel of sarcoplasmic reticulum from cardiac muscle. *J Biol Chem* 1995; **270**: 2074–2081.

32. Lokuta AJ, Rogers TB, Lederer WJ *et al.* Modulation of cardiac ryanodine receptors of swine and rabbit by a phosphorylation-dephosphorylation mechanism. *J Physiol* (Lond) 1995; **487**:609–622.
33. Györke S, Fill M. Ryanodine receptor adaptation: control mechanism of Ca^{2+}-induced Ca^{2+} release in heart. *Science* 1993; **260**:807–809.
34. Valdivia HH, Kaplan JH, Ellis-Davies GCR *et al.* Rapid adaptation of cardiac ryanodine receptors: Modulation by Mg^{2+} and phosphorylation. *Science* 1995; **267**:1997–2000.
35. Sitsapesan R, Montgomery RAP, Williams A. New insights into the gating mechanisms of cardiac ryanodine receptors revealed by rapid changes in ligand concentration. *Circ Res* 1995; **77**: 765–772.
36. Chu A, Diaz-Munoz M, Hawkes MJ *et al.* Ryanodine as a probe for the functional state of the skeletal muscle sarcoplasmic reticulum calcium release channel. *Mol Pharmacol* 1990; **37**: 735–741.
37. Meissner G, el-Hashem A. Ryanodine as a functional probe of the skeletal muscle sarcoplasmic reticulum Ca^{2+} release channel. *Mol Cell Biochem* 1992; **114**:119–123.
38. Yoshida A, Takahashi M, Imagawa T *et al.* Phosphorylation of ryanodine receptors in rat myocytes during β-adrenergic stimulation. *J Biochem* (Tokyo) 1992; **111**:186–190.
39. Patel JR, Coronado R, Moss RL. Cardiac sarcoplasmic reticulum phosphorylation increases Ca^{2+} release induced by flash photolysis of Nitr-5. *Circ Res* 1995; **77**:943–949.
40. Witcher DR, Strifler BA, Jones LR. Cardiac-specific phosphorylation site for multifunctional Ca^{2+}/calmodulin-dependent protein kinase is conserved in the brain ryanodine receptor. *J Biol Chem* 1992; **267**:4963–4967.
41. Lai FA, Dent M, Wickenden C *et al.* Expression of a cardiac Ca^{2+}-release channel isoform in mammalian brain. *Biochem J* 1992; **288**:553–564.
42. Yoshida A, Ogura A, Imagawa T *et al.* Cyclic AMP-dependent phosphorylation of the rat brain ryanodine receptor. *J Neurosci* 1992; **12**:1094–1100.
43. Airey JA, Grinsell MM, Jones LR *et al.* Three ryanodine receptor isoforms exist in avian striated muscles. *Biochemistry* 1993; **32**:5739–5745.
44. Suko J, Maurer-Fogy I, Plank B *et al.* Phosphorylation of serine 2843 in ryanodine receptor-calcium release channel of skeletal muscle by cAMP-, cGMP- and CaM-dependent protein kinase. *Biochim Biophys Acta* 1993; **1175**:193–206.
45. Hain J, Nath S, Mayrleitner M *et al.* Phosphorylation modulates the function of the calcium release channel of sarcoplasmic reticulum from skeletal muscles. *Biophys J* 1994; **67**: 1823–1833.
46. Mayrleitner M, Chandler R, Schindler H *et al.* Phosphorylation with protein kinases modulates calcium loading of terminal cisternae of sarcoplasmic reticulum from skeletal muscle. *Cell Calcium* 1995; **18**:197–206.
47. Varsányi M, Meyer HE. Sarcoplasmic reticular Ca^{2+} release channel is phosphorylated at serine 2843 in intact rabbit skeletal muscle. *Biol Chem Hoppe-Seyler* 1995; **376**:45–49.
48. Wang J, Best PM. Inactivation of the sarcoplasmic reticulum Ca^{2+} channel by protein kinase. *Nature* 1992; **359**:739–441.

49. Stein P, Palade P. Patch clamp pf sarcolemmal spheres from stretched skeletal muscle fibers. *Am J Physiol* 1989; **256**:C434–440.
50. Herrmann-Frank A, Varsányi M. Enhancement of Ca^{2+} release channel activity by phosphorylation of the skeletal muscle ryanodine receptor. *FEBS Lett* 1993; **332**:237–242.
51. Orr I, Gechtman Z, Shoshan-Barmatz V. Characterization of Ca^{2+}-dependent endogenous phosphorylation of 160000- and 150000-dalton proteins of sarcoplasmic reticulum. *Biochem J* 1991; **276**:89–96.
52. Gechtman Z, Orr I, Shoshan-Barmatz V. Involvement of protein phosphorylation in activation of Ca^{2+} efflux from sarcoplasmic reticulum. *Biochem J* 1991; **276**:97–102.
53. Damiani E, Margreth A. Subcellular fractionation to junctional sarcoplasmic reticulum and biochemical characterization of 170 kDa Ca^{2+}- and low-density-lipoprotein-binding protein in rabbit skeletal muscle. *Biochem J* 1991; **277**:825–832.
54. Shoshan-Barmatz V, Orr I, Weil S *et al.* The identification of the phosphorylated 150/160-kDa proteins of sarcoplasmic reticulum, their kinase and their association with the ryanodine receptor. *Biochim Biophys Acta* 1996; **1283**:89–100.
55. Hofmann SL, Brown MS, Lee E *et al.* Purification of a sarcoplasmic reticulum protein that binds Ca^{2+} and plasma lipoproteins. *J Biol Chem* 1989; **264**:8260–8270.
56. Damiani E, Picello E, Saggin L *et al.* Identification of triadin and of histidine-rich Ca^{2+}-binding protein as substrates of 60 kDa calmodulin-dependent protein kinase in junctional terminal cisternae of sarcoplasmic reticulum of rabbit fast muscle. *Biochem Biophys Res Comm* 1995; **209**:457–465.
57. Cala SE, Jones LR. Phosphorylation of cardiac and skeletal muscle calsequestrin isoforms by casein kinase II. Demonstration of a cluster of unique rapidly phosphorylated sites in cardiac calsequestrin. *J Biol Chem* 1991; **266**:391–398.
58. Airey JA, Beck CF, Tanksley S *et al.* Identification and localization of two triad junctional foot protein isoforms in mature avian fast twitch skeletal muscle. *J Biol Chem* 1990; **265**:14187–94.
59. Oyamada H, Murayama T, Takagi T *et al.* Primary structure and distribution of ryanodine-binding protein isoforms of the bullfrog skeletal muscle. *J Biol Chem* 1994; **269**:17206–17214.
60. Ottini L, Marziali G, Conti A *et al.* α and β isoforms of ryanodine receptor from chicken skeletal muscle are homologues of mammalian RyR1 and RyR3. *Biochem J* 1996; **315**:207–216.
61. Olivares E, Tanksley S, Beck CF *et al.* Multiple foot protein isoforms in amphibian, avian and piscine skeletal muscles. *Biophys J* 1991; **59**:1153–63.
62. Guerrini R, Menegazzi P, Anacardio R *et al.* Calmodulin binding sites of the skeletal, cardiac, and brain ryanodine receptor Ca^{2+} channels: Modulation by the catalytic subunit of cAMP-dependent protein kinase? *Biochemistry* 1995; **34**:5120–5129.
63. Netticadan T, Xu A, Narayanan N. Divergent effects of ruthenium red and ryanodine on Ca^{2+}/calmodulin-dependent phosphorylation of the Ca^{2+} release channel (ryanodine receptor) in cardiac sarcoplasmic reticulum. *Arch Biochem Biophys* 1996; **333**:368–376.
64. Meissner G, Rios E, Tripathy A. Regulation of skeletal muscle Ca^{2+} release channel (ryanodine receptor) by Ca^{2+} and monovalent cations and anions. *J Biol Chem* 1997; **272**:1628–1638.
65. Sutko JL, Willerson JT. Ryanodine alterations of the contractile state of rat ventricular myocardium. Comparison with dog, cat and rabbit ventricular tissues. *Circ Res* 1980; **46**:332–343.

66. Shattock MJ, Bers DM. Rat vs. rabbit ventricle: Ca flux and intracellular Na assessed by ion-selective microelectrodes. *Am J Physiol* 1989; **256**:C813–C822.
67. Brilliantes A-MB, Ondrias K, Scott A *et al.* Stabilization of calcium release channel (ryanodine receptor) function by FK506-binding protein. *Cell* 1994; **77**:513–523.
68. Timerman A, Jayaraman T, Weiderrecht G *et al.* The ryanodine receptor from canine heart sarcoplasmic reticulum is associated with a novel FK-506 binding protein. *Biochem Biophys Res Commun* 1994; **198**:701–706.
69. Cameron AM, Steiner JP, Roskams AJ *et al.* Calcineurin associated with the inositol 1,4,5-trisphosphate receptor-FKBP12 complex modulates Ca^{2+} flux. *Cell* 1995; **83**:463–72.
70. Krebs EG, Beavo JA. Phosphorylation-dephosphorylation of enzymes. *Ann Rev Biochem* 1979; **48**:923–959.
71. Manning DR, DiSalvo J, Stull JT. Protein phosphorylation: Quantitative analysis in vivo and in intact cell systems. *Mol Cell Endocrinol* 1980; **19**:1–19.
72. McPherson PS, Campbell KP. Characterization of the major brain form of the ryanodine receptor/Ca^{2+} release channel. *J Biol Chem* 1993; **268**:18785–19790.
73. Kim DH, Ikemoto N. Involvement of 60-kilodalton phosphoprotein in the regulation of calcium release from skeletal muscle sarcoplasmic reticulum. *J Biol Chem* 1986; **261**:11674–11679.

Chapter 9

Ca^{2+} Release Channel/Ryanodine Receptor - L-Type Ca^{2+} Channel/Dihydropyridine Receptor Interactions in Skeletal Muscle

Gerhard Meissner

OVERVIEW

Ryanodine receptors (RyRs) constitute a family of Ca^{2+} release channels that have an important Ca^{2+} signaling role in muscle and nonmuscle cells by releasing Ca^{2+} from intracellular stores. Because of their abundance and critical role in muscle contraction, the function of the RyRs has been most extensively studied in striated muscle. In skeletal and cardiac muscle, an action potential triggers the rapid release of Ca^{2+} from the sarcoplasmic reticulum (SR) by a mechanism referred to as excitation-contraction (E-C) coupling. Different E-C coupling mechanisms exist in mammalian skeletal and cardiac muscle. A distinguishing feature is that cardiac muscle E-C coupling is dependent on extracellular Ca^{2+}, whereas skeletal muscle E-C coupling is not. In cardiac muscle, a dihydropyridine receptor (DHPR)/Ca^{2+} channel (L-type) located in the surface membrane and tubular infoldings (T-tubule) of the surface membrane mediates the influx of Ca^{2+} during an action potential by functioning as a voltage-dependent Ca^{2+} channel.[1] The rise in intracellular Ca^{2+} concentration triggers the massive release of Ca^{2+} by opening closely apposed SR Ca^{2+} release channels in a process known as 'calcium-induced calcium release' (CICR).[2] (Fig. 9.1). SR Ca^{2+} release channels are also referred to as ryanodine receptors (to distinguish them from the IP_3 receptor/Ca^{2+} release channel, Ref. 3) because they bind the neutral plant alkaloid ryanodine with high affinity and specificity. In contrast to cardiac muscle, the removal of extracellular Ca^{2+} does not eliminate E-C coupling in skeletal muscle. This finding led to the formulation of the "mechanical coupling" mechanism which suggests that voltage-sensing T-tubule DHPR/Ca^{2+} channels (L-type) open SR Ca^{2+} release channels through direct protein-protein interactions (Fig. 9.1). A more recent view holds that vertebrate skeletal muscle E-C coupling is regulated by DHPR- and Ca^{2+}-dependent mechanisms.[4,5] According to this view, Ca^{2+} ions are initially released by SR Ca^{2+} release channels that are directly linked to DHPRs. The released Ca^{2+} ions then open Ca^{2+} release channels not directly linked to DHPR.

Below we will briefly review the DHPR and RyR. We will then focus on studies carried out to determine the structural and functional interaction between the DHPR and RyR in mammalian skeletal muscle.

DHPR/CA^{2+} CHANNEL

The mammalian skeletal and cardiac muscle DHPR/Ca^{2+} channels (L-type) are related oligomeric protein complexes which are made up of up to five different subunits (α_1, α_2, β, γ, δ).[6] The α_1 subunit alone can form a Ca^{2+} channel, binds Ca^{2+} channel antagonists, and belongs to a voltage-sensitive ion channel superfamily which includes Na^+ and K^+ channels. In skeletal muscle, a proteolytically processed 175 - 190-kD form of the 212-kD α_1 subunit predominates. The α_1 subunit co-purifies with a 145-kD glycosylated, α_2 subunit which is disulfide-linked to a 25-kD δ subunit, a cytoplasmic 54-kD β subunit, and a 30-kD glycosylated, γ subunit. The α_2, β, γ and δ subunits do not participate directly in the formation of a voltage-gated conductance pathway, however, expression studies have indicated that

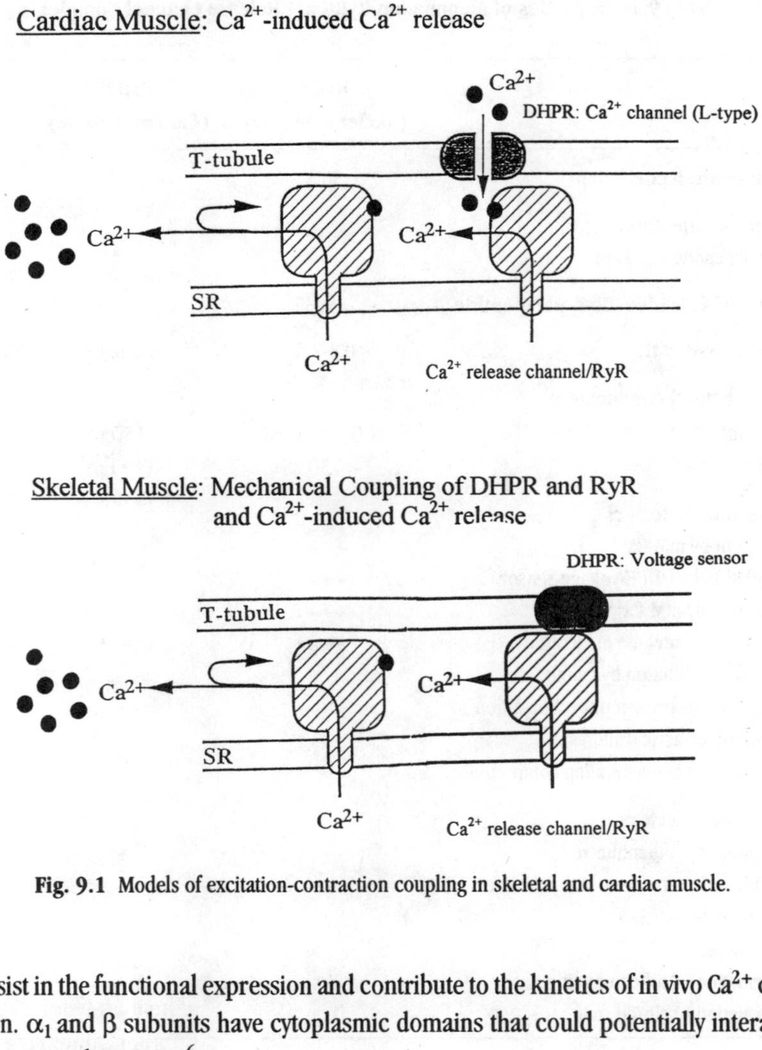

Fig. 9.1 Models of excitation-contraction coupling in skeletal and cardiac muscle.

they assist in the functional expression and contribute to the kinetics of in vivo Ca^{2+} channel function. α_1 and β subunits have cytoplasmic domains that could potentially interact with closely apposed SR RyRs.[6]

RyR/CA²⁺ RELEASE CHANNEL

Mammalian tissues express three structurally and functionally related M_r ~560,000 RyR polypeptides that are encoded by three different genes: ryr1, ryr2, and ryr3. RyR1, RyR2 and RyR3 are also called the skeletal, cardiac, and brain RyR because they were first identified

Table 9.1 Properties of Mammalian RyR/Ca^{2+} Release Channel Complexes

Type	RyR1 (Skeletal muscle)	RyR2 (Cardiac muscle)	RyR3 (Brain)
Sedimentation coefficient	30 S	30 S	30 S
Subunit compsition:			
RyR polypeptide and FKBP	4+4	4+4	
Amino acid residues/RyR polypeptide	~ 5035	~ 4970	~ 4870
DHPR to RyR ratio	0.3-2.0	0.1-0.3	
Single channel conductance			
in 50 mM Ca^{2+}	100 ~ 150 pS	~ 150 pS	
in 250 mM K^+	~ 750 pS	~ 750 pS	
Endogenous effectors			
Activation by mM Ca^{2+}	+++	+++	+++
Regulation by DHPR/voltage-sensor	+++	–	–
Inhibition by mM Ca^{2+} and Mg^{2+}	+++	+	+
Activation by adenine nucleotides	+++	+	
Activation/inhibition by calmodulin	+/+	–/+	
Regulation by protein phosphorylation	+	++	
Activation by acylcarnitines	+	–	
Activation by NO-generating compounds	+	+	
Exogenous effectors			
Inhibition by ruthenium red	+++	++	
Inhibition by tetracaine and procaine	++	++	
Activation by caffeine	+	+	+
Ryanodine:			
Formation of subconducting state	Yes	Yes	
Formation of closed state	Yes	Yes	

For Refs. see text.

in and isolated from skeletal muscle, cardiac muscle and brain, respectively (For a review see Refs. 7–11) (Table 9.1). The primary structure of the skeletal muscle (RyR1)[12,13] cardiac muscle (RyR2)[14,15] and brain and smooth muscle (RyR3)[16] RyRs was determined by cDNA cloning and sequencing. Each of the large polypeptides is comprised of ~5000 amino acids with a predicted molecular mass of ~560 kD. The predicted amino acid sequence between them is 65–70% identical.

The skeletal and cardiac muscle isoforms are primarily expressed in skeletal and cardiac muscle, respectively, and in brain and other tissues at lower levels.[17–19] RyR3 is expressed as a minor component in skeletal and cardiac muscles; however, apparently without being essential for striated muscle function.[20] Nonmammalian skeletal muscles express one RyR or two immunologically and functionally distinct RyR isoforms (initially referred to as α and βRyR).[11] The latter two are expressed in approximately equal amounts in amphibian and avian skeletal muscles, and slow (but not fast) fish skeletal muscles. The deduced amino acid sequence of the cloned α isoform from bullfrog and β isoforms from bullfrog and chicken reveals sequences highly homologous to those of the mammalian RyR1 and RyR3, respectively.

Analysis of the hydropathy of the predicted amino acid sequence has provided valuable clues regarding the disposition of the RyR in the SR membrane.[7–9] Four or twelve transmembrane segments were assigned to be located in about 500 or 1000 amino acid residues near the C-terminus which represents the most hydrophobic region of the RyR and accordingly has been proposed to form the Ca^{2+} channel pore. The remaining sequences are highly hydrophilic and were assigned to be in the cytoplasm. Studies with site-directed antibodies have suggested that the N- and C-termini of the skeletal RyR are cytoplasmically localized.[21] Evidence for two SR lumenal segments in the C-terminal region has been presented, and it was confirmed that the C-terminus is localized in the cytoplasm but is folded in an antibody-inaccessible conformation in the native receptor.[22]

The skeletal muscle and cardiac muscle RyRs have been purified as four-leaf clover-like (quatrefoil) 30 S protein complexes composed of four 560-kD polypeptides (RyR subunits)[8] (Fig. 9.2) in association with four 12-kD polypeptides (FK-506 binding protein, FKBP)[23] (Table 9.1). The structure of the skeletal muscle RyR has been resolved at 3 nm resolution by cryoelectron microscopy and three dimensional reconstruction.[24,25] The RyR showed a four-fold symmetry consisting of a large, loosely packed ‘cytoplasmic’ domain with overall dimensions of 29 × 29 × 12 nm and a smaller ‘transmembrane’ domain that extended 7 nm toward the SR lumen and likely forms the Ca^{2+} channel pore. Most interestingly, images of the open channel revealed a central “hole” in the transmembrane spanning domain which was not evident in the closed channel.[25] These studies have suggested that the RyR may undergo major conformational changes during channel opening and closing.

The RyRs are cation-selective channels which display a larger ion conductance for monovalent cations (~750 pS with 250 mM K^+ as current carrier) than for divalent cations (100–150 pS with 50 mM Ca^{2+}) [7–9] (Table 9.1). They are ligand-gated channels which are

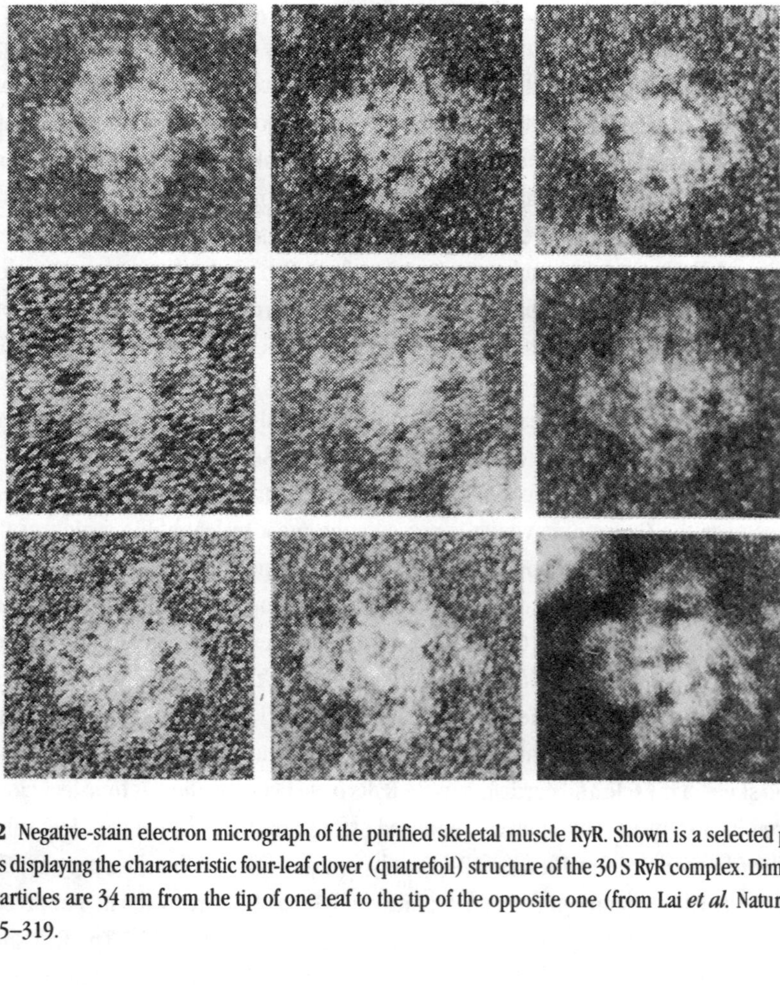

Fig. 9.2 Negative-stain electron micrograph of the purified skeletal muscle RyR. Shown is a selected panel of particles displaying the characteristic four-leaf clover (quatrefoil) structure of the 30 S RyR complex. Dimensions of the particles are 34 nm from the tip of one leaf to the tip of the opposite one (from Lai *et al.* Nature 1988; **334**:315–319.

activated by micromolar Ca^{2+} concentrations and by various adenine nucleotides.[7–9,26] Cytosolic Ca^{2+} ions are important, if not the principal, activators of the cardiac and brain isoforms. Regulation of the skeletal muscle isoform differs significantly from that of the two other isoforms. A unique property of the mammalian skeletal muscle RyR is that its activity is regulated by the DHPR acting as a voltage-sensing molecule and by Ca^{2+}. Millimolar concentrations of Ca^{2+} and Mg^{2+} inhibit channel opening. Activation of RyRs by micromolar cytosolic Ca^{2+} and inhibition by millimolar cytosolic Ca^{2+} suggest at least two classes of

Ca^{2+} binding sites, high-affinity activation sites and low-affinity inactivation sites. Skeletal and cardiac muscle RyRs are also sensitive to regulation by SR luminal Ca^{2+} which may bind to luminal channel sites[27] or have access to the cytosolic Ca^{2+} activation and Ca^{2+} inactivation sites.[28] Other possible Ca^{2+}-dependent mechanisms include regulation by SR luminal (calsequestrin[29]) and cytosolic (calmodulin[30,31]) Ca^{2+} binding proteins. Among the adenine nucleotides, cyclic ADP-ribose is most effective being a potent Ca^{2+}-releasing agent in sea urchin eggs[32], mammalian smooth muscle[33] and pancreatic islets.[34] However, whether this compound has a physiological role in striated muscle has remained controversial.[35–37] A complex regulation of skeletal muscle and cardiac muscle RyR activity by endogenous and exogenous kinases has been reported. [38–40] In addition, it has become apparent in recent years that other candidates must be considered to interact with Ca^{2+} release channels. These include lipid intermediates such as acylcarnitines or sphingosine[7–9] and NO.[41,42] Furthermore, a large number of drugs have been found to affect the RyRs.[6–8] These include caffeine which activates the RyR in the millimolar concentration range. Among the inhibitors, ruthenium red is often chosen because it inhibits at micromolar concentrations. Another extensively used drug is ryanodine, a plant alkaloid, which binds with high specificity and affinity to Ca^{2+} release channels giving them their second name. Ryanodine has a dual effect. Low concentrations open the channels by locking them into an open subconductance state. At concentrations above 10 μM, ryanodine completely closes the channels.

DHPR-RyR INTERACTIONS IN SKELETAL MUSCLE

Immunolabeling and electron microscopic studies show that the DHPR and RyR co-localize in the triad junction of skeletal muscle (For a review see Ref. 43). Tetrads, clusters of four particles presumed to represent four DHPRs, are located almost exactly opposite four subunits of every other RyR, which strongly suggests a specific interaction and has furthermore led to the concept of DHPR-linked and -unlinked RyRs in skeletal muscle. In support are the lack of tetrads in "dyspedic" (no feet) skeletal muscle fibers of a gene-knockout mouse model with a disrupted skeletal muscle RyR gene.[44] On the other hand, triad junctions were present, although these appeared to be narrower in the RYR1 knockout mouse than in normal muscle. In dyspedic myoblasts, DHPRs are clustered in peripheral couplings but do not form tetrads.[45] These observations suggest that the formation of DHPR tetrads requires the presence of the skeletal muscle RyR, although other proteins may play a role in the formation of the triad junction and/or the interaction between the DHPR and

RyR. One candidate protein is triadin, a M_r 90,000 SR junctional protein, which has been suggested to bind to the DHPR.[46]

In mammalian ventricle, RyR is located primarily near the surface and T-tubule, which provides a morphological basis for Ca^{2+}-induced Ca^{2+} release. In mammalian atrium,[47] as in avian heart,[48] many RyRs are located on nonjunctional SR away from DHPR/L-type Ca^{2+} channels, which suggests differences in the communication of DHPR and RyR in ventricle and atrium. Clusters but not tetrads of DHPRs are present in the junctional domains of avian cardiac muscle, in accordance with a less direct communication between the DHPR and RyR in cardiac E-C coupling.[49]

The expression of tissue-specific DHPR and RyR isoforms suggests that both may determine the mode of E-C coupling, i.e. mechanical coupling in skeletal muscle and Ca^{2+}-induced Ca^{2+} release in cardiac muscle. The first insights into the role of the DHPR in determining the mode of E-C coupling were obtained by Tanabe *et al.* using a dysgenic mouse model.[50] Muscular dysgenesis is a recessive lethal mutation in which skeletal muscle cells are incapable of E-C coupling as they lack the α_1 subunit of the DHPR. Microinjection of the skeletal muscle α_1 DHPR cDNA restored skeletal-type E-C coupling since it occurred in the absence of extracellular Ca^{2+}. Injection of the cardiac α_1 DHPR cDNA resulted in cardiac-type E-C coupling since it required the presence of extracellular Ca^{2+}.[51] Studies using skeletal and cardiac muscle chimeric cDNAs indicated that the putative II-III cytoplasmic loop region of the DHPR α_1 subunit is a major determinant for the type of E-C coupling (skeletal or cardiac) observed in dysgenic muscle cells.[52]

A DHPR β subunit gene-knockout mouse model has suggested that additional subunits of the DHPR are essential for skeletal muscle E-C coupling.[53] The dysgenic muscle cells were incapable of E-C coupling, although the contractile machinery was found to be normal. The β subunit of the DHPR binds to the cytoplasmic loop between transmembrane domains I and II of the α_1 subunit,[54] and the presence of the α_1 subunit is required for targeting the β subunit to triads.[55] A minor role of the I-II loop in skeletal muscle E-C coupling has been observed,[52] which suggests a regulation of RyR by DHPR through multiple interactions.

Two mutant animals have shown that a functional RyR1 is needed for skeletal muscle E-C coupling. A RyR1 gene-knockout mouse failed to show a contractile response to electrical stimulation under physiological conditions.[56] The mutant skeletal muscle fibers retained the ability to release Ca^{2+} in response to caffeine, suggesting the presence of functional RyR3 which is expressed as a minor component in skeletal muscle. Similarly, a lack of the skeletal muscle RyR isoform (αRyR) in chicken (Crooked Neck Dwarf, cn/cn) led to a loss of normal skeletal muscle E-C coupling.[11] The contributions of the αRyR could not be duplicated by a second RyR isoform (βRyR), which corresponds to the mammalian brain isoform (RyR3) and is expressed normally in dysgenic (cn/cn) skeletal muscle. That the non-skeletal RyR isoforms could not support skeletal muscle E-C coupling suggests that the skeletal muscle RyR is essential for skeletal muscle E-C coupling.

Whether the DHPR and RyR interact directly or whether other proteins are involved is unclear. Morphological evidence (discussed above) as well as co-immunoprecipitation of DHPR and RyR[57] suggest a close interaction between the two receptors. The presence of other proteins linking the DHPR and RyR or being required for a stable interaction between the two receptors could not, however, be ruled out in these studies. Indeed, the inability of chemical reagents to cross-link the DHPR and RyR suggested the presence of such proteins and led Caswell, Brandt and their coworkers to search for additional triadic proteins. A protein found by these investigators to support an interaction between the DHPR and RyR is triadin,[46] a M_r 90,000 SR junctional glycoprotein which co-localizes with the RyR in skeletal and cardiac muscle.[47,58] SR Ca^{2+} release was affected by an anti-triadin antibody[59] and exposure of SR vesicles to oxidizing reagents led to the formation of crosslinks between triadin and RyR.[60] These observations suggest a close association of triadin and RyR.

One skeletal muscle and three cardiac muscle isoforms of triadin have been cloned and their primary sequences have been determined.[61] The deduced amino acid sequences show that triadins are identical in their N-terminal sequences, whereas the C-terminal sequences are distinct for the skeletal and cardiac isoforms. Hydropathy analysis indicates that triadins have a small (47 amino acid residues) cytoplasmic N-terminal domain and a single transmembrane segment, with the remainder being found in the SR lumen. Triadin-GST fusion proteins interacted with RyR and calsequestrin, a SR lumenal Ca^{2+} binding protein, but not with DHPR.[62] These results have been challenged by Caswell, Brandt and co-workers who found that in protein overlays triadin bound to both the DHPR and RyR, whereas no direct interaction between the DHPR and RyR could be detected.[46] Based on immunological and biochemical evidence, these investigators proposed a unique membrane topology model for triadin.[63] A major portion of triadin was cytoplasmically placed, and triadin was proposed to contain, in addition to one α helix, three β sheets that span the SR membrane. Additional studies are required to clarify the protein's interaction with the DHPR and RyR and its role in organizing the E-C coupling components in the triad junction.

As discussed above, compelling evidence for a different function of DHPR in mammalian cardiac and skeletal muscle E-C coupling, i.e. as a Ca^{2+} channel in cardiac muscle and as a voltage-sensing molecule in skeletal muscle, has been obtained through the elegant work of Tanabe, Beam and their co-workers. These investigators showed that the putative II-III cytoplasmic loop of the α_1 subunit has a major role in determining the type of E-C coupling that exists in muscle. Lu *et al.*[64,65] expressed the skeletal and cardiac II-III cytoplasmic loop regions in E.coli and assessed their interaction with RyR in single channel, [^{3}H]ryanodine binding and protein overlay measurements. Both peptides bound to and activated the purified skeletal muscle RyR but not cardiac muscle RyR. These results provide strong evidence for a direct interaction between the skeletal muscle DHPR and RyR and imply that, in addition to the DHPR, the RyR has a critical role in determining the E-C coupling mode in muscle.

The putative cytoplasmic loop between transmembrane repeats II and III of skeletal muscle DHPR α_1 subunit is composed of 126 amino acids (Glu666-Leu791). Use of three smaller overlapping peptides in the loop region identified an amino acid region from Glu666 to Glu726 which bound to and activated the skeletal muscle RyR.[65] El-Hayek *et al.* further refined the activating region by synthesizing six peptides corresponding to the cytoplasmic loop.[66] One of these (peptide A, Thr671-Leu690) increased [^{3}H]ryanodine binding and Ca^{2+} release from rabbit skeletal muscle triads. Peptide A-induced channel activation was suppressed by another DHPR-derived peptide (Glu724-Pro760). Accordingly, the II-III cytoplasmic loop region of DHPR α_1 subunit appears to have a region that triggers and another that blocks SR Ca^{2+} release. Other subunits and parts of the DHPR α_1 subunit may contribute to the functional coupling of the two receptors. A synthetic peptide corresponding to a highly conserved sequence located between amino acids 1487 and 1506 in the carboxy terminus of the α_1 subunit inhibited [^{3}H]ryanodine binding to SR membranes.[67] Thus, a complex picture is emerging, which suggests multiple interactions between DHPR and RyR.

The α_1 and β subunits of the skeletal muscle DHPR contain several consensus phosphorylation sites including a serine residue, Ser687, in II-III cytoplasmic loop region of the α_1 subunit. Ser687 is located in the 20 amino acid loop region identified by El-Hayek *et al.* to activate RyR.[66] This site is rapidly phosphorylated by cAMP-dependent protein kinase in vivo and in vitro.[68,69] Phosphorylation of Ser687 resulted in the formation of an inactive II-III loop peptide (Fig. 9.3) without loss of binding to RyR1.[65] Replacement of Ser687 by Ala also yielded a peptide incapable of activating RyR1. A Ser→Ala substitution is generally not considered to greatly affect the structure of unphosphorylated protein forms. It therefore appears that the β-OH of Ser687 is essential for the formation of an active loop peptide. The cardiac DHPR II-III loop region lacks a corresponding phosphorylation site, which suggests that cAMP-dependent protein kinase-mediated phosphorylation of Ser687 may specifically regulate skeletal muscle E-C coupling.

Yamazawa *et al.*[70] cultured mutant skeletal myocytes lacking both RyR1 and RyR3 to probe for regions of RyR1 that interact with DHPR. They expressed several mutated RyR1 cDNAs focusing on an amino acid region (1303–1406) that is highly divergent between RyR1 and RyR2 and is lacking in RyR3. Expression of RyR1 but not RyR2 rescued skeletal

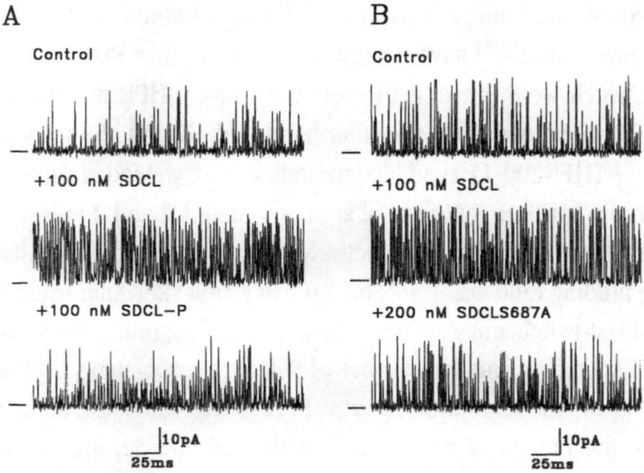

Fig. 9.3 Single channels of purified skeletal muscle RyR activated by peptide corresponding to putative cytoplasmic loop between transmembrane repeat II and III of the skeletal muscle DHPR α_1 subunit (SDCL), and inhibited by cytoplasmic loop peptides phosphorylated at Ser687 (SDCL-P) (A) and with Ser687→Ala substitution (SDCLS687A) (B). Single channel currents, shown as upward deflections, were recorded in symmetrical KCl media containing 9 µM free cytosolic Ca^{2+}. Top traces, control; middle traces, after the addition of 100 mM SDCL; bottom traces, after the further addition of 100 mM SDCL-P (A) or 200 mM (SDCLS687A). The addition of 100 mM SDCL caused ~4 fold increase in channel activity, whereas the further addition of the phosphorylated and mutant peptides decreased ~2 fold SDCL-activated channels (from Lu *et al.* Biol Chem 1995; **270**:18459–18464).

muscle E-C coupling in the mutant cells. Deletion of amino acids 1303–1406 or 1303–1367 resulted in loss of E-C coupling, whereas deletion of amino acids 1357–1406 or substitution of amino acids 1303–1406 by the corresponding sequence of RyR2 resulted in the formation of functional channels capable of skeletal muscle E-C coupling. An increase in intracellular Ca^{2+} on application of 25 mM caffeine indicated that Ca^{2+} release channel function was preserved. These results suggest that amino acid residues 1303–1356 of RyR1 are critical for E-C coupling in skeletal muscle. However, the observation that mutated RyR1 carrying the sequence of RyR2 exhibited skeletal muscle E-C coupling suggests that additional regions of RyR1 are essential for a functional coupling of DHPR and RyR.

DHPR-UNLINKED RyR

Although the direct control of RyR ion channel in skeletal muscle suggests a defined stoichiometric relationship, the ratio of DHPR to RyR in vertebrate skeletal muscle is uncertain. Morphological studies, initially with toadfish swimbladder[71] and more recently with mouse skeletal muscle,[44] have suggested that every other RyR is linked to a group of four particles, which were presumed to represent four DHPR molecules. These studies suggest a DHPR/RyR ratio of 2. On the other hand, radioligand binding with two receptor-specific ligands, [^{3}H]PN200-110 and [^{3}H]ryanodine, have yielded variable ratios. Margreth *et al.*[72] measured a DHPR to RyR ratio of approximately 1.2 and 2 in fast-twitch and slow-twitch muscles of rat and rabbit, respectively. In other studies, the [^{3}H]PN200-110 to [^{3}H]ryanodine binding ratio was 1.4[73] and 1.0[74] in rabbit skeletal muscle, and 0.3 and 0.9 in rat slow- and fast-twitch muscles, respectively.[75] Binding ratios of less than 2 imply that, at least in some species, a smaller number of RyRs are associated with the voltage sensor than predicted by the model of Block *et al.*[71] In rat skeletal muscle, some of these may be located outside the junctional area.[76] In frog skeletal muscle, the DHPR/RyR ratio was 1.1[72] and 0.6.[74] Two RyR isoforms (RyR1 and RyR3) are expressed in frog skeletal muscle.[11] It is therefore possible that only one isoform, the skeletal muscle isoform (RyR1), is linked to the DHPR. In favor of a linkage to the skeletal muscle isoform is that in chicken, which also expresses RyR1 and RyR3, the skeletal muscle RyR is essential for skeletal muscle E-C coupling.[11]

An only partial linkage of RyRs to DHPRs raises the question by which mechanism the unlinked RyRs are activated. Are they activated by Ca^{2+} released from DHPR-linked channels or are there other, perhaps unknown diffusible transmitter molecules that regulate the unlinked RyRs? Studies with skinned, cut, and intact skeletal muscle fibers favor the idea of a Ca^{2+}-dependent Ca^{2+} release mechanism.[4,5,10] Strong evidence in support of a Ca^{2+}-gated SR Ca^{2+} release has been obtained from studies with skeletal muscle fibers containing a fast complexing Ca^{2+} buffer such as BAPTA or Fura-2. In voltage-clamped intact and cut muscle fibers, depolarization of the surface membrane typically results in an intracellular Ca^{2+} transient comprised of an early peak and a quasi-steady component which is maintained until repolarization of the surface membrane.[4,5,10] The latter component is believed to be under the direct control of the T-tubule voltage sensor (DHPR). The microinjection of a relatively high concentration (~2–4 mM) of the fast complexing Ca^{2+} buffers BAPTA or Fura-2 into intact frog muscle fibers caused elimination of the early peak component of depolarization-induced Ca^{2+} release, while leaving the maintained quasi-steady level unaffected.[77] These results suggested that the initially released Ca^{2+} ions open additional (DHPR-unlinked) Ca^{2+} release channels by a Ca^{2+}-induced Ca^{2+} release mechanism. The idea of a Ca^{2+}-induced Ca^{2+} release process has been challenged by Jong *et al.*[78] who observed that Fura-2 concentrations of 0.5–3 mM increased both the amount of Ca^{2+} release and the rate of release, which was attributed to the removal by Fura-2 of Ca^{2+}-induced inactivation of Ca^{2+} release. In the hands of these investigators, Fura-2 concentrations as high as 8 mM did not prevent Ca^{2+} inactivation of Ca^{2+} release, which led them to propose that DHPR-linked and -unlinked channels may function as a singly gated unit.[79] In support of this idea, it has been shown that in cardiac and skeletal muscle intracellular Ca^{2+} transients may arise as the sum of localized Ca^{2+} release events (called 'Ca^{2+} sparks').[80–82] In cardiac muscle, the opening of a single T-tubule Ca^{2+} channel is apparently sufficient to evoke a Ca^{2+} spark by activating a Ca^{2+} release unit which may consist of one or several Ca^{2+}-gated Ca^{2+} release channels.[83] In frog skeletal muscle, the appearance of large-amplitude events consisting of two or more sparks has led Klein *et al.*[82] to propose that voltage-gated release channels may activate adjacent ligand-gated release channels.

T-tubule depolarization-induced SR Ca^{2+} release has been also examined using triad-enriched membrane fractions isolated from rabbit skeletal muscle. In an early study, Ikemoto *et al.*[84] observed that T-tubule depolarization induced by rapid ion replacement resulted in 3 phases of Ca^{2+} release: (1) a lag phase, (ii) the rapid release of a small amount of Ca^{2+} with a rate constant of $k = 100\ s^{-1}$, and (iii) the slow release of a larger amount of Ca^{2+} with $k \sim 1\ s^{-1}$. More recently, these investigators showed that a fluorescent probe (methylcoumarin acetate) attached to the RyR can be used to monitor the conformational states of RyR during T-tubule polarization and depolarization.[85] DHPR antagonists and antibodies have implicated the DHPR and the SR integral membrane protein, triadin, in mediating the slow release phase while a 28 kDa protein, not part of the DHPR, was implicated in mediating the rapid release phase.[59] These results have challenged the idea of DHPR-mediated control of rapid SR Ca^{2+} release during EC coupling. The magnitude and rate of rapid SR Ca^{2+} release were dependent on the magnitude of the T-tubule potential change, and Ca^{2+}, Mg^{2+} and ATP concentrations of T-tubule depolarizing media.[84,86,87] Anderson and Meissner[87] used a rapid filtration assay and the fast Ca^{2+} buffer BAPTA to distinguish in cell homogenates from rabbit skeletal muscle between T-tubule depolarization-dependent and Ca^{2+}- dependent SR Ca^{2+} release. T-tubule depolarization-induced SR Ca^{2+} release was inhibited by high [BAPTA] and 1 mM Ca^{2+} (Fig. 9.4). The results of Fig. 9.4 provide direct evidence for the idea that the released Ca^{2+} ions activate and inhibit the SR Ca^{2+} release channel during T-tubule depolarization, depending on the Ca^{2+} concentration in the junctional gap.

CONCLUDING REMARKS

In conclusion, skeletal muscle T-tubule depolarization-induced SR Ca^{2+} release appears to be controlled by T-tubule- and Ca^{2+}-dependent mechanisms. This model is supported by

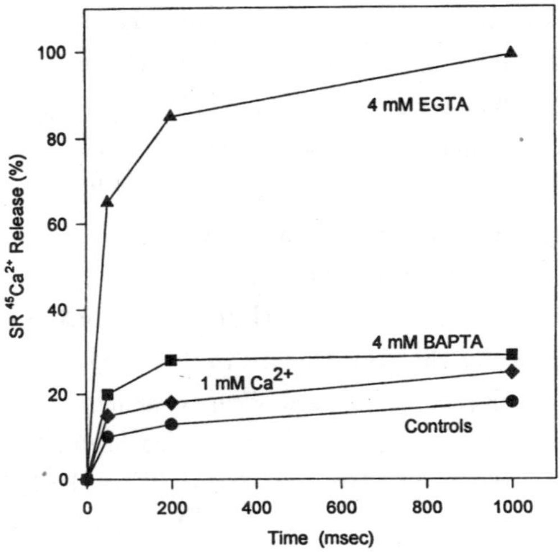

Fig. 9.4 Effects of Ca^{2+} and 'fast' Ca^{2+} buffer BAPTA on T-tubule depolarization-induced sarcoplasmic reticulum Ca^{2+} release. T-tubule depolarization-induced SR Ca^{2+} release was measured in rabbit skeletal muscle homogenates using an ion-replacement protocol. Triads in homogenates were actively loaded with $^{45}Ca^{2+}$ under T-tubule polarizing conditions, placed on filters, and rinsed with T-tubule depolarizing solution for the indicated times using a rapid filtration apparatus. T-tubule depolarizing solutions contained a 'slow' Ca^{2+} buffer (4 mM EGTA), a 'fast' Ca^{2+} buffer (4 mM BAPTA) or 1 mM Ca^{2+}, and in "Controls" in addition the DHPR antagonist D888. The maximal amount of $^{45}Ca^{2+}$ released (100%) corresponded to the caffeine-sensitive $^{45}Ca^{2+}$ pool (Adapted from Anderson and Meissner. J Gen Physiol 1995; 105:363–383).

morphological studies and is consistent with speculations that DHPR conformational changes, induced by T-tubule depolarization, trigger the opening of RyR/Ca^{2+} release channels through a physical linkage between the two receptors. The model is also consistent with recent observations of a direct functional interaction of the skeletal muscle Ca^{2+} release channel with peptides derived from the DHPR and the presence of RyR1 regions that appear to be essential for skeletal muscle E-C coupling. Major questions that remain to be resolved regarding the interaction of DHPR and RyR include: In which way does linkage to the DHPR alter the RyR's regulation by Ca^{2+} and other endogenous effectors? Which regions of the RyR and the DHPR are involved in the functional interaction of the two receptors? Are other SR and T-tubule proteins required for T-tubule mediated SR Ca^{2+} release?

ACKNOWLEDGMENTS

Support by United States Public Service grants AR18687 and HL27430 is gratefully acknowledged.

REFERENCES

1. Wier WG. Cytoplasmic [Ca^{2+}] in mammalian ventricle: Dynamic control by cellullar processes. *Ann Rev Physiol* 1990; **52**:467–485.
2. Fabiato A. Calcium-induced release of calcium from the cardiac sarcoplasmic reticulum. *Am J Physiol* 1983; **245**:C1–C14.
3. Berridge MJ. Inositol trisphosphate and calcium signalling. *Nature* 1993; **361**:315–325.
4. Rios E, Pizarro G. Voltage sensor of excitation-contraction coupling in skeletal muscle. *Physiol Rev* 1991; **71**:849–908.
5. Schneider MF. Control of calcium release in functioning skeletal muscle fibers. *Ann Rev Physiol* 1994; **56**:463–484.
6. Catterall WA. Structure and function of voltage-gated ion channels. *Ann Rev Biochem* 1995; **64**:493–531.
7. Coronado R, Morrissette J, Sukhareva M *et al.* Structure and function of ryanodine receptors. *Am J Physiol* 1994; **266**:C1485–C1504.
8. Meissner G. Ryanodine receptor/Ca^{2+} release channels and their regulation by endogenous effectors. *Ann Rev Physiol* 1994; **56**:485–508.
9. Ogawa Y. Role of ryanodine receptors. *Crit Rev Biochem Mol Biol* 1994; **29**:229–274.
10. Melzer W, Herrmann-Frank A, Luttgau HCh. The role of Ca^{2+} ions in excitation-contraction coupling of skeletal muscle fibres. *Biochim Biophys Acta* 1995; **1241**:59–116.
11. Sutko JL, Airey JA. Ryanodine receptor Ca^{2+} release channels: Does diversity in form equal diversity in function? *Physiol Rev* 1996; **76**:1027–1071.
12. Takeshima H, Nishimura S, Matsumoto T *et al.* Primary structure and expression from complementary DNA of skeletal muscle ryanodine receptor. *Nature* 1989; **339**:439–445.
13. Zorzato F, Fujii J, Otsu K *et al.* Molecular cloning of cDNA encoding human and rabbit forms of the Ca^{2+} release channel (ryanodine receptor) of skeletal muscle sarcoplasmic reticulum. *J Biol Chem* 1990; **265**:2244–2256.
14. Nakai J, Imagawa T, Hakamata Y *et al.* Primary structure and functional expression from cDNA of the cardiac ryanodine receptor/calcium release channel. *FEBS Lett* 1990; **271**:169–177.
15. Otsu K, Willard HF, Khanna VK *et al.* 1990. Molecular cloning of cDNA encoding the Ca^{2+} release channel (ryanodine receptor) of rabbit cardiac muscle sarcoplasmic reticulum. *J Biol Chem* 1990; **265**:13472–13483.
16. Hakamata Y, Nakai J, Takeshima H *et al.* Primary structure and distribution of a novel ryanodine receptor/calcium release channel from rabbit brain. *FEBS Lett.* 1992; **312**: 229–235.

17. Ledbetter MW, Preiner JK, Louis CF *et al.* Tissue distribution of ryanodine receptor isoforms and alleles determined by reverse transcription polymerase chain reaction. *J Biol Chem* 1994; **269**:31544–31551.
18. Giannini G., Conti A., Mammarella S *et al.* (1995) The ryanodine receptor/Ca^{2+} release channel genes are widely and differentially expressed in murine brain and peripheral tissues. *J Cell Biol* 1995; **128**:893–904.
19. Tunwell REA, Lai FA. Ryanodine receptor expression in the kidney and a non-excitable kidney epithelial cell. *J Biol Chem* 1996: **271**:29583–29588.
20. Takeshima H, Ikemoto T, Nishi M *et al.* Generation and characterization of mutant mice lacking ryanodine receptor type 3. *J Biol Chem* 1996; **271**:19649–19652.
21. Marty I, Villaz M, Arlaud G *et al.* Transmembrane orientation of the N-terminal and C-terminal ends of the ryanodine receptor in the sarcoplasmic reticulum of rabbit skeletal muscle. *Biochem J* 1994; **298**:743–749.
22. Grunwald R, Meissner G. Lumenal sites and C terminus accessibility of the skeletal muscle calcium release channel (ryanodine receptor). *J Biol Chem* 1995; **270**:11338–11347.
23. Timerman AP, Onoue H, Xin HB. Selective binding of FKBP12.6 by the cardiac ryanodine receptor. *J Biol Chem* 1996; **271**:20385–20391.
24. Radermacher M, Rao V, Grassucci, R *et al.* Cryo-electron microscopy and three-dimensional reconstruction of the calcium release channel/ryanodine receptor from skeletal muscle. *J Cell Biol* 1994; **127**:411–423.
25. Orlova EV, Serysheva II, van Heel M *et al.* Two structural configurations of the skeletal muscle calcium release channel. *Nature Struct Biol* 1996; **3**:547–551.
26. Murayama T, Ogawa Y. Properties of Ryr3 ryanodine receptor isoform in mammalian brain. *J Biol Chem* 1996; **271**:5079–5084.
27. Sitsapesan R, Williams AJ. The gating of the sheep skeletal sarcoplasmic reticulum Ca^{2+} release channel is regulated by luminal Ca^{2+}. *J Membrane Biol* 1995; **146**:133–144.
28. Tripathy A, Meissner G. Sarcoplasmic reticulum lumenal Ca^{2+} has access to cytosolic activation and inactivation sites of skeletal muscle Ca^{2+} release channel. *Biophys J* 1996; **70**:2600–2615.
29. Ikemoto N, Antoniu B, Kang JJ *et al.* Intravesicular calcium transient during calcium release from sarcoplasmic reticulum. *Biochemistry* 1991; **30**:5230–5237.
30. Tripathy A, Xu L, Mann G *et al.* Calmodulin activation and inhibition of skeletal muscle Ca^{2+} release channel (ryanodine receptor). *Biophys J* 1995; **69**:106–119.
31. Ikemoto T, Iino M, Endo M. Enhancing effect of calmodulin on Ca^{2+}-induced Ca^{2+} release in the sarcoplasmic reticulum of rabbit skeletal muscle fibres. *J Physiol* 1995; 487:573–582.
32. Galione A, Lee HC, Busa WB. Ca^{2+}-induced Ca^{2+} release in sea urchin egg homogenates: Modulation by cyclic ADP-ribose. *Science* 1991: **253**:1143–1146.
33. Kummerle JF, Makhlouf GM. Agonist-stimulated cyclic ADP ribose. Endogenous modulators of Ca^{2+}-induced Ca^{2+} release in intestinal longitudinal muscle. *J Biol Chem* 1995; **270**: 25488–25494.
34. Noguchi N, Takasawa S, Nata K *et al.* Cyclic ADP-ribose binds to FK506-binding protein 12.6 to release Ca^{2+} from islet microsomes. *J Biol Chem* 1997; **272**:3133–3136.

35. Meszaros LG, Bak J, Chu A. Cyclic ADP-ribose as an endogenous regulator of the non-skeletal type ryanodine receptor Ca^{2+} channel. *Nature* 1993; **364**:76–79.
36. Sitsapesan R, McGarry SJ, Williams AJ. Cyclic ADP-ribose competes with ATP for the adenine nucleotide binding site on the cardiac ryanodine receptor Ca^{2+} release channel. *Circ Res* 1994; **75**:596–600.
37. Guo X, Laflamme MA, Becker PL. Cyclic ADP-ribose does not regulate sarcoplasmic reticulum Ca^{2+} release in intact cardiac myocytes. *Circ Res* 1996; **79**:147–151.
38. Witcher DR, Kovacs RJ, Schulman H *et al.* Unique phosphorylation site on the cardiac ryanodine receptor regulates calcium channel activity. *J Biol Chem* 1991; **266**:11144–11152.
39. Suko J, Maurer-Fogy I, Plank B *et al.* Phosphorylation of serine2843 in ryanodine receptor-calcium release channel of skeletal muscle by cAMP, -cGMP- and CaM-dependent protein kinase. *Biochim Biophys Acta* 1993; **1175**:193–206.
40. Hain J, Onoue H, Mayrleitner M *et al.* Phosphorylation modulates the function of the calcium release channel of sarcoplasmic reticulum from cardiac muscle. *J Biol Chem* 1995; **270**: 2074–2081.
41. Stoyanovsky D, Murphy T, Anno PR *et al.* Nitric oxide activates skeletal and cardiac ryanodine receptors. *Cell Calcium* 1997; **21**:19–29.
42. Xu L, Meissner G, Eu P *et al.* S-Nitrosylation activates the cardiac Ca^{2+} release channel (ryanodine receptor). *Biophys J* 1997; **72**:A375.
43. Franzini-Armstrong C, Jorgensen AO. Structure and development of E-C coupling units in skeletal muscle. *Ann Rev Physiol* 1994; **56**:509–534.
44. Takekura H, Nishi M, Noda T *et al.* Abnormal junctions between surface membrane and sarcoplasmic reticulum in skeletal muscle with a mutation targeted to the ryanodine receptor. *Proc Natl Acad Sci U S A* 1995; 92:3381–3385.
45. Protasi F, Franzini-Armstrong C, Allen PD. Dihydropyridine receptors are clustered but do not form tetrads in peripheral couplings of dyspedic myotubes. *Biophys J* 1997; **72**:A13.
46. Caswell AH, Brandt NR, Brunschwig JP *et al.* Localization and partial characterization of the oligomeric disulfide-linked molecular weight 95,000 protein (triadin) which binds the ryanodine and dihydropyridine receptors in skeletal muscle triadic vesicles. *Biochemistry* 1991; **30**: 7507–7513.
47. Lewis SC, Felix K, Caswell AH *et al.* Immunolocalization of sarcolemmal dihydropyridine receptor and sarcoplasmic reticular triadin and ryanodine receptor in rabbit ventricle and atrium. *J Cell Biol* 1995; **129**:673–682.
48. Junker J, Sommer JR, Sar M *et al.* Extended junctional sarcoplasmic reticulum of avian cardiac muscle contains functional ryanodine receptors. *J Biol Chem* 1994; **269**:1627–1634.
49. Sun XH, Protasi F, Takahashi M *et al.* Molecular architecture of membranes involved in excitation-contraction coupling of cardiac muscle. *J Cell Biol* 1995; **129**:659–671.
50. Tanabe T, Beam KG, Powell JA *et al.* Restoration of excitation-contraction coupling and slow calcium current in dysgenic muscle by dihydropyridine receptor complementary DNA. *Nature* 1988; **336**:134–139.
51. Tanabe T, Mikami A, Numa S *et al.* Cardiac-type excitation-contraction coupling in dysgenic skeletal muscle injected with cardiac dihydropyridine receptor cDNA. *Nature* 1990; **344**: 451–453.

52. Tanabe T, Beam KG, Adams BA *et al.* Regions of the skeletal muscle dihydropyridine receptor critical for excitation-contraction coupling. *Nature* 1990; **346**:567–569.
53. Strube C, Beurg M, Powers PA *et al.* Reduced Ca^{2+} current, charge movement, and absence of Ca^{2+} transients in skeletal muscle deficient in dihydropyridine receptor β_1 subunit. *Biophys J* 1996; **71**:2531–2543.
54. DeWaard M, Witcher DR, Pragnell M *et al.* Properties of the a_1-b anchoring site in voltage-dependent Ca^{2+} channels. *J Biol Chem* 1995; **270**:12056–12064.
55. Chien AJ, Zhao X, Shirokov RE *et al.* Roles of a membrane-localized b subunit in the formation and targeting of functional L-type Ca^{2+} channels. *J Biol Chem* 1995; **270**:30036–30044.
56. Takeshima H, Yamazawa T, Ikemoto T *et al.* Ca^{2+}-induced Ca^{2+} release in myocytes from dyspedic mice lacking the type-1 ryanodine receptor. *EMBO J* 1995; **14**:2999–3006.
57. Marty I, Robert M, Villaz M *et al.* Biochemical evidence for a complex involving dihydropyridine receptor and ryanodine receptor in triad junctions of skeletal muscle. *Proc Natl Acad Sci USA* 1994; **91**:2270–2274.
58. Carl LS, Felix K, Caswell AH *et al.* Immunolocalization of triadin, DHP receptors, and ryanodine receptors in adult and developing skeletal muscle of rats. *Muscle & Nerve* 1995; **18**: 1232–1243.
59. Brandt NR, Caswell AH, Brunschwig JP *et al.* Effects of anti-triadin antibody on Ca^{2+} release from sarcoplasmic reticulum. *FEBS LETT* 1992; **299**:57–59.
60. Liu G, Pessah IN. Molecular interaction between ryanodine receptor and glycoprotein triadin involves redox cycling of functionally important hyperreactive sulfhydryls. *J Biol Chem* 1994; **269**:33028–33034.
61. Guo W, Jorgensen AO, Jones LR *et al.* Biochemical characterization and molecular cloning of cardiac triadin. *J Biol Chem* 1996; **271**:458–465.
62. Guo W, Campbell KP. Association of triadin with the ryanodine receptor and calsequestrin in the lumen of the sarcoplasmic reticulum. *J Biol Chem* 1995; **270**:9027–9030.
63. Fan H, Brandt NR, Peng M *et al.* Binding sites of monoclonal antibodies and dihydropyridine receptor alpha 1 subunit II-III loop on skeletal muscle triadin fusion peptides. *Biochemistry* 1995; **34**:14893–14901.
64. Lu X, Xu L, Meissner G. Activation of the skeletal muscle calcium release channel by a cytoplasmic loop of the dihydropyridine receptor. *J Biol Chem* 1994; **269**:6511–6516.
65. Lu X, Xu L, Meissner G. Phosphorylation of dihydropyridine receptor II-III loop peptide regulates skeletal muscle calcium release channel function. Evidence for an essential role of the b-OH group of Ser^{687}. *J Biol Chem* 1995; **270**:18459–18464.
66. El-Hajek R, Antoniu B, Wang J *et al.* Identification of calcium release-triggering and blocking regions of the II-III loop of the skeletal muscle dihydropyridine receptor. *J Biol Chem* 1995; **270**:22116–22118.
67. Wang JP, Seryshev A, Wu Y *et al.* The functional regulation site of SR Ca^{2+} release channel on the DHPR α_1 subunit. *Biophy J* 1995: **68**:A14.
68. Rohrkasten A, Meyer HE, Nastainczyk W *et al.* cAMP-dependent protein kinase rapidly phosphorylates serine-687 of the skeletal muscle receptor for calcium channel blockers. *J Biol Chem* 1988; **263**:15325–15329.

69. Lai Y, Seagar MJ, Takahashi M *et al.* Cyclic AMP-dependent phosphorylation of two size forms of a1 subunits of L-type calcium channels in rat skeletal muscle cells. *J Biol Chem* 1990; **265**: 20839–20848.
70. Yamazawa T, Takeshima H, Shimuta M *et al.* A region of the ryanodine receptor critical for excitation-contraction coupling in skeletal muscle. *J Biol Chem* 1997; **272**:8161–8164.
71. Block BA, Imagawa T, Campbell KP *et al.* Structural evidence for direct interaction between the molecular components of the transverse tubule/sarcoplasmic reticulum junction in skeletal muscle. *J Cell Biol* 1988; **107**:2587–2600.
72. Margreth A, Damiani E, Tobaldin G. Ratio of dihydropyridine to ryanodine receptors in mammalian and frog twitch muscles in relation to the mechanical hypothesis of excitation-contraction coupling. *Biochem Biophys Res Comm* 1993; **197**:1303–1311.
73. Bers DM, Stiffel VM. Ratio of ryanodine to dihydropyridine receptors in cardiac and skeletal muscle and implications for E-C coupling. *Am J Physiol* 1993; **264**:C1587–C1593.
74. Anderson K, Cohn AH, Meissner G. High-affinity [^{3}H]PN200-110 and [^{3}H]ryanodine binding to rabbit and frog skeletal muscle. *Am J Physiol* 1994; **266**:C462–C466.
75. Delbono O, Meissner G. Sarcoplasmic reticulum Ca^{2+} release in rat slow- and fast-twitch muscles. *J Membrane Biol* 1996; **151**:123–130.
76. Dulhunty AF, Junankar PR, Stanhope C. Extrajunctional ryanodine receptor in the terminal cisternae of mammalian skeletal muscle fibres. *Proc R Soc Lond. B. Biol Sci* 1992; **247**:69–75.
77. Jacquemond V, Csernoch L, Klein MG *et al.* Voltage-gated and calcium-gated calcium release during depolarization of skeletal muscle fibers. *Biophys J* 1991; **60**:867–873.
78. Jong DS, Pape PC, Chandler WK *et al.* Reduction of calcium inactivation of sarcoplasmic reticulum calcium release by Fura-2 in voltage-clamped cut twitch fibers from frog muscle. *J Gen Physiol* 1993; **102**:333–370.
79. Jong DS, Pape PC, Baylor SM *et al.* Calcium inactivation of calcium release in frog cut muscle fibers that contain millimolar EGTA or Fura-2. *J Gen Physiol* 1995; **106**:337–388.
80. Cheng H, Lederer WJ, Cannell MB. Calcium sparks: Elementary events underlying excitation-contraction coupling in heart muscle. *Science* 1993; **262**:740–744.
81. Tsugorka A, Rios E, Blatter LA. Imaging elementary events of calcium release in skeletal muscle cells. *Science* 1995; **269**:1723–1726.
82. Klein MG, Cheng H, Santana LF *et al.* Two mechanisms of quantized calcium release in skeletal muscle. *Nature* 1996: **379**:455–458.
83. Santana LF, Cheng H, Gomez AM *et al.* Relation between the sarcolemmal Ca^{2+} current and Ca^{2+} sparks and local control theories for cardiac excitation-contraction coupling. *Circ Res* 1996; **78**:166–171.
84. Ikemoto N, Antoniu B, and Kim DH. Rapid calcium release from the isolated sarcoplasmic reticulum is triggered via the attached transverse tubular system. *J Biol Chem* 1984; **259**: 13151–13158.
85. Ikemoto N, El-Hayek R. Reciprocal control of the conformational state of the sarcoplasmic reticulum calcium channel protein by polarization and depolarization in the transverse tubule. *FEBS Lett.* 1996; *394*:330–334.

86. Ritucci NA, Corbett AM. Effect of Mg^{2+} and ATP on depolarization-induced Ca^{2+} release in isolated triads. *Am J Physiol* 1995; **269**:C85–C95.
87. Anderson K, Meissner G. T-tubule depolarization-induced SR Ca^{2+} release is controlled by dihydropyridine receptor- and Ca^{2+}-dependent mechanisms in cell homogenates from rabbit skeletal muscle. *J Gen Physiol* 1995; **105**:363–383.

Chapter 10

Regulation of Ryanodine Receptor by Peptide Toxins

Georgina B. Gurrola and Hector H. Valdivia

INTRODUCTION

During excitation-contraction coupling of cardiac and skeletal muscle, the calcium release channel (also termed ryanodine receptor, RyR) constitutes the major pathway for Ca^{2+} release from the sarcoplasmic reticulum (SR).[1] RyRs are also found in neurons, exocrine cells, smooth muscle cells, epithelial cells, lymphocytes, and sea urchin eggs. In all of these cells, RyRs play a central role in the regulation of the intracellular free Ca^{2+} concentration, whose elevation triggers a cascade of events that culminates in muscle contraction, hormone secretion, lymphocyte activation, egg fertilization, etc.. To gain the functional flexibility necessary to respond to different triggering signals, at least three tissue-specific RyR isoforms are expressed. In mammals, RyR1 is expressed predominantly in fast- and slow-twitch skeletal muscle, while RyR2 is expressed predominantly in cardiac muscle. RyR3 appears to be localized to brain, smooth muscle, and epithelial cells, although low levels of expression of RyR1 and RyR2 are also found in some of these tissues.[2] Despite this different pattern of expression, there are functional and structural attributes common to all RyR isoforms: they are homotetramers of large molecular size (~2 million Da),[3] they form Ca^{2+}-gated Ca^{2+}-selective channels of large conductance,[4] and they are distinctively affected by the plant alkaloid ryanodine.[5]

RYANODINE AS A PROBE OF RyR: IS THERE NEED FOR OTHER LIGANDS?

Ryanodine has been an invaluable tool in defining the pharmacological profile of RyRs. The binding of [^{3}H]ryanodine is enhanced by activators of Ca^{2+} release (μM Ca^{2+}, ATP, caffeine) and decreased by inhibitors of Ca^{2+} release (Mg^{2+}, ruthenium red), suggesting that the alkaloid binds to a conformationally-sensitive domain on the RyR protein.[6] Therefore, [^{3}H]ryanodine has been used as a probe of the functional state of the channel. This approach has contributed to the identification, purification and cloning of the Ca^{2+} release channel,[7–9] and to the identification of endogenous regulators,[10] novel ligands,[11–12] and regulatory proteins.[14] With all this wealth of information provided by the use of ryanodine, is there a need for additional probes of RyR?

The use of ryanodine to dissect the contribution of RyRs to intracellular Ca^{2+} signals in situ has been hampered by several of its kinetic features: (1) at low concentrations (approximately 5 to 50 nM), ryanodine increases the mean open time of the channel without modifying its unitary conductance.[14,15] At intermediate concentrations (50 nM to 10 μM), ryanodine "locks" the channel in a subconducting state.[5] Higher concentrations (>100 μM) of the alkaloid fully and irreversibly close the channel.[15] Thus, ryanodine may act as an agonist and blocker of the RyR,[6] producing ambiguous effects. (2) The association rate of ryanodine is too slow[16,17] for experiments with intact cells. To accelerate the onset of effect, micromolar, instead of nanomolar levels of ryanodine are used. During the initial phase of incubation, ryanodine may open RyRs and deplete the SR Ca^{2+} content. At equilibrium, higher ryanodine concentrations may close RyRs. Therefore, although the effect of depleting the SR Ca^{2+} content or blocking its release may be macroscopically the same (paralysis), the dual action of ryanodine makes it difficult to assess the availability of a releasable Ca^{2+} pool at any experimental point. (3) Ryanodine is strictly dependent on Ca^{2+} for binding,[16] but Ca^{2+} itself adds its own component of channel activation through the virtually instantaneous mechanism of Ca^{2+}-induced Ca^{2+} release. Hence, it is difficult to discern the amount of Ca^{2+} mobilized by Ca^{2+} before the onset of the ryanodine effect. (4) The recovery of SR function after treatment with ryanodine is also difficult to assess because the slow dissociation rate of ryanodine[17] makes the pharmacological effect essentially irreversible.

SCORPION PEPTIDES: ALTERNATIVE LIGANDS TO PROBE RyR STRUCTURE AND FUNCTION

Scorpion venoms contain families of small basic proteins that modify the gating mechanism of Na^+ channels, or block with high affinity K^+ channels of excitable cells.[18] A Cl^--channel specific blocker peptide was also recently isolated from a scorpion venom.[19] These toxins have been invaluable tools in the identification, purification, structural mapping and functional characterization of the corresponding ionic channels. We thus searched for peptide toxins that could regulate RyRs in a rapid, reversible, and simpler manner than ryanodine.

The existence of peptide toxins specific for RyRs was first known from work conducted with venom from the African scorpion *Buthotus hottentota*.[20] A peptide fraction from this venom produced a rapid, specific and reversible activation of skeletal RyRs, engendering hopes that new probes of RyR that acted in a simpler manner than ryanodine were available. Shortly after, Valdivia *et al.* purified to homogeneity two peptide components from the scorpion *Pandinus imperator* that targeted RyRs with nanomolar affinity.[21] **Imperatoxin I** ($IpTx_i$), a ~15 kDa protein, inhibited [^{3}H]ryanodine binding to cardiac and skeletal RyRs with an apparent K_d = 50 nM. **Imperatoxin A** ($IpTx_a$), a ~4 kDa peptide, specifically increased [^{3}H]ryanodine binding to skeletal (but not cardiac) RyRs with a K_d = 6 nM.[22] At concentrations higher than those needed to regulate RyRs, neither $IpTx_i$ nor $IpTx_a$ affected other Ca^{2+} channels or ion transporters of muscle and brain.[21] Using these prototype toxins, other peptide toxins from scorpion[23] and lizard[24] venom have been characterized. Thus, there exist a vast array of natural peptides to probe the structure and function of RyRs. This chapter describes recent advances made towards the functional and structural characterization of $IpTx_a$ and $IpTx_i$.

PURIFICATION OF IMPERATOXINS

Whole *P. imperator* venom completely inhibits [^{3}H]ryanodine binding to cardiac and skeletal SR with a concentration of 15 μg/ml yielding the half-maximal effect (ED_{50}) (Fig. 10.1F). The presence of components that stimulate RyRs is only apparent after

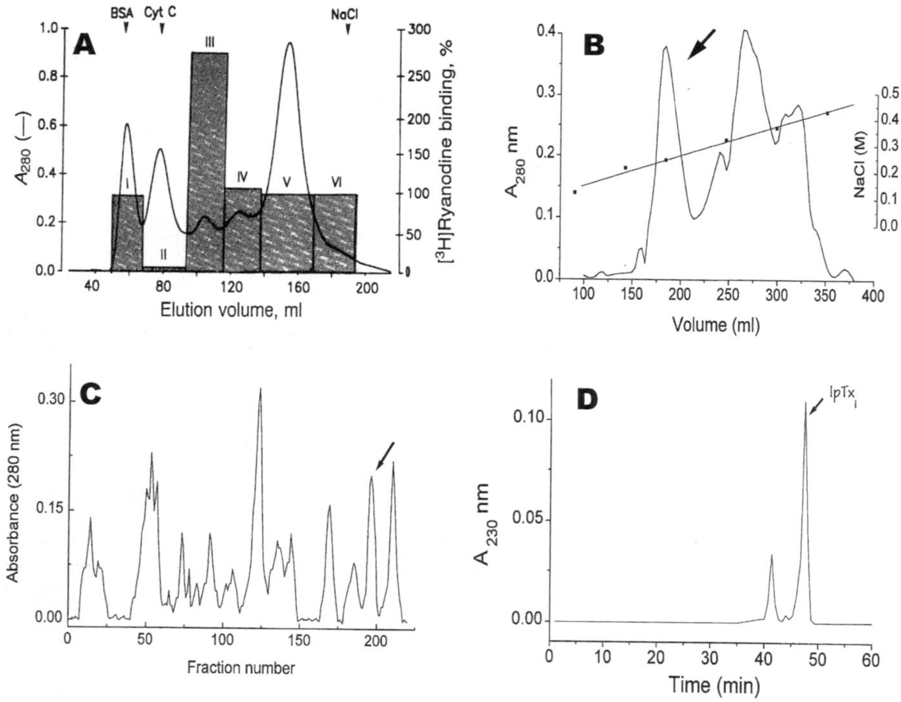

Fig. 10.1 Purification of IpTx$_i$ and IpTx$_a$. (A) *P. imperator* whole venom(120 mg per batch) was suspended in double-distilled water and centrifuged at 15,000 g for 30 min. The supernatant was applied onto a column (0.9 × 190 cm) of Sephadex G-50 super-fine (Pharmacia). Fractions were eluted with 20 mM NH_3OAc (pH 4.7) at a flow rate of 20 ml/h. (B) Fraction II containing IpTx$_i$ was applied to a column (0.9 × 30 cm) of carboxymethyl (CM)-cellulose 32 (Whatman) equilibrated with 20 mM NH_3OAc (pH 4.7). Peptides were eluted at a flow rate of 20 ml/h with a linear gradient of 250 ml of 20 mM NH_3OAc (pH 4.7) and 250 ml of the same buffer containing 0.55 M NaCl. (C) Fraction III containing IpTx$_a$ was run under conditions identical to those described for Fraction II. Peaks containing lpTx$_i$ and lpTx$_a$ were dialyzed against deionized water (3 × 30 min), concentrated by vacuum centrifugation, and injected into a C_8 reverse-phase HPLC column (Vydac). (D) IpTx$_i$ was eluted with a linear gradient of 0–60% acetonitrile in 0.1% trifluoroacetic acid running at 1 ml/min for 60 min. (E) IpTx$_a$ was chromatographed under conditions identical to those described for IpTx$_i$. Pure IpTx$_i$ and IpTx$_a$ are quantified either by amino acid analysis or by absorbance at 280 nm ($A_{280\ nm}$) using an extinction coefficient (ε) = 14,952 and 1,852 M^{-1} cm^{-1}, respectively. (F) Dose-response relation for each of the active venom components. 7 nM [^{3}H]ryanodine were incubated with 0.3 mg/ml rabbit skeletal SR in medium containing 0.2 M KCl, 10 mM Na-Hepes (pH 7.2), and 10 μM free Ca^{2+}. After 90 min at 36°C, the reaction was terminated by rapid filtration in Whatman GF/B filters. Nonspecific binding was determined in the presence of 10 μM ryanodine and has been subtracted from each sample.

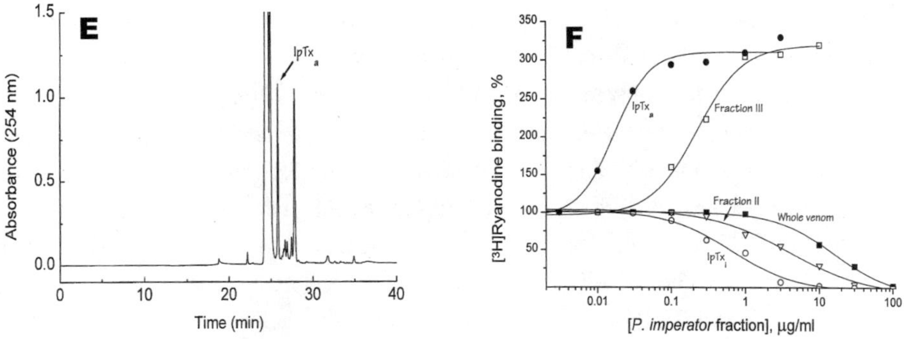

Fig. 10.1 *(Continued)*

fractionation of the whole venom. In a column of Sephadex G-50, which separates components based on molecular mass, whole *P. imperator* venom yields six fractions (Fig. 10.1A). Fraction I is composed of proteins in the range of 40–200 kDa, probably corresponding to hyaluronidases, collagenases and proteases that facilitate the penetration and diffusion of smaller toxic compounds.[25] Fraction II contains polypeptides in the range of 8 to 20 kDa, some of which correspond to phospholipases A_2 (PLA_2). $IpTx_i$ is present in this fraction and is the main toxin responsible for inhibition of [^{3}H]ryanodine binding. Fraction III and IV contain polypeptides in the range of 3 to 7 kDa. Fraction III increases [^{3}H]ryanodine binding (Fig. 10.1A), but since it accounts for ≤ 5% of the total venom, its effect is probably masked in the whole venom by that of Fraction II. The stimulation of [^{3}H]ryanodine binding by Fraction III is due to the presence of $IpTx_a$. Peptide blockers of several K^+ channels are also present in these fractions,[26,27] but they do not affect RyRs (see below). Fraction V and VI are composed of small peptides, free amino acids and neurotransmitters that contribute to the systemic effects caused by scorpion envenommation.[25]

After separation in Sephadex G-50, Fraction II (Fig. 10.1B) and Fraction III (Fig. 10.1C) are further fractionated in a cation-exchange column (carboxymethyl-cellulose). A peptide from Fraction II containing >90% of the inhibitory binding activity ($IpTx_i$) and a peptide from Fraction III containing >90% of the stimulatory activity ($IpTx_a$) (arrow in Figs. 10.1B and 10.1C) are eluted when the NaCl concentration at the top of the column reaches 65 and 340 mM, respectively. $IpTx_i$ and $IpTx_a$ are then dialyzed against deionized water, concentrated and injected separately into a C_8 reverse-phase HPLC column. $IpTx_i$ (Fig. 10.1D) and $IpTx_a$

(Fig. 10.1E) elute as single sharp peaks in 60 min gradients of 0–100% acetonitrile with average retention times of 48 and 27 min, respectively.

Dose-response relationship for each of the active venom components were made to assess their potency to modulate RyRs (Fig. 10.1F). Whole *P. imperator* venom and Fraction II inhibit [^{3}H]ryanodine binding to cardiac and skeletal SR with an ED_{50} = 20.2 and 4.3 μg/ml, respectively. ED_{50} for $IpTx_i$ is 0.7 μg/ml. Based on this value and the molecular mass of $IpTx_i$ (~15 kDa), the estimated apparent dissociation constant (K_d) is 46 nM.[28] On the other hand, Fraction III and $IpTx_a$ increase [^{3}H]ryanodine binding to skeletal SR with an ED_{50} = 0.19 and 0.016 μg/ml, respectively. Given the molecular mass of $IpTx_a$ (3.7 kDa), its apparent K_d is ~5 nM.[29]

PRIMARY STRUCTURE AND PROPOSED THREE-DIMENSIONAL STRUCTURE OF $IpTx_a$

Fig. 10.2 shows the complete amino acid sequence of $IpTx_a$ as determined by direct automated microsequencing.[29] $IpTx_a$ is composed of 33 amino acids with a calculated M_r = 3,765. As are most short scorpion toxins, $IpTx_a$ is a basic peptide containing three pairs of cysteine residues that stabilize the three-dimensional conformation by forming disulfide bridges. Fig. 10.2 also shows a comparison of the amino acid sequence of $IpTx_a$ with those of Pi-1 and Pi-3, two peptide blockers of K^+ channels from the same *P. imperator* venom,[26,27] reportedly the smallest ionic channel-targeted scorpion peptides. Pi-3 is representative of a group of homologous toxins with three disulfide bridges such as noxiustoxin, charybdotoxin, iberiotoxin, etc.[30] Pi-1 forms a K^+ toxin subgroup on its own by having the remarkable characteristic of possessing 4 disulfide bridges.[27] No significant similarity is observed between $IpTx_a$ and any of these toxins, even when gaps (–) are introduced at the level of Cys_{10} to maximize homology. Even less homology is found when the comparison involves the larger Na^+ channel-selective peptides from scorpion venoms (not shown). Thus, $IpTx_a$ constitutes a novel class, the smallest yet, of scorpion toxins targeted against ionic channels.

A higher sequence homology is found when the comparison is made with Tx2-9 of the spider *P. nigriventer*,[31] and agelenin of the spider *A. asperta*,[32] two peptide blockers of neuronal P-type calcium channels (Fig. 10.2). If gaps are introduced to maximize homology, Tx2-9 is the most similar with 45% sequence identity. Agelenin has only limited resemblance on a one-to-one amino acid comparison (21% sequence identity) however, higher homology may be ascribed if two-residue gaps are introduced. Regardless of the extent of similarity, it is clear that $IpTx_a$ is more related to these two spider toxins than to other toxins present

Pi-1	L V K C R G T S D - C G R P C Q Q Q T G C P N S K C I N R M C K C Y G C
Pi-3	T I S C T N P K Q - C Y P H C K K E T G Y P N A K C M N R K C K C F G R
$IpTx_a$	G D C L P H L K R C K A D N D C C G K K C K R R G T N A E K R C R
Tx2-9	S F C I P - F K P C K S D E N C C - K K F K C K T T G I V K L C R W
Agelenin	G G C L P H N R F C N A L S G P R C C S G L K C K E L S I W D S R C L-NH_2

Fig. 10.2 Amino acid sequence of $IpTx_a$ and comparison with other scorpion and spider toxin sequences. The amino acid sequence of $IpTx_a$ was obtained by direct Edman degradation of the reduced and alkylated $IpTx_a$.[29] Gaps (–) in the amino acid sequence of Pi-1 and Pi-3, two K^+ channel blockers from the scorpion *P. imperator.*[27,28] as well as in Tx2-9, a P-type Ca^{2+} channel blocker from the spider *P. nigriventer*,[31] have been introduced to maximize homology. Agelenin, also a spider toxin that blocks P-type Ca^{2+} channel,[32] is presented without gaps. Taken, with permission, from Zamudio, *et al.* (1997).[29]

in scorpion venoms. Thus, $IpTx_a$ adds to the emerging notion that peptide toxins found in venomous animals of different phyla contain similar structural motifs.[33]

Recently, the three-dimensional structures of numerous peptide toxins specifically targeted against different ion channels have been reported. A salient finding from these studies is that all peptide toxins so far described, although different in their animal origin and ion channel specificity, may be grouped in a small number (3) of structural subsets that share the same global folding.[33] The major criterion for classification of peptide toxins within a subset is not their linear amino acid sequence but the residues that appear in their β-sheet strands. The conserved residues appearing in the core β-sheet include the cysteines, which are critical for maintenance of the three-dimensional structures. The distinctive arrangement of the six cysteines in the sequence of $IpTx_a$ is similar to that of other spider (μ-agatoxin IVA) and snail (ω-conotoxin GVIA) toxins whose three-dimensional structures are known. This allowed us to build a three-dimensional model of $IpTx_a$ (Fig. 10.3) based on the three-dimensional structure of these two closely related toxins. The central structural elements of this subset of toxins are a three-stranded β-sheet with a +2x, –1 topology and a long loop connecting strands II and III. In the case of $IpTx_a$, the three β sheets are formed by residues Leu_7-Cys_{10} (I) Gln_{14}-Cys_{17} (II) and Glu_{29}-Cys_{32} (III) and the long loop is comprised of residues Gly_{18}-Glu_{29} (Fig. 10.3). Within this long loop there is an hydroxyl-containing amino acid (Thr_{26}) preceded by a cluster of five basic amino acids (Lys_{19}-Arg_{24}). This special amino acid arrangement is a unique feature of $IpTx_a$ and we hypothesized that it participates in the binding of $IpTx_a$ to the RyR. Other $IpTx_a$-related toxins with equally long loop but different amino acid sequence such as ω-conotoxin GVIA lack the capacity to modulate [^{3}H]ryanodine binding (unpublished).

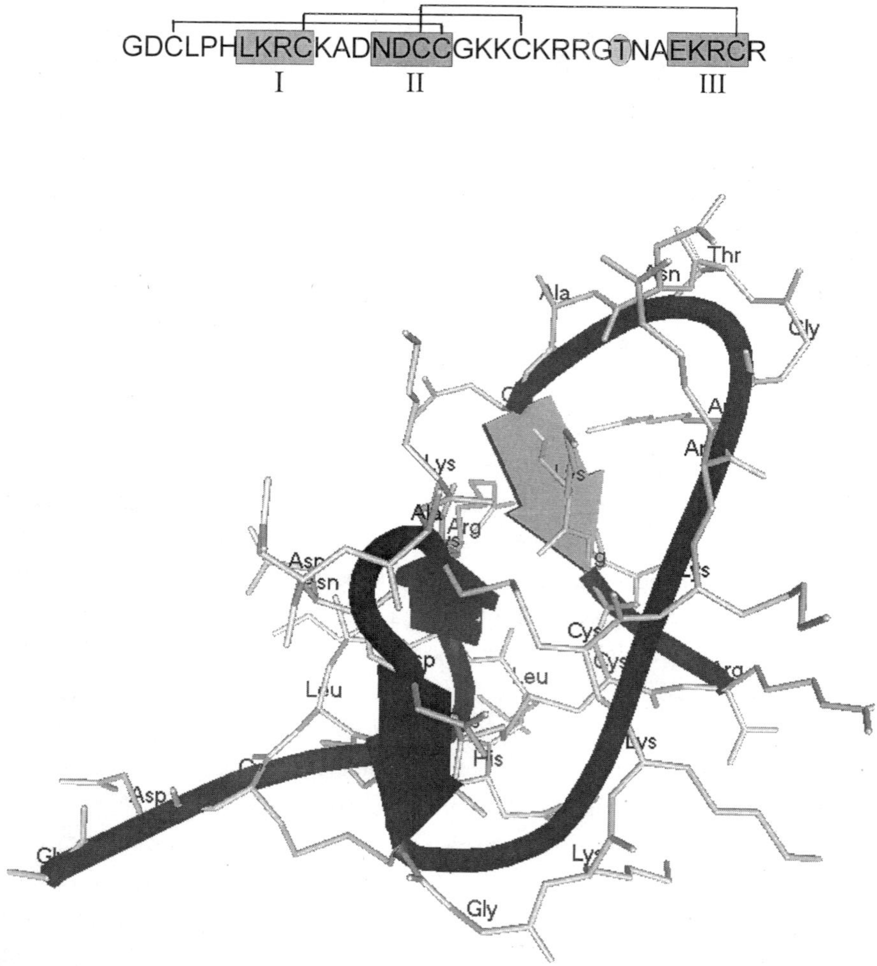

Fig. 10.3 Structural elements and three-dimensional model of $IpTx_a$. Top: disulfide bonding pattern and position of β-sheet strands. Thr_{26} is encircled to gain notoriety. Bottom: Energy-minimized model structure of $IpTx_a$. The amino acid sequence of $IpTx_a$ was entered into the program Insight II (Molecular Simulations) and then homology-modeled using the atomic coordinates for μ-Agatoxin IV and ω-conotoxin GVIA, two related toxins. The resulting structure was subjected to cycles of descent energy minization until convergence was reached. In this view, the amino-terminus (corresponding to Gly_1) is in the left arm of the toxin and Thr_{26} is at the top of the long arm extending from β-sheet II to β-sheet III.

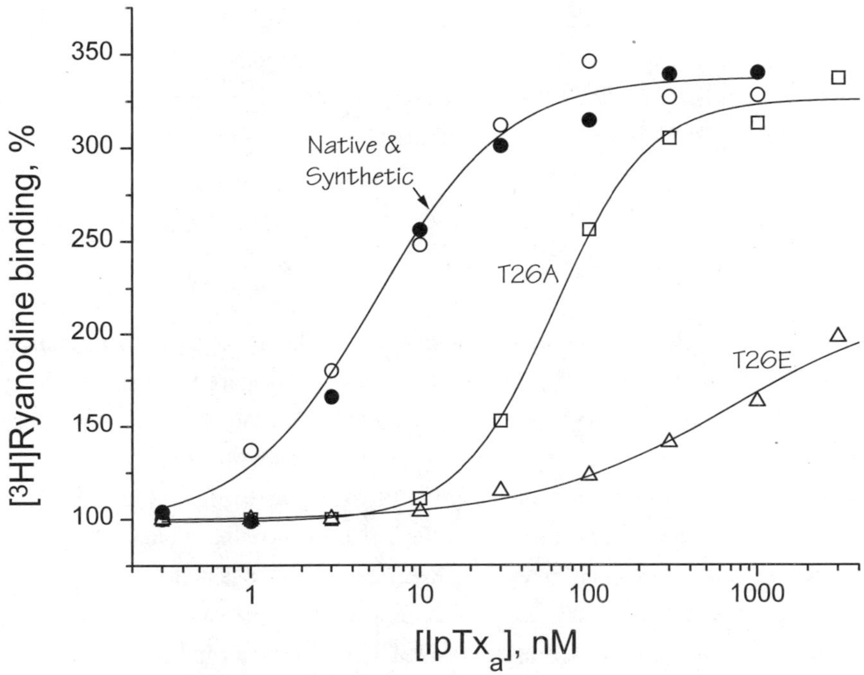

Fig. 10.4 Mutating Thr_{26} affects the affinity of $IpTx_a$ for RyRs. Peptides with amino acid sequence identical to that of native $IpTx_a$ (filled circles) or those in which Thr_{26} was substituted by Ala (T26A, squares) or Glu (T26E, triangles), were synthesized using solid-phase methodology and HPLC-purified before testing on [^{3}H]ryanodine binding assays. Fitting of data points was performed with the computer program Origin 4.0 using the equation described in the text. [^{3}H]Ryanodine was carried out as described in the legend to Fig. 10.1.

To test the hypothesis above, we synthesized a peptide with amino acid sequence identical to that of $IpTx_a$ (synthetic $IpTx_a$) and two others in which Thr_{26} was substituted by Ala (T26A) or Glu (T26E). Synthetic $IpTx_a$ increased [^{3}H]ryanodine binding to skeletal SR with the same potency and affinity as native $IpTx_a$ (Fig. 10.4). Both sets of data points were overlapping and could be fitted with the same line. We used the function: $B = B_{max} / 1 + (K_d + [IpTx_a])^n$ to fit the data, where B is the specific binding of [^{3}H]ryanodine, B_{max} is the maximal activation of [^{3}H]ryanodine binding evoked by $IpTx_a$ (325%), K_d is the apparent dissociation constant of $IpTx_a$ (5 nM), and n is the Hill number (1.58). Thus, synthetic $IpTx_a$ retains the same biological activity as native $IpTx_a$. By contrast, the mutated peptides decreased their affinity and potency to stimulate [^{3}H]ryanodine binding. K_d for T26A and for T26E were 62 and 715 nM, respectively. In addition, B_{max} for T26E was reduced to 221%. These results suggest that Thr_{26} is an important component of $IpTx_a$'s binding site; removal of the hydroxyl group (T26A) increases K_d 12-fold and substitution by a negatively-charged group (T26E) decreases it by 143-fold.

SELECTIVITY OF $IpTx_a$

Among ligands of RyRs, a unique characteristic of $IpTx_a$ is its capacity to stimulate [^{3}H]ryanodine binding to the RyR1 (skeletal-type) isoform.[22] Figure 10.5A shows the Ca^{2+}-dependence of [^{3}H]ryanodine binding to skeletal RyRs and the effect of $IpTx_a$. Specific binding in the absence of $IpTx_a$ (control, open circles) has a threshold for detection at ~100 nM [Ca^{2+}] (pCa 7) and is maximal at 10–100 mM [Ca^{2+}]. Higher [Ca^{2+}] decreases binding. This dual effect of Ca^{2+} gives rise to a bell-shaped curve that is similar to the Ca^{2+}-dependence of open probability for skeletal RyRs.[4] In the presence of $IpTx_a$ (filled circles) the binding curve also has a bell-shape but is dramatically augmented in absolute values. Furthermore, the threshold for activation is shifted by $IpTx_a$ to pCa 8 and the optimal binding to pCa 5. These results suggest that $IpTx_a$ exerts its stimulatory effect by sensitizing RyRs to Ca^{2+}.

In contrast to the dramatic effect on skeletal RyRs, the effect of $IpTx_a$ on cardiac RyRs is complex, showing no effect at pCa 8 and 7, a modest but statistically significant stimulation at pCa 6, and a modest inhibition at pCa 5 to 3 (Fig. 10.5B). Therefore, the activation induced by $IpTx_a$ on cardiac RyRs occurs at one [Ca^{2+}] only and constitutes a small fraction of that observed on skeletal RyRs. Due to these restrictions, it is reasonable to assume that, in systems where intracellular [Ca^{2+}] fluctuates rapidly due to the concurrent action of several RyR isoforms (as it does in non-mammalian skeletal muscle), $IpTx_a$ is likely to target skeletal-type RyRs and produce a substantial modification of the parameters directly controlled by these isoforms.

In summary, since $IpTx_a$ fails to increase binding of [^{3}H]ryanodine to cardiac RyRs (as well as cerebrum and liver RyRs),[22] this in all likelihood reflects the existence of a distinct modulatory site for $IpTx_a$ on the skeletal RyR. However, we cannot rule out that the $IpTx_a$-binding site is absent in cardiac (RyR2) isoforms because the Ca^{2+}-dependence of [^{3}H]ryanodine binding to cardiac RyRs is, albeit modestly, altered by $IpTx_a$. Thus, the use of $IpTx_a$ as a selective modulator of skeletal RyR needs to be investigated further.

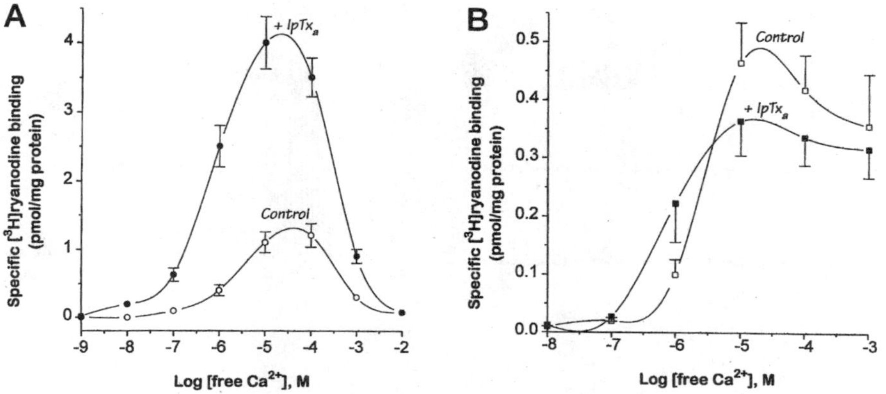

Fig. 10.5 Effect of IpTx$_a$ on the Ca^{2+}-dependence of [^{3}H]ryanodine binding to skeletal and cardiac RyRs. The binding of [^{3}H]ryanodine to rabbit skeletal (A) and porcine cardiac (B) SR was performed as described in the legend to Fig. 10.1, with the exception that the incubation medium was supplemented with 1 mM EGTA and several $CaCl_2$ concentrations to yield the indicated [free Ca^{2+}]. IpTx$_a$ (100 nM) was present throughout the incubation period. Data points are the mean (± SEM) of 4 independent determinations. Smooth lines linking data points have no theoretical meaning. Taken, with permission, from El-Hayek, *et al.* (1995).[22]

EFFECT OF IpTx$_a$ ON SINGLE RyRs

If the binding of [^{3}H]ryanodine may be used as an index of the activity of the channel,[6,17] the simplest interpretation for the IpTx$_a$-evoked stimulation of [^{3}H]ryanodine binding is that the peptide increases the channel open probability (P_o). We reconstituted rabbit skeletal RyRs in planar lipid bilayers to test this hypothesis.[22,28] Figure 10.6 shows traces of single skeletal RyR channel taken before ("Control") and 1 min after addition of 10 nM IpTx$_a$ ("10 nM IpTx$_a$"). Holding potential is +40 mV and current flows from cis (cytosolic) to trans (lumenal) side. In addition to decreasing the mean closed time,[22] IpTx$_a$ evokes the appearance of a subconducting state that represents ~30% of the full conducting level. Although of small amplitude, the subconductance state has a mean open time which is ~100 to 200-fold longer than that of unmodified channels. Thus, ion flow is greater for an IpTx$_a$-modified channel despite its lower conducting level. Both the decrease of mean closed time[22] and the stabilization of a long-lived subconductance state are consistent with the view that the stimulation of [^{3}H]ryanodine binding produced by IpTx$_a$ results from an overall increase in P_o, which in turn favors the binding of the alkaloid.

The stabilization of channel openings in a subconductance state seems to be a common feature of scorpion peptide modulators of RyRs. It was first noted by Valdivia, *et al.* (1991)[20]

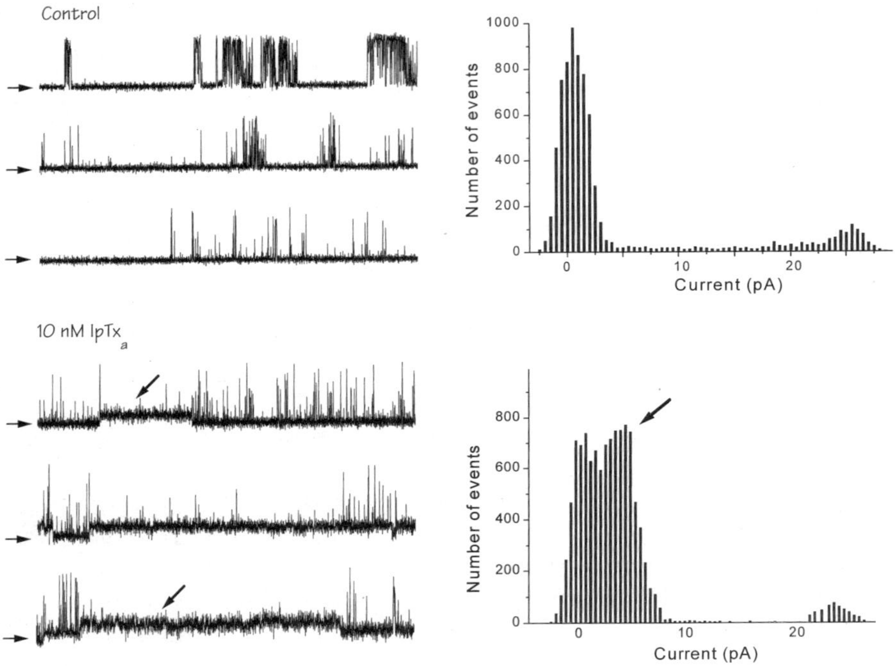

Fig. 10.6 Modification of RyR kinetics by $IpTx_a$. Top panel: Cs^+ currents through a single skeletal RyR activated by 10 μM cytosolic Ca^{2+}.[22,28] Holding potential: 40 mV. Recording solutions were 300 mM cesium methanesulfonate and 10 mM Na-Hepes (pH 7.2) in both chambers. Bottom panel: the same channel ~1 min after addition of 10 nm $IpTx_a$ to the cytosolic face of the channel. Arrows point to a subconductance state that was virtually absent in control. All traces are 1 sec long. Right panels: current histograms obtained from 30 sec recordings of the channel under control (above) or in the presence of the toxin (bottom).

from effects produced by *Buthotus bottentota* venom on skeletal RyRs. The whole venom increased the frequency of open events and converted a fraction of full-conducting openings into subconducting states.[20] The activity of the whole venom could be reproduced by a peptide fraction of M_r ~4,000 to 7,000.[20] A single peptide from a similar fraction of *Buthotus judaicus* was later identified by Morrissette *et al.* (1996)[23] and christened **ryanotoxin** to note the similarity between the peptide toxin and ryanodine in inducing the appearance of a subconductance state in single RyR. Although both ryanodine and the scorpion peptides stabilize RyR openings in a subconductance state, there exist fundamental differences that point to different mechanism of action by each of these ligands. First, under identical conditions of channel recording, the amplitude of the subconductance level induced by ryanodine and ryanotoxin (or $IpTx_a$) is different. Second, competition experiments reveal that ryanodine and $IpTx_a$ share different binding sites (see below), suggesting that the kinetic effects produced by each of these ligands occur by modification of different regulatory sites in the RyR protein.

DIRECT BINDING OF $[^{125}I]IpTx_a$ TO RyRs

The results presented thus far do not allow us to safely ascertain if the binding of $IpTx_a$, like that of ryanodine, is dependent on the conformational state of the receptor. Although the effect of $IpTx_a$ is potentiated by ligands that open RyRs such as Ca^{2+} (Fig. 10.5) and caffeine,[22,29] an unequivocal demonstration that $IpTx_a$ binds to the open receptor must be provided by experiments where $IpTx_a$ is directly used as the radiolabelled ligand. Using the Bolton-Hunter technique, we radiolabelled $IpTx_a$ and tested the binding properties of $[^{125}I]IpTx_a$. Chinese hamster ovary (CHO) cells lack endogenous RyR and the capacity to bind $[^{125}I]IpTx_a$. However, when transfected with a vector containing the full-length gene enconding for the skeletal RyR (RyR1), binding of $[^{125}I]IpTx_a$ appears parallel to the binding of $[^3H]$ryanodine.[34] This suggests that $[^{125}I]IpTx_a$ retains its ability to recognize RyR. Fig. 10.7A shows that the binding of $[^{125}I]IpTx_a$ to skeletal SR is inhibited by ruthenium red with an IC_{50} similar to that shown for inhibition of $[^3H]$ryanodine binding (~4 μM).[8] Ryanodine shows a more complex effect, stimulating $[^{125}I]IpTx_a$ binding at low concentration but decreasing it at high concentration (Fig. 10.7B). Thus, ligands that are specific for RyRs also modify the binding of $[^{125}I]IpTx_a$. Binding of $[^{125}I]IpTx_a$ is as high at $[Ca^{2+}]$ that keep the RyR channel closed (~10–100 nM) as it is at $[Ca^{2+}]$ that open RyRs maximally ~10–100 μM) (Fig. 10.7C). However, high (millimolar) $[Ca^{2+}]$ inhibit $[^{125}I]IpTx_a$ binding. In all, the Ca^{2+}-dependence of $[^{125}I]IpTx_a$ binding is essentially different to the bell-shaped curve for $[^3H]$ryanodine binding (Fig. 10.7C, dotted line), suggesting that the peptide is able to bind to several conformational states of the RyR.

IMPERATOXIN I ($IpTx_i$), A HETERODIMERIC PROTEIN WITH PHOSPHOLIPASE A_2 ACTIVITY

$IpTx_i$ is another protein from the scorpion P. imperator that affects the activity of RyRs with nanomolar affinity (Fig. 10.1),[21,28] but with a mechanism of action radically different from

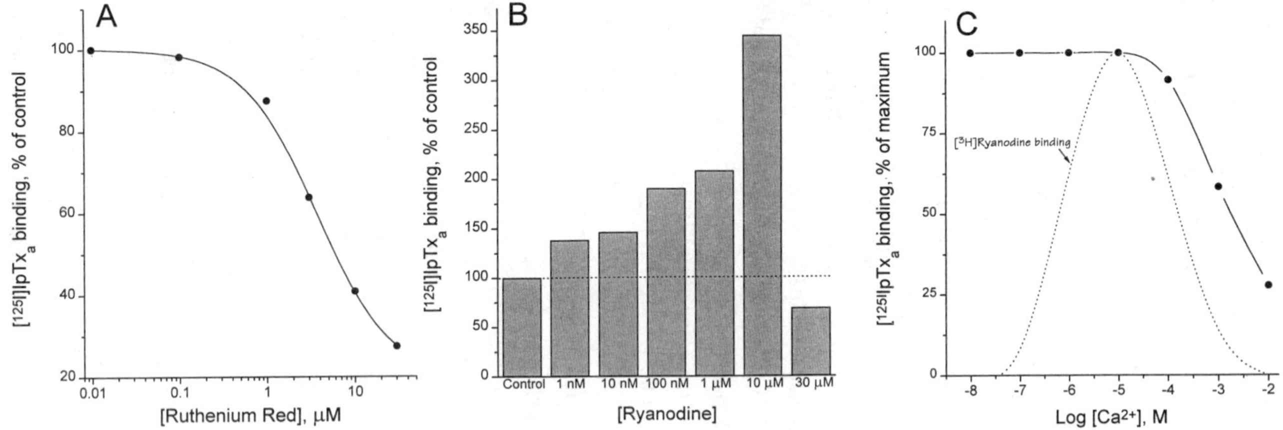

Fig. 10.7 Effect of RyR regulators on the binding of [^{125}I]IpTx$_a$ to skeletal SR. Bolton-Hunter-labeled [^{125}I]IpTx$_a$ was incubated 90 min at 36°C with rabbit skeletal SR (0.2 mg/ml) in medium containing 0.2 M KCl, 10 mM Na-Hepes (pH 7.2) and 10 μM $CaCl_2$. The reaction was terminated by rapid filtration as described for [^{3}H]ryanodine binding assays.[22,28,29] For experiments were ryanodine was tested (panel B), the indicated concentrations of ryanodine were added to SR vesicles 30 min before addition of [^{125}I]IpTx$_a$. (C) Effect of [Ca^{2+}] on the binding of [^{125}I]IpTx$_a$. The Ca^{2+}-dependence of [^{3H}]ryanodine binding obtained in the same vesicle population is included for comparison.

that described for $IpTx_a$. Knowledge of the primary structure of $IpTx_i$ was essential to unravel the mechanism by which it causes inhibition of cardiac and skeletal RyRs. A cDNA library was prepared from the venomous glands of this scorpion and used to clone the gene encoding $IpTx_i$.[28] $IpTx_i$ consists of a large subunit (104 amino acid residues) with amino acid sequence 38% homologous to that of phospholipase A_2 (PLA_2) from honey bee (*Apis mellifera*) venom and 35% homologous to PLA_2 from lizard (*Heloderma horridum*) venom (Fig. 10.8A). This large subunit is covalently linked by a disulfide bond to a smaller (27 amino acid residues), structurally unrelated subunit (Fig. 10.8B). Thus, $IpTx_i$ is a heterodimeric protein with lipolytic action, a property that is only shared with β-bungarotoxins, a group of neurotoxins from snake venoms.[35] The small subunit has no significant similarity to other peptides deposited in the protein data base. Hence, this small subunit constitutes a new class of scorpion peptides.

The presence in $IpTx_i$ of a PLA_2 subunit covalently linked to a small, structurally unrelated peptide is reminiscent of the bipartite arrangement of β-bungarotoxins, a group of snake neurotoxins that inhibit neurotransmitter release. Each β-bungarotoxin dimer has a PLA_2 subunit covalently bound to a smaller subunit related to Kunitz-type protease inhibitors. Members of the Kunitz superfamily bind to and block voltage-dependent K^+ and Ca^{2+} channels.[36] Thus, although the small subunit of $IpTx_i$ showed no homology to members of the Kunitz superfamily, its distinctive arrangement within the toxin suggested that it might act as a blocker. We synthesized a peptide with amino acid sequence identical to that of the small subunit of $IpTx_i$, but it failed to inhibit [^{3}H]ryanodine binding, as did two others in which Cys_4 was substituted by Ala or Met to avoid interpeptide disulfide bonding.[28] Hence, the target site for the small subunit, or its biological function, remain to be determined.

Given our failure to detect inhibition by the small subunit, we turned our attention to the large subunit of $IpTx_i$. In a reduced and carboxymethylated form, the large subunit of $IpTx_i$ failed to inhibit RyRs (not shown). However, this was not surprising given that disulfide bridges are essential to maintain the toxin's three-dimensional structure. It is the treatment of $IpTx_i$ with *p*-bromophenacyl bromide (*p*BPB), a covalent modifier of His residues that inhibits PLA_2 activity without affecting disulfide bridges,[37] that decreases substantially the capacity of $IpTx_i$ to inhibit RyRs (Fig. 10.9). This suggests that a lipid product of PLA_2 activity, more than a direct $IpTx_i$-RyR interaction, is responsible for RyR inhibition.

PLA_2's are abundant components of snakes, scorpions and bee venoms, and often constitute the main toxic component to mammals. Although all PLA_2's induce a variety of pathological symptoms including neurotoxicity and myotoxicity, they differ in their mechanism of action and in their molecular targets. For example, the snake neurotoxin **crotoxin** is

A

Large subunit:

```
-3   1                                                 50
        TMWGTKWCGSGNEATDISELGYWSNLDSCCRTHDHC-DNIPS-GQTKYGL
       IIYPGTLWCGHGNKSSGPNELGRFKHTDACCRTHDMCPDVM-SAGESKHGL
     GAFIMPGTLWCGAGNAASDYSQLGTEKDTDMCCRDHDHC-ENWISALEYKHGM


     51                                                100
        TNEGKYTMMNCKCETAFEQCLRNVTGGMEGPAAGFVRKTYFD-LYGNGCY
        TNTASHTRLSCDCDDKFYDCLKNSADTISSY---FVGKMYFN-LIDTKCY
        RNYYPSTISHCDCDNQFRSCLMKLKDGTADY----VGQTYFNVLKIP-CF


     101                                               150
        NVQCPSQ ..........................................
        KLEHPVTGCGERTEGR-CLHYTVDKSKPK-VYQWFDLRKYXXNSRG-GSX
        ELEEGE-GCVDWNFWLECTESKIMPVAKLVSAAPYQAQAETQSGEG ...
     151
        ........ (IpTxi)
        EYHFEIVY (Bee phospholipase)
        ........ (Heloderma phospholipase)
```

B

Small subunit:

```
1         10         20
SEECPDGVATYTGEAGYGAWAINKLNG
```

Fig. 10.8 Amino acid sequence of $IpTx_i$ and relation to PLA_2´s from bee and lizard venoms. (A) The amino acid sequence of the large subunit of $IpTx_i$ (first row) was aligned with that of bee (*Apis millifera*, second row) and lizard (*Heloderma suspectum*, third row) venom PLA_2. A few gaps (dashes) were introduced in order to enhance similarities. Bold letters indicate exact homology for the three PLA_2´s. Dots indicate end of the sequence, whereas X indicate unknown residues, as indicated in the protein data base. (B) Amino acid sequence of the small subunit of $IpTx_i$. No significant similarity with other proteins was found. Cys_4 from this subunit and Cys_{101} of the large subunit are believed to be involved in disulfide brige formation.

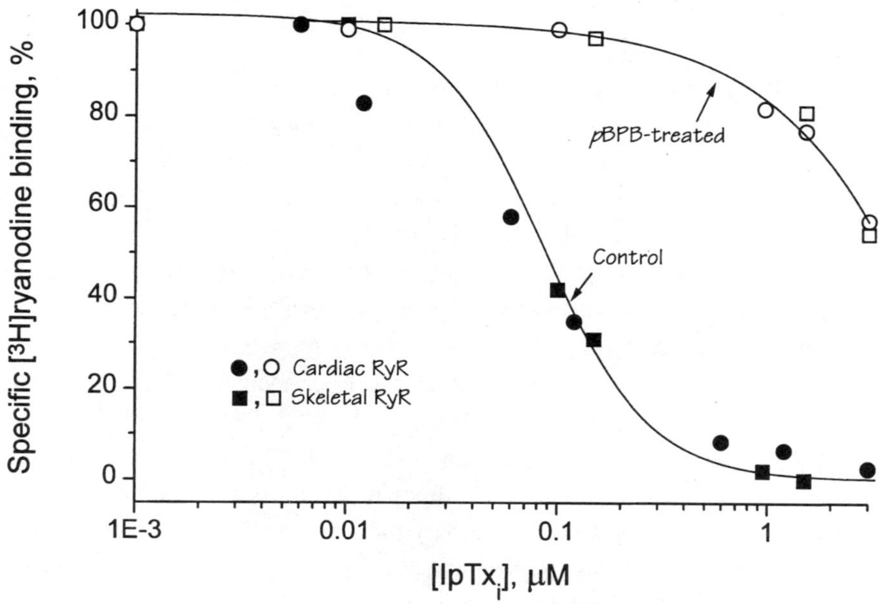

Fig. 10.9 Treatment with a PLA_2 inhibitor lowers the potency of IpTx$_i$ to inhibit RyR. IpTx$_i$ (200 μM) was incubated for 16 h at room temperature with 2 mM *p*BPB (open symbols), or with acetone (filled symbols), the drug vehicle, as control. Binding of [^{3}H]ryanodine to cardiac (circles) or skeletal (squares) SR was performed in the absence (defined as 100%) or in the presence of indicated concentrations of IpTx$_i$.

composed of a PLA_2 subunit and an inhibitory subunit, which keeps the phospholipase inactive until binding to a presynaptic receptor triggers the dissociation of the inhibitory subunit.[38] In contrast, **notexin** from *Netechis scutatus scutatus* is a PLA_2 without an associated subunit that produces muscle paralysis by binding to a specific receptor in the neuromuscular junction.[39] IpTx$_i$ is the first example of a scorpion toxin in which a PLA_2 is found chaperoned by a smaller, structurally unrelated subunit. In the first study of IpTx$_i$,[22] we found that several transporters and ion channels of striated muscle including the inositol trisphosphate receptor, the muscarinic receptor, voltage-sensitive Na^+ channels and the Ca^{2+}-ATPase of SR were insensitive to micromolar concentrations of IpTx$_i$. These results argue against an indiscriminate effect on key molecules of excitation-contraction coupling by the lipolytic action of IpTx$_i$. IpTx$_i$ thus offers an alternative way to block RyR distinct from ryanodine and other ligands that require a physical interaction with the RyR protein.

SUMMARY

Peptide toxins are emerging as useful probes of RyR structure and function. Their intrinsic properties confer to these peptides a unique set of attributes that may both favor and discourage their use as substitute ligands for ryanodine. $IpTx_a$ is a small basic peptide that binds directly to the RyR and acts rapidly and reversibly. In this respect, $IpTx_a$ may be more advantageous to physiologists than ryanodine, which is too slow to bind and dissociate from its effector site to attain equilibrium during the course of experiments with intact cells. As a natural peptide, $IpTx_a$ is suitable for iodination, and for attachment of crosslinkers for covalent binding to the receptor protein, thus offering the prospect of receptor tagging and mapping. As a high-affinity ligand, $IpTx_a$ may contribute considerably to the understanding of the structure-function relationship of the RyR in isolated SR vesicles or with the purified receptor, but its usefulness in intact cells may be limited by its incapacity to permeate the external membrane. In this respect, the lipophylic nature of ryanodine is clearly advantageous. Finally, the heterodimeric protein $IpTx_i$, by releasing lipid products of PLA_2 activity, may block RyRs without a physical interaction with the channel protein. However, the lack of a bona fide ligand-receptor interaction precludes the use of $IpTx_i$ in a variety of studies designed to count, map or tag RyRs. As a corollary, it may be clear by now that our experimental needs are too diverse to be satisfied by the use of a single ligand. This repertoire of natural probes should expand our opportunities to explore more integrally the structure and function of RyRs.

ACKNOWLEDGEMENTS

We thank Barbara Shelisko for her instrumental work in the modeling of $IpTx_a$, and to Lourival Possani, Fernando Zamudio, Carolina Arévalo, Jeff Walker and Raghava Sreekumar for providing excellent intellectual feedback and material support. This work was supported by Grant HL-55438 from the National Institutes of Health and a Grant-in-Aid from the American Heart Association.

REFERENCES

1. Bers DM. In Excitation-contraction coupling and cardiac contractile force. Boston, Kluwer Academic Publishers: 1991.
2. Ogawa Y. Role of ryanodine receptors. *Crit. Rev. Biochem. Mol. Biol.* 1994; **29**:229–274.

3. Takeshima H, Nishimura S, Matsumoto T, *et al.* Primary structure and expression from complementary DNA of skeletal muscle ryanodine receptor. *Nature (Lond.)* 1989; **339**: 439–445.
4. Smith JS, Imagawa T, Ma J, *et al.* Purified ryanodine receptor from skeletal muscle is the calcium-release channel of sarcoplasmic reticulum. *J. Gen. Physiol.* 1988; **92**:1–26.
5. Lindsay ARG, Tinker A, and Williams AJ. How does ryanodine modify ion handling in the sheep cardiac sarcoplasmic reticulum Ca^{2+} release channel. *J. Gen. Physiol.* 1994; **104**:425–447.
6. Meissner G, and El-Hashem A. Ryanodine as a functional probe of the skeletal muscle sarcoplasmic reticulum Ca^{2+} release channel. *Mol Cell Biochem.* 1992; **114**:119–123.
7. Inui M, Saito A, and Fleischer S. Purification of the ryanodine receptor and identity with feet structures of junctional terminal cisternae of sarcoplasmic reticulum from fast skeletal muscle. *J. Biol. Chem.* 1987; **262**:1740–1747.
8. Imagawa T, Smith JS, Coronado R, *et al.* Purified ryanodine receptor from skeletal muscle sarcoplasmic reticulum is the Ca^{2+}-permeable pore of the Ca^{2+} release channel. *J. Biol. Chem.* 1988; **262**:16636–16643.
9. Lai FA, Erickson HP, Rousseau E, *et al.* Purification and reconstitution of the calcium release channel from skeletal muscle. *Nature (Lond.)* 1988; **331**:315–319.
10. El Hayek R, Valdivia C, Valdivia HH, *et al.* Activation of the Ca^{2+} release channel of skeletal muscle sarcoplasmic reticulum by palmitoyl carnitine. *Biophys. J.* 1993; **65**:779–789.
11. Nakamura, Y, Kobayashi J, Gilmore J, *et al.* Bromo-eudistomin D, a novel inducer of calcium release from fragmented sarcoplasmic reticulum that causes contractions of skinned muscle fibers. *J. Biol. Chem.* 1986; **261**:4139–4142.
12. Mack MM, Molinski TF, Buck ED, *et al.* Novel modulators of skeletal muscle FKBP12/calcium channel complex from Iantella basta. Role of FKBP12 in channel gating. *J. Biol. Chem.* 1994; **269**:23236–23249.
13. Jayaraman T, Brillantes AM, Timerman AP, *et al.* FK506 binding protein associated with the calcium release channel (ryanodine receptor). *J. Biol. Chem.* 1992; **267**:9474–9477.
14. Bull R, Marengo JJ, Suarez-Isla B, *et al.* Activation of calcium release channels in sarcoplasmic reticulum from frog muscle by nanomolar concentrations of ryanodine. *Biophys. J.* 1989; **56**: 749–756.
15. Buck E, Zimany I, Abramson J A *et al.* Ryanodine stabilizes multiple conductance states of the skeletal muscle calcium release channel. *J. Biol. Chem.* 1992; **267**:23560–23567.
16. Pessah I, Stambuk RA, and Casida JE. Ca^{2+} activated ryanodine binding: mechanism of activation and intensity modulation by Mg^{2+}, caffeine and adenine nucleotides. *Mol. Pharmacol.* 1987; **31**:232–238.
17. Chu A, Díaz-Muñoz M, Hawkes MJ, *et al.* Ryanodine as a probe of the functional state of the skeletal muscle sarcoplasmic reticulum calcium release channel. *Mol. Pharmacol.* 1990; **37**: 735–741.
18. Ménez A, Bontems F, Roumestand C, *et al.* Structural basis for functional diversity of animal toxins. *Proc. R. Soc. Edinburgh* 1992; **99B**:83–103.

19. DeBin JA, Maggio JE, and Strichartz GR. Purification and characterization of chlorotoxin, a chloride channel ligand from the venom of the scorpion. *Am. J. Physiol.* (Cell Physiol 33) 1993; **264**: 361–369.
20. Valdivia HH, Fuentes O, El-Hayek R, *et al.* Activation of SR Ca^{2+} release channels by a novel scorpion venom. *J. Biol. Chem.* 1991; **266**:19135–19138.
21. Valdivia HH, Kirby MS, Lederer WJ, *et al.* Scorpion toxins targeted against the sarcoplasmic reticulum Ca^{2+} release channel of skeletal and cardiac muscle. *Proc. Natl. Acad. Sc. (USA)* 1992; **89**:12185–12189.
22. El-Hayek R, Lokuta AJ, Arévalo C, *et al.* Peptide probe of ryanodine receptor function. *J. Biol. Chem.* 1995; **270**:28696–28704.
23. Morrissette J, Beurg M, Sukhareva M, *et al.* Purification and characterization of ryanotoxin, a peptide with actions similar to those of ryanodine. *Biophys. J.* 1996; **71**:707–721.
24. Morrissette J, Kratzschmar J, Haendler B, *et al.* Primary structure and properties of helothermine, a peptide toxin that blocks ryanodine receptors. *Biophys. J.* 1995; **68**:2280–2288.
25. Tu, AT. In Handbook of natural toxins. Insect poisons, allergens, and other invertebrate venoms. New York: Marcel Dekker, Inc. 1983.
26. Gómez-Lagunas F, Olamendi-Portugal F, Zamudio FZ, *et al.* Two novel toxins from the venom of the scorpion Pandinus imperator show that the N-terminal amino acid sequence is important for their affinities towards Shaker B K^+ channels. *J. Membr. Biol.* 152, 49–56.
27. Olamendi-Portugal T, Gómez-Lagunas F, Gurrola GB, *et al.* A novel structural class of K^+ channel-blocking toxin from the scorpion Pandinus imperator. *Biochem. J.* 1996; **315**:977–981.
28. Zamudio FZ, Conde R, Arévalo C, *et al.* The mechanism of inhibition of ryanodine receptor channels by Imperatoxin I, a heterodimeric protein from the scorpion Pandinus imperator. *J. Biol. Chem.* 1997; **272**:11886–11894.
29. Zamudio, FZ, Gurrola GB, Arévalo C, *et al.* Primary structure and synthesis of Imperatoxin A ($IpTx_a$), a peptide activator of Ca^{2+} release channels/ryanodine receptors. *FEBS Letters* 1997; **405**:385–389.
30. Miller, C. The charybdotoxin familiy of K^+ channel-blocking peptides. *Neuron* 1995; **15**:5–10.
31. do Nascimiento-Cordeiro M, Ribeiro-Diniz C, do Carmo-Valentim A, *et al.* The purification and amino acid sequences of four Tx2 neurotoxins from the venom of the Brazilian "armed" spider Phoneutria nigriventer (Keys). *FEBS Letters* 1992; **310**:153–156.
32. Hagiwara K, Sakai T, Miwa A, *et al.* Complete amino acid sequence of a new type of neurotoxin from the venom of the spider Agelena opulenta. *Biomed. Res.* 1990; **1**:181–186.
33. Omecinsky DO, Holub KE, Adamns ME, *et al.* Three-dimensional structure analysis of μ-agatoxins: further evidence for common motifs among neurotoxins with diverse ion channel specificities. *Biochemistry* 1996; **35**:2836–2844.
34. Ma J, personal communication.
35. Kwong PD, McDonald NQ, Sigler PB, *et al.* Structure of β-bungarotoxin: potassium channel binding by Kunitz modules and targeted phospholipase action. *Structure* 1995; **3**:1109–1119.
36. Schweitz H, Herteaux C, Bois P, *et al.* Calcicludine, a venom peptide of the Kunitz type protease inhibitor family, is a potent blocker of high-threshold Ca^{2+} channels with a high affinity for L-type channels in cerebellar granule cells. *Proc. Natl. Acad. Sc. (USA)* 1994; **91**:878–882.

37. Glaser KB, Mobilio D, Chang JY, *et al.* Phospholipase A_2 enzymes: regulation and inhibition. *Trend Pharmacol. Sc.* 1993; **14**:92–97.
38. Delot E, and Bon C. Model for the interaction of crotoxin, a phospholipase A_2 neurotoxin, with presynaptic membranes. *Biochemistry* 1993; **32**:10708–10713.
39. Harris JB, Karlsson E, and Thesleff S. Effects of an isolated toxin from Australian Tiger snake (Notechis scutatus scutatus) venom at the mammalian neuromuscular junction. *Br. J. Pharmacol.* 1973; **47**:141–146.

Chapter 11

Chemical Modification of Ryanodine Receptor/Ca^{2+} Release Channel activity

Varda Shoshan-Barmatz and Wei Feng

PERSPECTIVES AND SUMMARY

The purpose of this review is to present the contribution of chemical probes to the study of the structure and function of the ryanodine receptor (RyR)/Ca^{2+} release channel.

In the absence of molecular structure, the control mechanism of the RyR channel gating and its modulation by ligands, such as calmodulin, ATP and Ca^{2+}, can be studied using a variety of techniques. One can manipulate the genes to create point mutations, truncated forms, chimeric molecules, or manipulate the proteins by means of proteolytic enzymes and chemical reagents. Nevertheless, it is often difficult to distinguish between direct and indirect effects of mutations or chemical modifications. Direct evidence is crucial, therefore, for building a model of ligand binding sites, transport path via the protein, or the control mechanism of channel gating.

Chemical modification of proteins by various covalent probes has been used widely for establishing the functional importance of amino acid side chains in protein activity and as an approach to elucidation of structure-function relations in various proteins.

In the present review, the application of covalent probes (group-directed, specific site-directed reagents, and cross-linking reagents) to the study of the control mechanisms of RyR/Ca^{2+} release channel, its structure-function relationship and the location of some ligand binding sites will be discussed. Some mechanistic aspects of the RyR/Ca^{2+} release channel activity have been characterized using chemical probes, whereas, others remain to be elucidated. On the basis of results obtained with different amino group modifying reagents, highly reactive lysyl residues are postulated in the gating mechanism of the RyR channel. This group may be the charged group which is sensitive to a voltage-driven conformational change of the α_1 subunit of the DHPR which activates Ca^{2+} release by the RyR. Using various carboxyl modifying reagents, a carboxyl group was located at a hydrophobic region of the RyR, and it is proposed that this group is involved directly in the channel closure or in Ca^{2+} binding to the low-affinity site(s) that leads to the inhibition of RyR activities. Thiol groups have been implicated in the binding of ryanodine and in the gating of the RyR channel. Histidine-directed reagents were found to inactivate ryanodine binding and to activate Ca^{2+} release. Based on the protection against DEPC (diethyl pyrocarbonate) inactivation of ryanodine binding by divalent cations, it has been suggested that a hystidyl residue(s) in the RyR is involved in Ca^{2+} binding to the low-affinity sites. By using various photoaffinity analogs of RyR modulators, ATP and ryanodine binding sites were localized at the carboxyl-terminal portion of the RyR, while the multiple sites for calmodulin were found both in the C-terminus and the central regions of the RyR. Cross-linking studies of RyR by several bifunctional reagents support its tetrameric organization and indicate that ryanodine binding to the low-affinity but not to the high-affinity sites is inhibited by RyR cross-linking. Thus, chemical studies of the RyR/Ca^{2+} release channel have provided new information regarding the involvement of specific carboxyl, hystidyl, lysyl and thiol residues in channel activation and inactivation and in the localization of some of the modulator binding sites within the RyR.

INTRODUCTION

Sarcoplasmic reticulum (SR) is an intracellular membrane-bound network in cardiac and skeletal muscles which regulates myoplasmic Ca^{2+} concentration, thereby regulating the contraction and relaxation of muscle.[1,2] Muscle contraction is elicited by Ca^{2+} released from the SR, while muscle relaxation results from Ca^{2+} sequestration in the SR by the (Ca^{2+}/Mg^{2+})ATPase.[3,4] Ca^{2+} is released into the sarcoplasm during excitation-contraction coupling via the ryanodine receptor (RyR)/Ca^{2+} release channel.[5–7]

The RyR from skeletal and cardiac muscle SR has been purified as a large homotetramer with an apparent molecular mass of 450 kDa per subunit (in SDS-PAGE).[8,9] RyR cDNA from skeletal muscle has been cloned and sequenced, and a subunit molecular mass of 565 kDa determined.[10,11] Purified RyR incorporated in a planar lipid bilayer (PLB) was found to form Ca^{2+}-conducting channels. The channel activity depends on Ca^{2+} and is modified by ryanodine, which stabilizes the channel in a sub-conductance state.[5–9,12] In addition to ryanodine-binding sites, RyR also has other sites for binding Ca^{2+}, ATP, calmodulin and Mg^{2+} all of which modulate its activity.[5]

A well accepted hypothesis for excitation-contraction (E-C) coupling in skeletal muscle suggests that a voltage-sensor molecule, the dihydropyridine receptor, located on the transverse tubules, undergoes conformational changes in response to depolarization. These changes are transmitted in turn to the RyR/Ca^{2+} release channel, and modulate the opening of this channel.[13,14]

An understanding of RyR structure-function requires the identification and location of the amino acid residues essential for RyR activity as well as the location of the multiple recognition sites for the various modulating ligands. Chemical modification of proteins has been used widely for establishing the functional importance of amino acid side chains for protein activity. Covalent probes can be subdivided into group-directed reagents, specific site-directed reagents and cross-linking reagents. Group-directed reagents are relatively selective chemical modifiers of amino acid side chains. Binding site-directed reagents, comprising affinity and photoaffinity labels, have been used for mapping specific binding sites for calmodulin, ATP, and ryanodine. Cross-linking reagents have been applied to the RyR to investigate the structure-function relationship. Table 11.1 summarizes the effect of chemical modification of RyR, by group-directed reagents, on ryanodine binding, Ca^{2+} release, and single channel activities.

I. Group-Directed Reagents

Group-directed reagents are chemicals that bind covalently to side chains of amino acid residues in a protein with some degree of specificity. The protection against inactivation by ligands, such as Ca^{2+}, ryanodine, ATP and calmodulin, is indicative of a modification of the ligand binding site by the reagent.

Table 11.1 Chemical Modification of RyR Activities by Amino Acid Side Chain-Directed Reagents

Reagents	Type of group modified	Effect on the following activites			References
		Ca^{2+} release	Ryanodine binding	Single channel	
DIDS	Lysine ε-NH_2	Stimulation	N.D.	Stimulation	22–24
DFDNB	N-terminal-α-amino	Stimulation	Inhibition	N.D.	25
DFNPS		N.D.	Inhibition	N.D.	25
DNFB		N.D.	Inhibition	Inhibition	25, 26
FITC		Stimulation	Inhibition	Stimulation	16
NBD-C1		N.D.	Stimulation & Inhibition[a]	Stimulation & Inhibition[a]	19
DCCD	Glu α- or Asp β-carboxyl group	Inhibition	Inhibition	Inhibition	28–32
EDC		No effect	Inhibition	Inhibition[b]	28, 32, 33
WRK		No effect	Inhibition	Inhibition	28, 32
EEDQ		No effect	Inhibition	N.D.	28, 32
DEPC	Histidyl group	Stimulation	Inhibition	Stimulation	42, 46, Fig. 1
Rose-Bengal		Stimulation	Inhibition	Stimulation/Substate	43, 44
Coumarin maleimide (nM)	SH group	Inhibition	N.D.	Inhibition	52, 53
Cu^{2+} + Mercaptants		Stimulation	N.D.	N.D.	50
2, 2′-dithiodipyridine		Stimulation	Stimulation & Inhibition[d]	N.D.	50, 88
4, 4′-dithiodipyridine		Stimulation	Stimulation & Inhibition[d]	N.D.	50, 88
Doxorubicin		Stimulation	Stimulation	N.D.	47
Heavy metals (Hg^{2+}, Ag^{+}, Cu^{2+}, Cd^{2+})		Stimulation	N.D.	N.D.	50
H2O2		Stimulation	Stimulation & Inhibition[d]	Stimulation & Inhibition[a,d]	51
Phthalocyanine dyes		Stimulation	N.D.	N.D.	50
N-succinimidy1-3(2-pyridyldithiol) propionate		Stimulation	Inhibition	N.D.	48, 88
Thimerosal		Stimulation	Stimulation & Inhibition[d]	Stimulation & Inhibition[d]	54

N.D. - Not determined

a - time-dependent

b - under the conditions used inhibition is suggested to be due to neutralization of negative charges in the cardiac RyR

c - cardiac RyR channel converted into subconductanct state

d - concentration-dependent

i. Amino group-directed reagents

The ε-amino group of lysine and N-terminal α-amino groups are able to react with a number of reagents such as FITC (fluorescein isothiocyanate), DNFB (2,4-dinitrofluorobenzene), NBD-Cl (7-chloro-4-nitrobenzo-2-oxa-1,3,-diazole) and DIDS (4,4'-diisothiocyano-stilbene-2,2'-disulfonic acid). The effect of these reagents on the RyR activity is summarized here.

Modification by FITC

Chemical modification of the RyR/Ca^{2+} release channel protein by FITC, which reacts specifically with the ε-amino groups of lysine residues,[15] markedly decreases ryanodine binding, and dramatically alters the behavior of single Ca^{2+}-release channel gating.[16] FITC was found to inhibit ryanodine binding in a time- and concentration-dependent manner by eliminating some ryanodine binding sites with no effect on the binding affinity of the remaining sites. Kinetic analysis of the inactivation of ryanodine binding by FITC suggests that the binding of two or more moles of FITC per mole of active site prevents ryanodine binding. The binding stoichiometry of FITC indicates that the modification of three to four lysyl residues in each RyR subunit is required for a complete inhibition of ryanodine binding. The FITC moiety was found bound to the 160, 96, 76 and 60 kDa RyR tryptic fragments, the FITC site apparently being on the 21.5, 18 and 17 kDa fragments formed by the V_8 protease.[16]

Covalent modification by FITC also dramatically modifies the activity of single Ca^{2+} release channels incorporated in PLB. It causes a marked increase in the probability of the channel being in the open state with a noisy ~60% subconductance. FITC-modified channels were no longer affected by ryanodine, as expected from the elimination of ryanodine binding site(s) by FITC modification. However, the FITC-modified channel activity was abolished by Mg^{2+} and ruthenium red.[16]

The ε-amino groups of lysyl side chains typically exhibit pKa values of 10–10.5. The pH dependence of the inactivation of ryanodine binding suggests that FITC reacts with an ε-amino group with a pKa of about 7.5.[16] This low pKa indicates that the group is very reactive, and may reflect the microenvironment of the ε-amino group such that its location is in close proximity to a carboxyl group. Fill *et al.*[17] have demonstrated that the gating of the RyR is strongly dependent on both transmembrane voltage and proton activity, rising steeply with their increase. They also measured non-linear capacitance changes (charge movement), which may be involved in the gating. It has been postulated[16] that the amino group(s) modified by FITC is involved in channel activation, possibly through its sensitivity to t-tubule voltage changes via a structural change in the channel which converts it from a closed to an open conformational state.

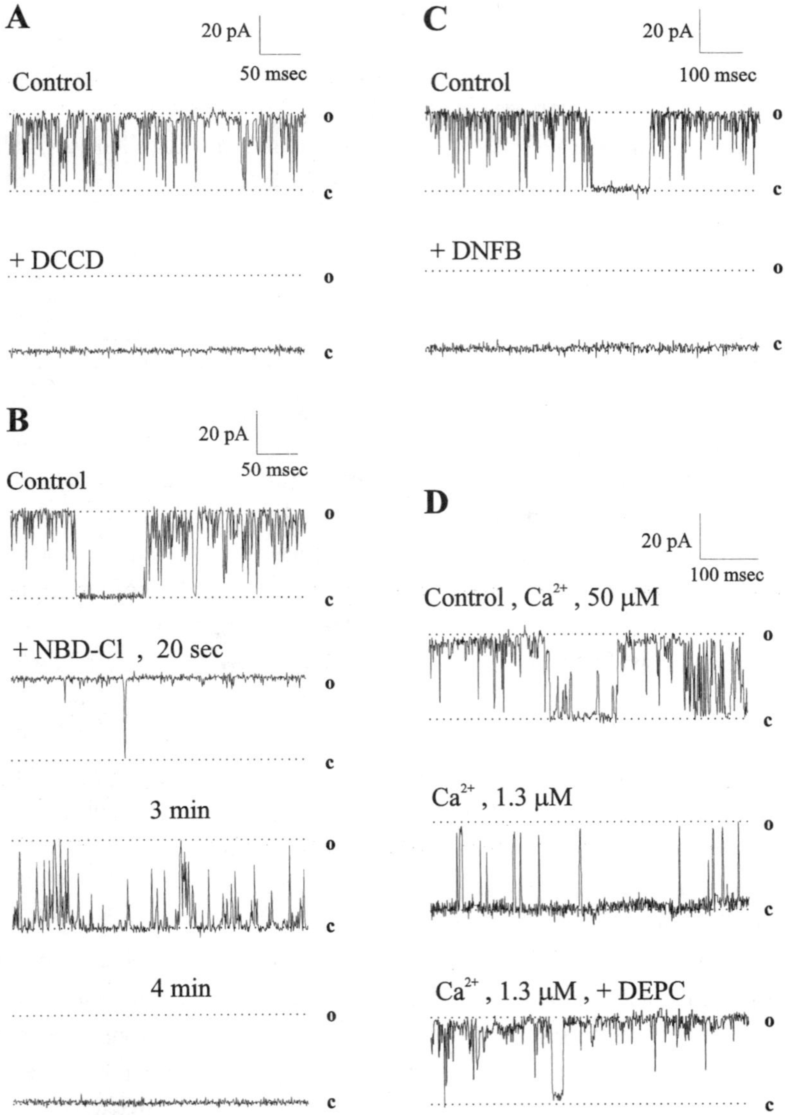

Fig. 11.1 Chemical modification of RyR single channel activity by NBD-Cl, DEPC, DNFB and DCCD. SR vesicles were reconstituted into planar lipid bilayer and single channel analysis was carried out as described previously.[32] Following the appearance of channel activity and its recording, the channel was exposed to the indicated modifying reagent and the channel activity was monitor as a function of time. A- DCCD (100 μM), B- NBD-ce (50 μM) C- DNFB (100 μM) and D- DEPC (150 μM) was added to the *cis* chamber, to where SR and voltage potential of +30 mV were applied. O, C indicate current levels representing open and closed states of the channel, respectively.

Modification by NBD-Cl

The fluorescent reagent NBD-Cl has the potential to react with the OH group of tyrosine, SH group of cysteine and ε-NH_2 of lysine.[18] This reagent has been found to lead to time-dependent activation and inactivation of RyR ryanodine binding and single channel activities[19] (Fig. 11.1B). RyR treated with NBD-Cl for a short period of time (5–20 sec) displayed an enhanced ryanodine binding and significant increased channel open probability. This NBD-Cl enhancement of ryanodine binding results from an increased site binding affinity. However, after a prolonged incubation (5–20 min), NBD-Cl was found to strongly inhibit ryanodine binding by decreasing the total binding sites. A prolonged incubation of the RyR channel with NBD-Cl also causes an irreversible inhibition of the channel activity.

The products of the reaction of NBD-Cl with ε-NH_2 of lysine, but not with either the SH group of cysteine or the OH group of tyrosine, are stable against treatment with SH-reducing agents such as dithiothreitol.[20,21] NBD-Cl-modified RyR is resistant to thiolysis by DTT, suggesting that a lysyl residue(s) is modified by NBD-Cl. This conclusion is supported by the absorption, and the fluorescence excitation and emission spectra[18,20,21] of the NBD-Cl-modified RyR.[19]

The interaction of RyR lysyl residues with NBD-Cl is summarized in the proposed model presented in Fig. 11.2A. This model suggests the involvement of two distinct lysine ε-amino groups in RyR channel gating, as we have previously proposed.[19] The interaction with the highly-reactive group takes place within a few seconds, after which the channel adopts the largely open conformation. The subsequent reaction with the second group of less reactive lysyl residues leads to complete channel inactivation.

Modification by DIDS and DNFB

DIDS (4,4'-diisothiocyano-stilbene-2,2'-disulfonic acid) has two SCN groups by which it can bind covalently to amino groups. DIDS, a known anion-channel blocker, was found to activate Ca^{2+} release in isolated SR vesicles and in reconstituted RyR channels in a planar bilayer.[22] However, RyR was not labeled with [$^3H_2$]DIDS, which labeled instead a 30 kDa protein,[24] suggesting an indirect interaction of DIDS with the RyR. In cardiac Ca^{2+} release channels reconstituted in PLB, DIDS was found to increase the open probability by inducing a long-lived open state.[23] It has been suggested that DIDS modifies an amino group residing in the gating structure of the channel.[23]

DNFB (2,4-dinitrofluorobenzene) reacts with amines to form dinitrophenyl derivatives. Depending on the incubation conditions, DNFB either inhibits or stimulates ryanodine binding

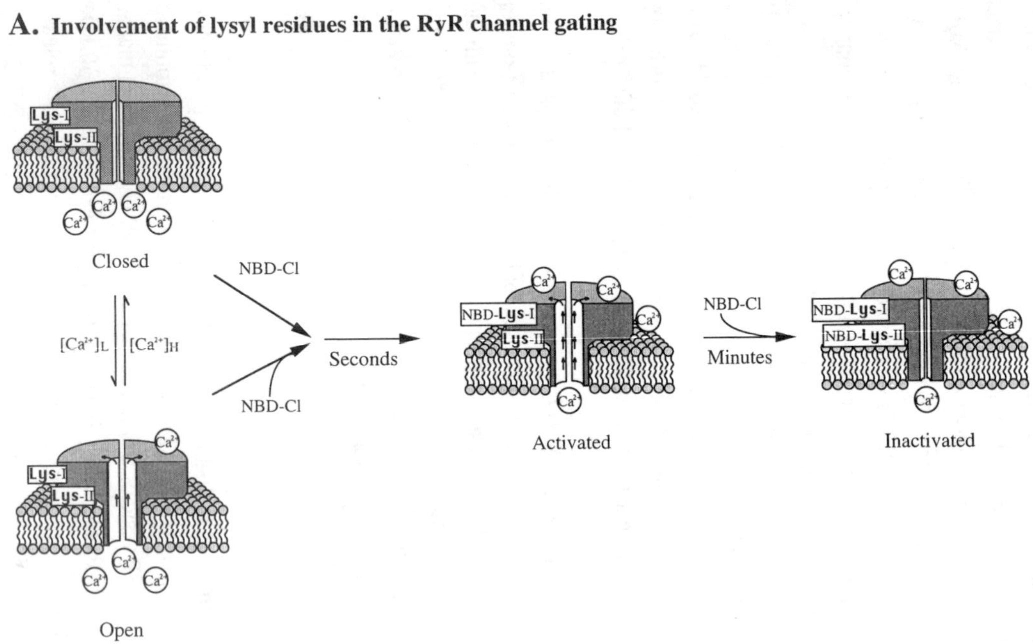

(a)

Fig. 11.2 Schematic model of RyR activation and inactivation by chemical modification with NBD-Cl and DCCD. The model shows the transmembrance topology of RyR, where one half of the homotetrameric structure is depicted. The involvement of two different lysyl residues (Lys I, Lys II) in the activation and inactivation of RyR/Ca^{2+} release channel is presented in A. The inveolvement of carboxyl group(s) in the channel closure is presented in B. Modification of a carboxyl group, located in a hydrophobic region of the channel, either by DCCD or by binding of Ca^{2+} results in channel closure. DCCD interacts with either closed (a) or open (b) state of the channel. Ca^{2+} binds, in a time-dependent manner, to the carboxyl group at the open state of the channel (b) and induces channel closure (c). Δ V indicates voltage change, and NH_3^+, NH_2 and COO^- indicate amino acid chain residues.

B. Involvement of carboxyl residues in channel closure

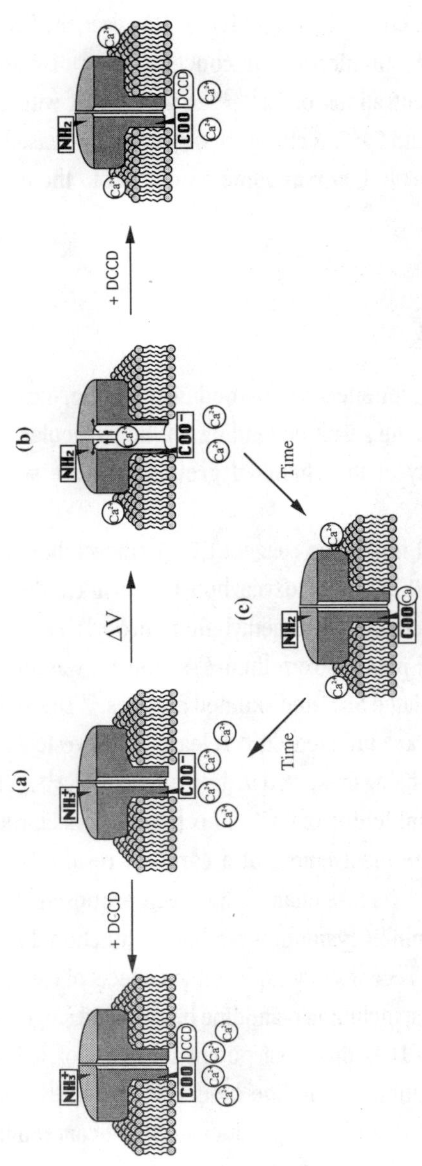

Fig. 11.2 (Continued)

to SR.[25,26] The stimulatory effect is due to an increase in the Ca^{2+} and ryanodine binding affinities. Exposure of RyR to DNFB under conditions which do not inhibit ryanodine binding leads to complete closure of the channel[26] (Fig. 11.1C). However, the DNFB-modified closed channel can be reactivated by submicromolar concentrations of ryanodine even in the presence of nanomolar concentrations of Ca^{2+},[26] in agreement with the increase in the binding affinity for ryanodine and Ca^{2+}. A closed channel with increased ryanodine binding affinity argues against the concept that ryanodine binds only to the open state of the RyR channel.[5]

ii. Carboxyl-modifying reagents

Carbodiimides have been used for site-specific modification of proteins to identify critical reactive carboxyl groups, with the result depending on the particular carbodiimide used, the accessibility and reactivity of the chemical groups, and the specific experimental conditions.[27]

The hydrophobic carboxyl modifying reagent DCCD (dicyclohexylcarbodiimide), but not the amphipathic reagent EEDQ (N-ethoxycarbonyl-2-ethyl-1,2-dihydro-quinoline), or hydrophilic reagents EDC [1-ethyl-3-(3-dimethyl-aminopropyl) carbodiimide] and WRK (Woodward's reagent K, N-ethyl-phenylisoxazolium-3'-sulfonate) was found to block alkaline pH-induced Ca^{2+} release in isolated SR[28] and skinned muscles.[29] DCCD also inhibited Ca^{2+}-activated choline influx and Ca^{2+}-induced Ca^{2+} release in SR vesicles from skeletal and cardiac muscles, this inhibition being prevented in the presence of Ca^{2+}.[30] DCCD modification of SR was found to abolish the inhibition of Ca^{2+} efflux by high concentrations of Ca^{2+}, Mg^{2+} and ruthenium red. Hence, the involvement of a carboxyl residue(s) in the low-affinity cationic binding site of the Ca^{2+} release channel has been proposed.[31]

DCCD was also found to inhibit ryanodine binding to junctional SR membranes, and was found to label the RyR.[28] A recent study explored the effects of carboxyl group reagents on skeletal muscle RyR activities, including ryanodine binding and single channel activities.[32] The hydrophobic reagent DCCD is the most potent inhibitor of RyR activity, while the amphipathic reagent EEDQ inhibits ryanodine binding at relatively high concentrations, and the hydrophilic reagents EDC and WRK produce no significant inhibition of ryanodine binding. Similar results were obtained when Ca^{2+} release from SR vesicles was tested.[28] DCCD was found to alter the binding affinity (Kd) and the total binding sites (Bmax) for ryanodine as well as the Ca^{2+} affinity of both the high- and low-affinity sites of the RyR.[32] It was suggested that a carboxyl residue(s) is linked to ryanodine binding directly or via the Ca^{2+} binding sites. This group is presumably located within a hydrophobic region of the protein, and therefore is accessible to the hydrophobic reagent DCCD but not to the water-soluble reagents EDC and WRK. However, it becomes accessible to these latter reagents under specific conditions such as at alkaline pH, or when an isolated RyR is used.[32] The DCCD-modified non-active channel was found to be re-activated and stabilized in a sub conducting state by addition of ryanodine in the presence of 250 μM Ca^{2+}.[32]

DCCD and WRK were found to block the skeletal muscle RyR channel opening completely[32] (see also Fig. 11.1A). EDC was found to reduce the cardiac RyR channel conductance (by 36%) only when applied to the luminal side of the channel and at low ionic strengths. It has been suggested that fixed negative surface charges on the luminal mouth of the RyR channel potentiate conductance.[33] Martinez-Azorin *et al.*[31] reported that DCCD derivatization of the carboxyl group prevents the blocking capacity of di- and polyvalent cations, while Feng and Shoshan-Barmatz[32] suggest that DCCD modification of a carboxyl residue(s) by itself induces channel closure (see Fig. 11.2B).

Studies with hydrophobic and hydrophilic carboxyl reagents have thus led to the conclusion that a carboxyl group located at a hydrophobic region of the RyR is involved in the channel closure and/or Ca^{2+} binding. This group may be involved in the inactivation by high Ca^{2+} concentrations of Ca^{2+} release,[34–36] ryanodine binding,[5–7,36] and single channel activity[37–41] (see Fig. 11.2B). In single channel measurements with RyR/Ca^{2+} release channels, an increase in luminal Ca^{2+} from micromolar to millimolar levels was found to decrease channel activity.[37–41] Sitsapesan and Williams[40] suggested that the effects of luminal Ca^{2+} were most likely mediated by specific Ca^{2+} binding sites on the luminal face of the channel and that these sites were exposed only when cytosolic ATP was bound to the channel. However, it has been suggested that luminal Ca^{2+} exerts its regulating effect by acting on Ca^{2+} binding sites accessible from the cytoplasmic side of the channel.[37] It has been pointed out that Ca^{2+} can move from the luminal to the cytosolic side of the channel through the channel pore and thus bind to the cytosolic Ca^{2+}-activating sites.[41] The suggestion that luminal calcium regulates Ca^{2+} release via the RyR/Ca^{2+} release channel is now generally accepted.[34–41] In view of the hydrophobic nature of the DCCD binding site, it has been postulated that the reactive carboxyl group is located in the ion conductance pathway and is involved in the channel closure by Ca^{2+} binding. This hypothesis is in agreement with the suggestion of Fill *et al.*[39] that the Ca^{2+} inactivation site occupies the interior of the ion-conducting pathway, since it is accessible from both sides of the SR membranes.

iii. Histidine-directed reagents

Exposure of SR to light in the presence of the dye Rose Bengal, known to induce photo-oxidation of histidyl residues, caused an activation of Ca^{2+} release, inhibition of ryanodine binding[42] as well as an activation of reconstituted Ca^{2+} release channels from skeletal muscle SR.[43] Similarly, illumination of Ca^{2+} release channels, isolated from sheep cardiac SR, in the presence of Rose Bengal resulted in the loss of ryanodine binding and the appearance of channels with subconductance states.[44]

Diethylpyrocarbonate (DEPC) reacts with histidyl residue in proteins to yield the N-carbethoxyhistidyl derivative.[45] This histidine modification can be specifically reversed by the addition of hydroxylamine.[45] DEPC (in mM range) was shown to induce Ca^{2+} release from SR vesicles.[42]

Exposure of junctional SR or purified RyR to DEPC also led to concentration- and time-dependent inactivation of ryanodine binding.[46] Based on the pH dependence of the inactivation of ryanodine binding by DEPC and the reversal of this inactivation by hydroxylamine, it has been suggested that the modified group is a histidyl residue(s).[46] According to an analysis of the time course of inactivation of ryanodine binding by DEPC, the inactivation resulted from modification of a single class of histidine residue per ryanodine binding site. The high-binding affinities for ryanodine and Ca^{2+} were not altered by DEPC-modification, but the total ryanodine binding sites were decreased.[46] On the other hand, a variety of divalent cations were found to prevent the DEPC-inactivation of ryanodine binding in a series of decreasing efficiency : $Mn^{2+} > Ba^{2+} > Mg^{2+} > Ca^{2+}$, similar to their effectiveness in inhibiting ryanodine binding, suggesting modification of the low-affinity Ca^{2+} binding sites.

As found for other histidine modifying reagents,[42,44] DEPC-modification dramatically increased the Ca^{2+} permeability of the SR vesicles.[46] Furthermore, DEPC activated the RyR/Ca^{2+} release channel incorporated in PLB at very low Ca^{2+} concentrations (see Fig. 11.1D).

It has been suggested that a histidyl residue(s) in the RyR is involved either directly in the binding of Ca^{2+} or in a conformational change that may be required for Ca^{2+} binding to the low-affinity sites. DEPC modification of the RyR resulted in <u>diminished</u> ryanodine binding and activation of Ca^{2+} release from SR, which resulted from the <u>opening</u> of the RyR/Ca^{2+} release channel (see Fig. 11.1D). The relationship between these effects is not clear, since ryanodine binds to the open channel.[5] These results suggest that DEPC modification stabilizes an open conformational state of the RyR that does not bind ryanodine (see discussion in Concluding remarks).

iv. Thiol group-directed reagents

Oxidation of protein SH groups to disulfides increased the Ca^{2+} permeability of SR, the increase being reversed by the reduction of the disulfide.[47–54] Oxidation of sulfhydryl groups to disulfide results in: (a) contraction of skinned muscle fibers; (b) stimulation of Ca^{2+} release from SR vesicles; (c) modification of the gating characteristics of single Ca^{2+} channels reconstituted into a PLB; (d) and alteration of the high-affinity ryanodine binding site of the RyR. The fluorescent sulfhydryl reagent coumarin maleimide was found to associate with RyR during activation of the Ca^{2+} release channel. This complex dissociated upon reduction leading to closure of the RyR/Ca^{2+} release channel.[52]

The interaction of the sulfhydryl reagent thimerosal, with the RyR was demonstrated recently.[54] Thimerosal (nM), which has been shown to increase the intracellular Ca^{2+} concentration and to induce Ca^{2+} spikes in several cell types,[55] was found to stimulate Ca^{2+} release, to inhibit ryanodine binding and to stimulate single channel activity. At high concentrations (mM) the reagent first activated the channel and then inhibited it in a time-dependent manner.[54] It has been postulated that oxidation and reduction of critical sulfhydryl groups is the mechanism underlying the gating of the Ca^{2+} release channel.[50,54]

II. Binding Site-Directed Reagents

The activity of RyR/Ca^{2+} release channels is modulated by a variety of ligands, including ATP, Ca^{2+}, calmodulin and Mg^{2+}. However, the properties and the location of the binding sites for these modulators and the relationship between them and RyR activities are as yet not understood.

For an understanding of the mechanism of ligand modulation, the identification of the amino acid residues in the binding site is desirable. The use of specific affinity labeling or photoaffinity labeling has been the classical approach to acquiring such information. An affinity label is characterized by two moieties: one of these has a structure closely related to that of the substrate/ligand and is therefore complementary to the binding site; the other is a group, generally having an electrophilic nature, which reacts with a nucleophilic amino acid residue to form a covalent bond. In the photoaffinity labels the modifying group is activated by light to generate a highly reactive intermediate that binds covalently to a neighboring group. The chemical modification of a protein with photoaffinity analogs of substrate/ligand can provide valuable structural information about the location of the binding site.

i. Ryanodine binding sites

Ryanodine binds to RyR, in a Ca^{2+}-dependent manner, to multiple binding sites with high- and low-affinities.[5–9,38,56,57] It is accepted that there is only one high-affinity site for [^{3}H]ryanodine per tetrameric receptor.[9] Binding of ryanodine to the high-affinity site locks the channel in the open, low-conductance state. Occupation of the low-affinity sites by ryanodine stabilizes the channel in the closed state.[5–9]

A ratio of 1:3 between high affinity (Kd = 8 nM) and low affinity (Kd = 5 μM) binding sites for ryanodine in SR has been reported. However, the ratio is dependent on the free Ca^{2+} concentration,[58] suggesting that the sites are interconvertible. Furthermore, a co-operativity between the ryanodine low-affinity binding sites has been suggested.[58–60]

The photoactive derivative of ryanodine ABRy {10-O-[3-(4-azidobenzamido) propionyl]ryanodine} was used for characterizing and photolabeling of the ryanodine binding sites of skeletal, cardiac, and brain membranes.[61] This derivative was found to specifically displace [^{3}H]ryanodine from its binding site(s). Photoaffinity labeling of triads with [^{3}H]ABRy resulted in specific and covalent incorporation of [^{3}H]ABRy in a 565 kDa protein (the skeletal RyR monomer). The [^{3}H]ABRy binding site has been located in a 76 kDa tryptic fragment that was identified by use of antibody directed against the carboxyl terminal of the RyR. It should be noted that the ATP binding site was also located in a 76k Da tryptic fragment of the RyR.[62]

ii. Nucleotide binding sites

ATP is a known modulator of RyR activity. ATP has been shown to enhance Ca^{2+} release from SR vesicles,[1,2,35] to increase ryanodine binding,[5–7,36,63] and to stimulate the activity of a single Ca^{2+} release channel reconstituted into PLB.[6,64]

Based on the amino acid sequence of skeletal muscle RyR, deduced from the cDNA, a potential ATP binding domain (containing the GXGXXG motif)[65] was predicted to lie in residues 4449–4454 or 4452–4457.[10] However, potential nucleotide binding motifs were also identified in other positions such as at amino acids 699–706, 1195–1200,[11] and 2370–2375, 4602–4607.[66]

Using the photoreactive derivative of ATP, 8-azido-ATP, an ATP binding site has been shown to be present within the ryanodine receptor.[9]

The photoreactive ATP analog 3'-O-(4-benzoyl)benzoyl-ATP (Bz^2ATP) has been used extensively as a probe for the identification of ATP binding sites and for their location within the polypeptide chain of various proteins, such as in the mitochondrial ATPase,[67] and the

SR Ca^{2+}-ATPase.[68] The ATP binding site of the RyR was characterized using Bz_2ATP.[62] The Bz_2ATP binding site was located in the 76 kDa tryptic fragment and in the 27 and 13 kDa fragments obtained by digestion with the V_8 protease.

The ATP analog, 2′,3′-dialdehyde ATP (oATP), was shown to enhance the open probability of the RyR/Ca^{2+} release channel incorporated in PLB.[69] The effect of the analog has a Hill coefficient of about 2, suggesting co-operativity between the ATP binding sites in the homotetramer. oATP binds to RyR forming a stable Schiff base between the aldehyde groups of oATP and ε-amino groups of lysines. Reduction by borohydride allows covalent incorporation of [α-^{32}P] oATP into the RyR.[69]

The ATP binding domain was located in the 76 kDa tryptic fragment[62] identified as the C-terminal part of the protein.[61] This fragment also contains the FITC binding site.[16] The RyR ATP binding domain most probably contains lysyl residue(s), as indicated by the inhibition of the photoaffinity labeling of the RyR by [α-^{32}P]Bz_2ATP in FITC-modified SR membranes[16] and by the RyR labeling by oATP.[69] Lysyl residues were proposed to be a part of the ATP binding site of six cation-transporting ATPases, including the SR Ca^{2+}-ATPase, Na,K-ATPase, and H,K-ATPase.[70] Location of the functional ATP binding site(s) to a specific region of the channel protein may help to locate a particular gating region within the protein primary sequence.

Brain RyR/Ca^{2+} release channels were found to be activated by adenine nucleotides.[71,72] The amino acid sequence of a brain-specific RyR has been deduced from a rabbit brain cDNA sequence.[73] It contained four nucleotide-binding consensus sequences.[73] A type 2 RyR was found to be labeled with [α-^{32}P]Bz_2ATP.[74]

The three different ATP analogs used for direct labeling of the nucleotide binding site(s) on the RyR/Ca^{2+} release channel, i.e. 8-azido-[α-^{32}P]ATP,[9] 3-O-(4-benzoyl) benzoyl ATP,[62,74] and 2′,3′-dialdehyde ATP[69] differ with respect to the modification introduced. 8-Azido-ATP identifies amino acid residues in the vicinity of a purine base. Bz_2ATP most probably labels a hydrophobic area within the nucleotide binding pocket. oATP specifically labels lysine residues within the ribose binding region. Thus, identification of the nucleotide binding domain of the RyR using these three analogs may provide insight to the structure of the nucleotide binding site(s) on the RyR.

iii. Calmodulin binding sites

Calmodulin (CaM) has been shown to affect Ca^{2+} release from SR.[75–77] Smith *et al.*[78] showed that CaM can interact directly with the Ca^{2+} release channel in a Ca^{2+}-dependent way. This CaM-RyR interaction reduced the mean duration of single-channel open events of the RyR without affecting channel conductance. The CaM effect was exerted via the interaction with the myoplasmic side of the RyR.[76]

Based on the deduced amino acid sequence of the RyR, two to three putative CaM binding sites at the C-terminal half of the protein have been proposed. Takeshima *et al.*[10] predicted two CaM binding sites in residues 3614–3637 and 4295–4325, whereas Zorzato *et al.*[11] predicted three different sites at residues 2807–2840, 2909–2930 and 3031–3049. Possible sites for CaM binding were identified based on the observation that CaM inhibits calpain digestion of RyR.[79] Analysis of the digestion pattern of the RyR in the presence and absence of CaM, coupled with the use of computer algorithms to identify consensus sequences for calpain digestion and CaM binding, led to the identification of residues 1383–1400, 1974–1996 and 3358–3374 as candidate sites for CaM.[79]

Ligand overlay experiments with skeletal muscle RyR fusion peptides have indicated the presence of up to seven different CaM binding regions.[80] Affinity labeling of RyR by CaM was demonstrated using CaM iodinated and derivated with the photoaffinity label benzophenon-4-maleimide {[^{125}I]Bz-CaM}.[81] CaM binding sites of skeletal, cardiac, and brain RyR/Ca^{2+} release channels were identified by following the binding of dansyl-calmodulin to synthetic peptides corresponding to the calmodulin binding sites present in the skeletal muscle RyR.[82]

Calmodulin interaction with the skeletal muscle RyR was studied by fluorescence anisotropy by using rhodamine maleimide-labeled wheat germ calmodulin (Rh-CaM).[81] These studies indicated that CaM binds with high affinity to RyR in the absence and presence of Ca^{2+}. Multiple calmodulin-binding sites on each channel protein subunit (229 pmol Rh-CaM and 10.6 pmol ryanodine were bound per mg of protein) were determined. However, the affinities of these calmodulin-binding sites were dependent on Ca^{2+} and Mg^{2+} concentration.[81] [^{125}I]CaM binding to SR and purified RyR indicated that the tetrameric complex binds with high affinity (K_D = 5-25 nM) 16 CaM at < 0.1 μM Ca^{2+} and 4 CaM at 100 μM Ca^{2+}.[83] Electron microscopic studies with purified RyRs and gold cluster labeled CaM indicated that one CaM binding site is at the surface that faces the cytoplasm.[84] The role of CaM in the physiological function of RyR/Ca^{2+} channel is still unclear. It has been suggested that CaM may play a role in E-C coupling by modulating both activation and inactivation of RyR/Ca^{2+} release channel.[83]

III. Cross-Linking Reagents

Cross-linking reagents comprise a class of modifiers which can bind to two amino acid side chains by virtue of their bifunctional activity. Cross-linking of protein subunits requires not only contact between them, but also the accessibility and suitable positioning of reactive amino acid residues.

Important biophysical characteristics of the RyR/Ca^{2+}-release channel are its organization in an oligomeric complex and the relationship between this organization and its activities, both as a ryanodine-binding protein and a Ca^{2+} release channel. One approach to study these relationships is the use of chemical cross-linking. Cross-linking experiments have demonstrated that close contacts exist between the 4 subunits of RyR.[25,58]

Cross-linking of RyR was carried out using several bifunctional reagents differing in their linear dimensions and flexibility, as well as in the reactivity of the active groups.[25] Cross-linking of SR membranes with DFDNB, DFNPS, DMS, EDC and glutaraldehyde was found to result in cross-linking of the RyR and of almost no other SR protein. Cross-linking of the RyR takes place both in membrane bound or purified soluble forms of the protein.[25] According to monomer copy numbers analysis,[85] the new cross-linked protein bands, which cross react with antibodies against the RyR, closely corresponded to the dimer and tetramer of the RyR subunits. Glutaraldehyde was shown to react with the purified RyR, and it was proposed to form dimers, trimers and tetramers.[58] In contrast, no trimer was detected[24] when either the membrane-bound and the purified RyR were reacted with glutaraldehyde at a concentration 100 times lower than that used for cross-linking of the purified RyR.[58]

Cross-linking of RyR was also observed with EDC, a water-soluble carbodiimide that forms zero-length covalent bonds between carboxyl groups and nucleophilic groups, usually amino groups.[86] Comparison of cross-linking using a zero length cross-linker, such as EDC, and DFDNB, in which the distance between the two reactive groups is about 5Å,[87] indicates a tight contact between the subunits in the tetramer. Cross-linking is also observed with DFNPS and with DMS, which have effective lengths of 9Å and 11Å, respectively.[87]

Although RyR was completely cross-linked to form tetramers by the various reagents, ryanodine binding to the high-affinity site was insignificantly affected, whereas the binding to the low-affinity site was inhibited.[25] RyR possesses two to three binding sites of lower affinity, in addition to the high-affinity ryanodine binding site.[56–60] It has been suggested that the ryanodine receptor exhibits "positive co-operativity" whereby binding at the low-affinity sites decreases the dissociation rate of ryanodine from the high-affinity site.[57,58] A model for the allosteric negative co-operativity between the four ryanodine binding sites in the homotetramer RyR has also been proposed, whereby the binding of ryanodine to one of the four sites (with high affinity) triggers the conversion of the other sites to the low-affinity binding sites.[59] This model suggests that in order for RyR to manifest binding co-operativity, there must be two or more different states of the receptor with different affinities. In addition, each subunit of the protein must undergo a conformational change(s). In the light of the allosteric interaction between the RyR subunits,[57–60] the observation, that the "fixation" of the RyR by cross-linkers in an irreversible tetrameric structure(s) does not affect ryanodine binding to the high-affinity site, is of interest. Ryanodine binding depends on Ca^{2+} and the complex [Ca^{2+}. RyR] is a very stable complex from which ryanodine is dissociated very slowly ($t_{1/2}$ of 14.4 h).[63] Hence, it seems that RyR undergoes conformational changes upon binding of Ca^{2+} and/or ryanodine. In the "new" conformation, ryanodine is physically occluded and hindered from free diffusion from its binding site. The cross-linking of two or more residues in each monomer, which leads to fixation of dimers or tetramers, can not, however, prevent the conformational changes involved in the binding and occlusion of ryanodine at the high-affinity site. On the other hand, cross-linking of the ryanodine receptor results in inhibition of ryanodine binding to the low-affinity binding sites.[25]

CONCLUDING REMARKS

Since the identification of RyR as the Ca^{2+} release channel of SR, much has been learned about the biochemical and molecular basis of its activation and inactivation, as reflected in ryanodine binding, Ca^{2+} release from SR and single channel activities.

Although the molecular mechanism of activation and inactivation of RyR/Ca^{2+} release channel is far from being elucidated, a number of new findings concerning the involvement of specific amino acid residues in the control of channel opening and closure have recently arisen from the use of chemical probes (see Fig. 11.2). On the basis of results with different amino group modifying reagents, highly-reactive lysyl residues are postulated in the gating mechanism of the RyR channel. This group may be the charged group sensitive to t-tubule voltage changes and associated with the structural changes which convert the channel from a stable closed conformation to the open conformation. Based on the modification of RyR by the various carboxyl-modifying reagents including DCCD, a carboxyl group located at a hydrophobic region of the RyR is proposed to be involved directly in the channel closure or in Ca^{2+} binding that leads to the inhibition of Ca^{2+} release, ryanodine binding and single channel activities by high Ca^{2+} concentrations (see Fig. 11.2B). Using a variety of SH reagents, it has been suggested that critical sulfhydryl groups in the RyR, via their oxidation and reduction, play an important role in the gating of the RyR/Ca^{2+} release channel. Modification of histidyl residues in RyR was found to inhibit ryanodine binding and to stimulate Ca^{2+} release, most probably via modification of the RyR low-affinity Ca^{2+} binding site(s).

It has been generally accepted that ligands known to open the channel and stimulate Ca^{2+} release, such as μM Ca^{2+}, mM ATP or mM caffeine, stimulate [^{3}H]ryanodine binding to a high-affinity site. Ligands known to close the channel and inhibit Ca^{2+} release, such as mM ruthenium red, mM Ca^{2+} and mM Mg^{2+}, inhibit ryanodine binding.[5] However, this correlation between ryanodine binding and the gating state of the channel was not found for RyR modification by several agents, such as Rose-Bengal, DEPC, FITC, thimerosal or peroxide, which were found to stimulate Ca^{2+} release and single channel activities, but to inhibit ryanodine binding (see Table 11.1). DNFB modification of RyR, on the other hand, resulted in channel closure, but this DNFB-modified channel has higher affinities for Ca^{2+} and ryanodine binding. Furthermore, the DCCD-modified closed channel was reactivated by ryanodine, suggesting ryanodine binding to closed channel.[32] These findings indicate that various modifications stabilize an open conformational state of the RyR that does not bind ryanodine, or *vice versa*, that some modifications induced a closed channel that binds ryanodine with a high affinity, thus suggesting a more complex relationship. Studies involving site-directed and group-directed reagents have provided a new insight into the structure and function of the RyR. These covalently bound probes have been used as a tool for the location of the ligand binding sites and of functional domains involved in the control mechanism of channel activity. These studies demonstrated that the carboxyl-terminal portion of the RyR includes both the channel-forming region and the modulator-binding sites for ATP, ryanodine, calmodulin and also the FITC-reactive group involved in channel gating. These putative modulator binding sites are located at the cytoplasmic surface of the RyR close to the predicated transmembrane segments, thus their modulating effects on the RyR channel might be via the regulation of the pore forming transmembrane segments of the RyR.

Cross-linking data indicate that a close contact exists between the 4 subunits of the RyR. The cross-linking studies also reveal that the linkage between the RyR subunits to form irreversible dimers or tetramers does not affect its ryanodine binding to the high-affinity site, but inhibits the binding to the low-affinity sites. This may suggest that cross-linking of RyR does not prevent the conformational changes involved in binding and occlusion of ryanodine at the high-affinity site, but interferes with the changes involved in the channel closure induced by ryanodine binding to the low-affinity binding sites.

ACKNOWLEDGEMENTS

The work was supported by the fund for basic research administered by the Israel Academy of Science and Humanities and by a grant from the Chief Scientist's Office, Ministry of Health, Israel.

REFERENCES

1. Endo M. Calcium release from sarcoplasmic reticulum. *Physiol Rev* 1977; **7**:71–108.
2. Ebashi S. Excitation-contraction coupling and mechanism of muscle contraction. *Ann Rev Physiol* 1991; **53**: 1–16.
3. Inesi G. Mechanism of calcium transport. *Ann Rev Physiol* 1985; **47**: 573–601.
4. Martonosi AN, Istvan J, Molnar E, *et al.* Emerging views on the structure and dynamics of the Ca^{2+} -ATPase in sarcoplasmic reticulum. *FEBS Lett* 1990; **268**: 365–370.
5. Coronado R, Morrissette J, Sukhareva M, *et al.* Structure and function of ryanodine receptors. *Am J Physiol* 1994; **266**: C1485–C1504.
6. Lai FA, Meissner G. The muscle ryanodine receptor and its intrinsic Ca^{2+} channel activity. *J Bioenerg Biomembr* 1989; **21**: 227–245.
7. Fleischer S, Inui, M. Biochemistry and Biophysics of excitation-contraction coupling. *Ann Rev Biophys Biophys Chem* 1989; **18**: 333–364.
8. Williams AJ. Ion conduction and discrimination in the sarcoplasmic reticulum ryanodine receptor/calcium release channel. *J Muscle Res Motil* 1992; **13**: 7–26.
9. Lai FA, Erickson HP, Rousseau E, *et al.* Purification and reconstitution of the Ca release channel from skeletal muscle. *Nature* 1988; **331**: 315–319.
10. Takeshima H, Nishimura S, Matsumoto T, *et al.* Primary structure and expression from complementary DNA of skeletal muscle ryanodine receptor. *Nature* 1989; **339**: 439–445.
11. Zarzato F, Fuyjii J, Otsu K, *et al.* Molecular cloning of cDNA encoding human and rabbit forms of the Ca^{2+} release channel (ryanodine receptor) of skeletal muscle sarcoplasmic reticulum. *J Biol Chem* 1990; **265**: 2244–2256.
12. Hymel L, Inui M, Fleischer S, *et al.* Purified ryanodine receptor of skeletal muscle sarcoplasmic reticulum forms Ca^{2+}-activated oligomeric Ca^{2+} channels in planar bilayers. *Proc Natl Acad Sci* 1988; **85**: 441–445.
13. Rios E, Brum G. Involvement of dihydropyridine receptors in excitation-contraction coupling in skeletal muscle. *Nature* 1987; **325**: 717–720.
14. Block BA, Imagawa T, Campbell KP, *et al.* Structural evidence for direct interaction between the molecular components of the transverse tubule/sarcoplasmic reticulum junction in skeletal muscle. *J Cell Biol* 1988; **107**:2587–2600.
15. Segel IH. *Enzyme Kinetics*, pp. 885, New York1975: Wiley-Interscience.
16. Orr I, Martin C, Ashley R, *et al.* The interaction of fluorescein isothiocyanate with the ryanodine receptor/Ca^{2+} release channel of sarcoplasmic reticulum. *J Biol Chem* 1993; **268**: 1376–1382.
17. Fill M, Ma J, Knudson CM, *et al.* Role of the ryanodine receptor of skeletal muscle in excitation-contraction coupling. *Ann New York Academy Science* 1989; **560**: 155–162.
18. Randall K, Sze H. Probing the catalytic subunit of the tonoplast H^+-ATPase from oat roots. *J Biol Chem* 1987; **262**: 7135–7141.
19. Feng W, Shoshan-Barmatz V. Involvement of lysine residues in the gating of the ryanodine receptor/Ca^{2+} release channel of skeletal muscle sarcoplasmic reticulum. *Eur J Biochem* 1997;**247**: 955–962.

20. Aboderin AA, Boedefeld E, Luisi PL. Reaction of chicken egg white lysozme with 7-chloro-4-nitrobenz-2-oxa-1,3-diazole. *Biochim Biophys Acta* 1973; **328**:20–30.
21. Ferguson SJ, Lloyd WJ, Radda GK. The mitochondrial ATPase. Evidence for a single essential tyrosine residue. *Eur J Biochem* 1975; **54**: 117–126.
22. Kawaski T, Kasai M. Disulfonic stilbene derivatives open the Ca^{2+} release channel of sarcoplasmic reticulum. *J Biochem* 1989; **106**: 401–405.
23. Zahradnikova A, Zahradnik I. Modification of cardiac Ca^{2+} release channel gating by DIDS. *Pflugers Arch* 1993; **425**: 555–557.
24. Yamaguchi N, Kawasaki T, *et al.* DIDS binding 30-kDa protein regulates the calcium release channel in the sarcoplasmic reticulum. *Biochem Biophys Res Commu* 1995; **210**: 648–653.
25. Shoshan-Barmatz V, Hadad-Halfon N, Ostersetzer O. Cross-linking of the ryanodine receptor/ Ca^{2+} release channel from skeletal muscle. *Biochim Biophys Acta* 1995; **1237**: 15511–161.
26. Feng W, Hadad N, Shoshan-Barmatz V. *Modification of ryanodine receptor/ Ca^{2+} release channel activities by DNFB.* 1998 (submitted).
27. Solioz M. Dicyclohexylcarbodiimide as a probe for proton translocating enzymes. *Trends Biochem Sci* 1984; 7: 309–312.
28. Argaman A, Shoshan-Barmatz V. DCCD interaction with sarcoplasmic reticulum: Inhibition of Ca^{2+} efflux. *J Biol Chem* 1988; **263**: 6315–6321.
29. Shoshan V, MacLennan DH, Wood DD. A proton gradient controls a Ca^{2+} release channel in sarcoplasmic reticulum. *Proc Natl Acad Sci USA* 1981; **78**: 4828–4832.
30. Yamamoto N, Kasai M. Mechanism and function of the Ca^{2+}-gated cation channel in sarcoplasmic reticulum vesicles. *J Biochem* (Tokyo) 1982;**92**: 485–496.
31. Martinez-Azorin F, Gomez-Fernandez JC, Fernandez-Belda F. Limited carbodiimide derivatization modifies some functional properties of the sarcoplasmic reticulum Ca^{2+} release channel. *Biochemistry* 1993; **32**: 8553–8559.
32. Feng W, Shoshan-Barmatz V. The action of carboxyl modifying reagents on the ryanodine receptor/ Ca^{2+} release channel of skeletal muscle sarcoplasmic reticulum. *Molec Membr Biol* 1996; **13**: 85–93.
33. Tu Q, Velez P, Cortes-Gurierrez M, *et al.* Surface charge potentiates conduction through the cardiac ryanodine receptor channel. *J Gen Physiol* 1994; **103**: 853–867.
34. Fabiato A. Time and calcium dependence of activation and inactivation of Ca^{2+}-induced release of calcium from the sarcoplasmic reticulum of a skinned canine purkinge cell. *J Gen Phsiol* 1985; **254**: 249–289.
35. Meissner G, Darling E, Eveleth J. Kinetics of rapid Ca^{2+} release by sarcoplasmic reticulum. Effects of Ca^{2+}, Mg^{2+} and adenine nucleotides. *Biochemistry* 1986; **25**: 236–244.
36. Michalak M, Dupraz P, Shoshan-Barmatz V. Ryanodine binding to sarcoplasmic reticulum membrane; comparison between cardiac and skeletal muscle. *Biochem Biophys Acta* 1988; **939**: 587–594.
37. Herrmann-Frank A, Lehmann-Horn F. Regulation of the purified Ca^{2+} release channel/ryanodine receptor complex of skeletal muscle sarcoplasmic reticulum by luminal calcium. *Pflug Arch Eur J Physiol* 1996; **432**: 155–157.

38. Ma JJ, Fill M, Knudson M, *et al.* Ryanodine receptor of skeletal muscle is gap - Junction type channel. *Science* 1988; **242**: 99–102.
39. Fill M, Coronado R, Mickelson JR, *et al.* Abnormal ryanodine receptor channels in malignant hyperthermia. *Biophys J* 1990: **50**: 471–475.
40. Sitsapesan R and Williams AJ. The gating of the sheep skeletal sarcoplasmic reticulum Ca^{2+}-release channel is regulated by luminal Ca^{2+}. *J Membr Biol* 1996; **146**: 133–144.
41. Tripathy A and Meissner G. Sarcoplasmic reticulum luminal Ca^{2+} has access to cytosolic activation and inactivation sites of skeletal muscle Ca^{2+} release channel. *Biophys J* 1996; **70**: 2600–2615.
42. Stuart J, Pessah IN, Favero TG, *et al.* Photooxidation of skeletal muscle sarcoplasmic reticulum induced rapid calcium release. *Arch Biochem Biophys* 1992; **292**: 512–521.
43. Xiong H, Buck E, Stuart J, *et al.* Rose bengal activates the Ca^{2+} release channel from skeletal muscle sarcoplasmic reticulum. *Arch Biochem Biophys* 1992; **292**: 522–528.
44. Holmberg SRM, Cumming D, Kusama Y, *et al.* Reactive oxygen species modify the structure and function of the cardiac sarcoplasmic reticulum calcium release channel. *Cardioscience* 1991; **2**: 19–25.
45. Miles, E.W. Modification of histidyl residues in protein by diethylpyrocarbonate. *Methods Enz* 1977; **47**: 431–442.
46. Shoshan-Barmatz V, Weil S. Diethylpyrocarbonate modification of ryanodine receptor from skeletal muscle sarcoplasmic reticulum. *Biochem J* 1994; **299**: 177–181.
47. Abramson JJ, Buck E, Salama G, *et al.* Mechanism of anthraquinone-induced calcium release from skeletal muscle sarcoplasmic reticulum. *J Biol Chem* 1988; **263**: 18750–18758.
48. Salam G, Abramson JJ, Silver ions trigger Ca^{2+} release by acting at the apparent physiological release site in sarcoplasmic reticulum. *J Biol Chem* 1984; **259**: 13363–13369.
49. Nagasaki K, Fleischer S. Modulation of the calcium release channel of sarcoplasmic reticulum by adriamycin and other drugs. *Cell Calcium* 1989;**10**: 63–70.
50. Abramson JJ, Salama G. Critical sulfhydryls regulate calcium release from sarcoplasmic reticulum. *J Bioenerg Biomembr* 1989; **21**: 283–294.
51. Favero T, Zable AC, Abramson JJ. Hydrogen peroxide stimulates the Ca^{2+} release channel from skeletal muscle sarcoplasmic reticulum. *J Biol Chem* 1995; **270**: 25557–25563.
52. Liu G, Abramson JJ, Zable A, *et al.* Direct evidence for the existence and functional role of hyperactive sulhydryls on the ryanodine receptor triadin complex selectivity labeled by the coumarin maleimide 7-diethyamino-3-(4'-maleimidylphenyl)-4-methylcoumarin. *Mol Pharmacol* 1994; **45**: 189–200.
53. Liu G, Pessah IN. Molecular interaction between ryanodine receptor and glycoprotein triadin involved redox cycling of functionally important hyperreactive sulfhydryls. *J Biol Chem* 1994; **269**: 33028–33034.
54. Abramson JJ, Zable AC, Favero TG, *et al.* Thimerosal interacts with the Ca^{2+} release channel ryanodine receptor from skeletal muscle sarcoplasmic reticulum. *J Biol Chem* 1995; **270**: 29644–29647.
55. Bootman MD, Taylor CW, Berridge MJ. The thiol reagent, thimerosal, evokes Ca^{2+} spikes in HeLa cells by sensitizing the inositol 1,4,5-trisphosphate receptor. *J Biol Chem* 1992; **267**: 25113–25119.

56. Meissner G, El Hashem A. Ryanodine as a functional probe of the skeletal muscle sarcoplasmic reticulum Ca^{2+} release channel. *Mol Cell Biochem* 1992;**114**:119–123.
57. Wang JP, Needleman DH, Hamilton SL. Relationship of low affinity [^{3}H] ryanodine binding sites to high affinity sites on the skeletal muscle Ca^{2+} release channel. *J Biol Chem* 1993; **268**: 20974–20982.
58. Lai FA, Misra M, Xu L, *et al.* The ryanodine receptor-Ca^{2+} release channel complex of skeletal muscle sarcoplasmic reticulum. Evidence for a cooperatively coupled, negatively charged homotetramer. *J Biol Chem* 1989; **264**: 16776–167785.
59. Pessah IN, Zimanyi I. Characterization of multiple [^{3}H] ryanodine binding sites on the calcium release of sarcoplasmic reticulum from skeletal and cardiac muscle: Evidence for a sequential mechanism in ryanodine action. *Mol Pharmacol* 1991; **39**: 679–689.
60. McGrew SG, Wolleben C, Siegl P, *et al.* Positive cooperativity of ryanodine binding to the calcium release channel of sarcoplasmic reticulum from heart and skeletal muscle. *Biochemistry* 1989; **28**: 1686–1691.
61. Witcher DR, McPherson PS, Kahl SD, *et al.* Photoaffinity labeling of the ryanodine receptor Ca^{2+} release channel with an azido derivative of ryanodine. *J Biol Chem* 1994; **269**: 13076–13079.
62. Zarka A, Shoshan-Barmatz V. Characterization and photoaffinity labeling of the ATP binding site of the ryanodine receptor from skeletal muscle. *Eur J Biochem* 1993; **213**: 147–154.
63. Pessah IN, Stambuk RA, Casida JE. Ca^{2+}-activated ryanodine binding: mechanisms of sensitivity and intensity modulation by Mg^{2+}, caffeine and adenine nucleotides. *Mol Pharmacol* 1987; **31**: 232–238.
64. Meissner G, Rousseau E, Lai FA. Structural and functional correlation of the trypsin-digested calcium release channel of skeletal muscle sarcoplasmic reticulum. *J Biol Chem* 1989; **264**: 1715–1722.
65. Wierenga RK, Hol WG. Predicted nucleotide-binding properties of p21 protein and its cancer-associated variant. *Nature* 1983; **302**: 842–844.
66. Brandt NR, Caswell AH, Brandt T, *et al.* Mapping of the calpain proteolysis products of the junctional foot protein of the skeletal muscle triad junction. *J Membr Biol* 1992; **127**: 35–47.
67. Williams N, Ackerman SH, Coleman PS. Benzophenone-ATP: A photoaffinity label for the active site of ATPases. *Methods Enzymol* 1986; **257**: 2834–2841.
68. Cable MB, Briggs FN. Labeling the adenine nucleotide binding domain of the sarcoplasmic reticulum Ca, Mg-ATPase with photoaffinity analogs of ATP. *J Biol Chem* 1984; **259**: 3612–3615.
69. Hohenegger M, Hermann-Horn F. Activation and labeling of the purified skeletal muscle ryanodine receptor by an oxidized ATP analogue. *Biochem J* 1995; **308**: 119–125.
70. Taylor WR, Green NM. The predicated secondary structure of the nucleotide-binding sites of six cation-transporting ATPase lead to a probable tertiary fold. *Eur J Biochem* 1989; **179**: 241–248.
71. Ashley RH. Activation and conductance properties of ryanodine-sensitive calcium channels from brain microsomal membranes incorporated into planar lipid bilayers. *J Membr Biol* 1989; **111**: 179–189.

72. Lai FA, Dent M, Wickenden C, *et al.* Expression of a cardiac Ca^{2+} release channel isoform in mammalian brain. *Biochem J* 1992; **288**: 553–564.
73. Hakamata Y, Nakai J, Takeshima H, *et al.* Primary structure and distribution of a novel ryanodine receptor/calcium release channel from rabbit brain. *FEBS Lett* 1992; **312**: 229–235.
74. Hadad N, Martin C, Ashley R, *et al.* Photoaffinity labeling of ATP binding sites identifies a brain microsomal membrane ryanodine receptor and brain dynein. *Biochem J* 1998; (submitted).
75. Meissner G. Evidence of a role for calmodulin in the regulation of Ca^{2+} release from skeletal muscle sarcoplasmic reticulum. *J Biol Chem* 1986;**261**: 6300–6306.
76. Plank B, Wyskovsky W, Hohenegger M *et al.* Inhibition of calcium release from skeletal muscle sarcoplasmic reticulum. *Biochim. Biophys Acta* 1988; **938**: 79–88.
77. Meissner G, Henderson JS. Rapid calcium release from cardiac sarcoplasmic reticulum vesicles is dependent on Ca^{2+} and is modulated by Mg^{2+}, adenine nucleotide and calmodulin. *J Biol Chem.* 1987; **262**: 3065–3073.
78. Smith JS, Rousseau E, Meissner, G. Calmodulin modulation of single sarcoplasmic reticulum Ca^{2+}-release channels from cardiac and skeletal muscle. *Circ Res* 1989; **64**: 352–359.
79. Seiler S, Wegener AD, Whang DD, *et al.* High molecular weight proteins in cardiac and skeletal muscle junctional sarcoplasmic reticulum vesicles bind calmodulin, are phosphorylated, and are degraded by Ca^{2+}-activated protease. *J Biol Chem* 1984; **259**: 8550–8557.
80. Chen SRW, MacLennan DH. Identification of calmodulin- Ca^{2+}- and ruthenium red-binding domains in the Ca^{2+} release chanel (ryanodine receptor) of rabbit skeletal muscle sarcoplasmic reticulum. *J Biol Chem* 1994; **269**: 22698–22704.
81. Yang H-C, Reedy MM, Burke LC, *et al.* Calmodulin interaction with the skeletal muscle sarcoplasmic reticulum calcium channel protein. *Biochemistry* 1994; **33**:518–525.
82. Guerrini R, Menegazzi P, Anacardio R, *et al.* Calmodulin binding sites of the skeletal, cardiac, and brain ryanodine receptor Ca^{2+} channels; Modulation by the catalytic subunit of cAMP-dependent protein kinase? *Biochemistry* 1995; **34**: 5120–5129.
83. Tripathy A, Xu L, Mann J, *et al.* Calmodulin activation and inhibition of skeletal muscle Ca^{2+} release channel (ryanodine receptor). *Biophys J* 1995; **69**: 106–119.
84. Wagenknecht T, Berkowitz J, Grassucci R, *et al.* Localization of calmodulin- binding sites on the ryanodine receptor from skeletal muscle by electron microscopy. *Biophys J* 1994; **67**: 2286–2295.
85. Hames BD. in Gel Electrophoresis of Proteins, A Practical Approach (Hames, B.D. and Rickwood, D. Eds) Second Edition 1990; pp 1–139.
86. Thomas JO, Sternberg N, Weisberg R. Altered arrangement of the DNA in injection-defective lambda bacteriophage. *J Mol Biol* 1978; **123**: 149–161.
87. Peters K, Richards FM. Chemical cross-linking reagents and problems in studies of membrane structure. *Ann Rev Biochem* 1977; **46**: 523–551.
88. Zaidi NF, Lagenaur CF, Abramson JJ, *et al.* Reactive disulfides trigger Ca^{2+} release from sarcoplasmic reticulum via an oxidation reaction. *J Biol Chem* 1989; **264**: 21725–21736.

Chapter 12

The Role of the Ryanodine Receptor in Inherited Muscle Diseases

Charles F. Louis and James R. Mickelson

INTRODUCTION

It is now clear that a number of inherited diseases of skeletal muscle result from mutations in membrane channel proteins. Mutations have been identified in the sarcolemmal sodium channel gene resulting in two different muscle diseases, hyperkalemic periodic paralysis and paramyotonia congenita, while mutations in the sarcolemmal chloride channel are associated with myotonia.[1–4] Numerous biochemical and genetic mapping studies in both the pig and human have now implicated a defect in the skeletal muscle ryanodine receptor (RyR1) as being responsible for the well-described inherited skeletal muscle disorders Malignant Hyperthermia (MH) and Central Core Disease (CCD).[5–7] In this article we will review the functional effects of naturally occurring and targeted mammalian RyR1 mutations on skeletal muscle Ca^{2+} regulation and excitation-contraction coupling.

CLINICAL FEATURES OF MH AND CCD

Human MH is a potentially fatal pharmacogenetic disorder that can occur when susceptible individuals are exposed to halogenated general anesthetics and/or non-depolarizing muscle relaxants.[8] It is estimated that approximately 1 in 12,000 to 1 in 40:000 individuals exhibit MH.[9,10] The essentially identical syndrome was identified in pigs in the late 1960s,[11] and has provided an invaluable animal model for studying the molecular basis of MH. MH-like syndromes have also been reported in a number of other species, most notably dogs.[12,13] The strong association of the porcine MH syndrome with various stressors such as exercise, fighting, reproduction, and transport has led to this disorder also being called porcine stress syndrome (PSS);[14] the association of stress with human MH remains controversial.

During an MH episode there is a rapid and sustained rise in body temperature (as much as 1°C/5 min in MHS humans),[15] a striking increase in aerobic and anaerobic metabolism, tachycardia, elevated arterial CO_2 and lactic acid, and usually the contracture of skeletal muscles, accompanied by muscle breakdown.[16] The patient may die if the early clinical symptoms are not recognized and therapy with the muscle relaxant dantrolene is not initiated immediately.[17] Triggering of anesthetic-induced MH is characteristic of most halogenated volatile anesthetics including halothane, enflurane, isoflurane, desflurane and sevoflurane.[18] There is now considerable evidence indicating that in addition to their anesthetic-inducing action in the central nervous system, these anesthetic agents also act on the sarcoplasmic reticulum in muscle.[19,20]

There is little or no consistently observed ultrastructural abnormality in most MHS muscle. However, the relatively rare autosomal dominant CCD myopathy is characterized by skeletal muscle wasting and weakness.[21] CCD patients are diagnosed on the basis of histological examination of skeletal muscle biopsy samples which exhibit amorphous central cores that are depleted of mitochondria in type 1 muscle fibers. A number of studies have shown that CCD is associated with susceptibility to MH,[22,23] and a number of genetic linkage studies have shown that the RyR1 gene is tightly linked to CCD.[24,25] This close association with the RyR1 suggested that MH and CCD can both result from RyR1 mutations, and are thus allelic.[23,26]

MYOPLASMIC CA^{2+} IN MHS AND NORMAL MUSCLES

Muscle rigidity is commonly encountered during the MH episode and the hyperthermia arises as a result of this muscle activity.[27] It has been argued that a defect in sarcoplasmic

Ca^{2+} regulation could be responsible both for the initiation and maintenance of MH episodes. Muscle contractile activity requires increased sarcoplasmic Ca^{2+}, while the hyperthermia can be explained by a Ca^{2+}-dependent activation of both aerobic and anaerobic metabolism, resulting in metabolic and respiratory acidosis. To date two different laboratories have performed experiments to measure MHS and normal muscle sarcoplasmic Ca^{2+} concentration under a variety of conditions.

Lopez and colleagues have used Ca^{2+} microelectrodes to measure the effect of halothane and/or dantrolene on myoplasmic Ca^{2+} concentration of both human and pig MHS and control muscles. In a number of studies, this group reported resting Ca^{2+} values of normal human and pig muscles of approximately 0.11 μM.[28,29] However, the resting Ca^{2+} concentrations in MHS human and MHS pig fibers were in the range 0.33–0.48 μM, approximately 3.5 fold greater than that of normal fibers.[28,29] Administration of halothane to MHS pig muscle fibers increased myoplasmic Ca^{2+} concentration to over 6–8 μM and initiated an MH episode.[30] Dantrolene reversed the MH episode and lowered myoplasmic Ca^{2+} from 8.4 μM to 0.17 μM.[30] Dantrolene decreased the resting Ca^{2+} concentration of both MHS human and MHS pig muscle fibers to levels similar to that of normal fibers when administered either prior to the biopsy,[31] or during the course of the experiment in vivo.[31,32]

An alternative approach for measuring myoplasmic Ca^{2+} concentration in muscle fibers of MHS and normal pigs and humans has been to use the fluorescent Ca^{2+} specific indicator dye Fura-2. In contrast to the results obtained using Ca^{2+} selective microelectrodes, resting Ca^{2+} concentrations of MHS and normal human or pig fibers estimated using Fura-2 were not significantly different (approximately 0.10 μM).[33,34] 1.5% halothane elevated myoplasmic Ca^{2+} in both MHS pig and MHS human fibers (up to 0.6 μM in pigs), and was accompanied by muscle contracture.[33,34] In contrast, up to 4% halothane had no effect on the myoplasmic Ca^{2+} concentration of normal human and pig muscle fibers.[33,34] Contracture was evident in MHS pig fibers at caffeine concentrations as low as 2 mM, which resulted in a myoplasmic Ca^{2+} concentration of 0.29 μM.[33] In contrast, contracture in normal pig fibers required 4 mM caffeine which resulted in a myoplasmic Ca^{2+} concentration of 0.16 μM.

Muscle bundle biopsy specimens are now routinely tested for MH susceptibility by standardized contracture protocols.[35–37] In this test, a segment of biopsied skeletal muscle is exposed to incremental doses of either caffeine or halothane, or a combination of these agents, in a tissue bath at 37°C to determine the contracture threshold of the muscle. Muscle bundles obtained from MHS individuals are more sensitive to the contractures induced by caffeine/halothane, and thus have lower contracture thresholds than do normal muscles. This increased sensitivity logically arises from an increased resting myoplasmic Ca^{2+} in MHS muscles induced by a given concentration of halothane or caffeine.

Thus, there is considerable evidence indicating that Ca^{2+} homeostasis is altered in MHS muscle, and that an inability to regulate myoplasmic Ca^{2+} is responsible for the initiation and propagation of MH episodes. A number of studies have now identified that these changes result from RyR1 mutations and the functional effect of these mutations on RyR1 Ca^{2+} channel activity.

IDENTIFICATION OF RyR1 MUTATIONS IN MH AND CCD

The gene responsible for porcine MH, synonymous with the "halothane gene" (HAL) or the "PSS gene", was well known for its deleterious effects on pork production, in addition to the unusual response of such pigs to halothane anesthesia. Haplotype analysis of linked marker loci facilitated the mapping of the HAL gene to porcine chromosome 6,[38] and physical mapping placed the HAL gene linkage group in the 6p11– q22 region.[39] After establishment of the RyR1 as a candidate gene for porcine MH, it was mapped to the same region of chromosome 6 that contained the HAL linkage group,[40] specifically band 6q12.[41]

Publication of the cDNA sequences of the MHS and normal pig RyR1 (coding for a 5036 amino acid protein)[5] identified the substitution of arginine615 in the normal sequence with a cysteine in the MHS sequence. It has now been clearly shown that pigs that develop MH episodes are homozygous for the Cys615 allele.[42] Heterozgotes display mild clinical abnormalities but do not develop full-blown MH episodes.[43] That all RyR1 sequence polymorphisms are inherited as a single MHS haplotype indicates that the MHS pig RyR1 mutation arose in a single founder animal.[5]

Studies of the porcine model clearly indicated that a mutation in the RyR1 gene could also be responsible for MH in humans, and physical and genetic linkage mapping studies of the RyR1 gene and MH trait were soon initiated. MacKenzie *et al.*[44] demonstrated that the RyR1 gene was located on the proximal long arm of human chromosome 19 (19q13.1) which is syntenic to the RyR1 and the halothane linkage group on porcine chromosome 6.[39,40] MH in some families was shown to be linked to markers in the region of chromosome 19 containing the RyR1 gene,[45] and linkage between the RyR1 gene and MH was demonstrated in several families using restriction fragment length polymorphisms.[46–48] Following the identification of the Arg615 Cys porcine RyR1 mutation,[5] the homologous Arg614Cys human mutation was soon identified in several families.[49,50]

Although it is clear that there is linkage of MH to the RyR1 and chromosome 19q13.1 in up to 50% of MHS families,[51–53] identifying additional RyR1 mutations has been slow. The pace of identification of RyR1 mutations increased rapidly, however, with the analysis of

Table 12.1 RyR1 Mutations in Human MH and CCD

Site of Substitution		Phenotype	Reference
Cys to Arg	35	MH	*McCarthy *et al.*
Arg to Cys	163	MH and CCD	Quane *et al.*[7]
Gly to Arg	248	MH	Gillard *et al.*[50]
Gyl to Arg	341	MH	Quane *et al.*[7]
Ile to Met	403	CCD	Quane *et al.*[7]
Tyr to Ser	522	MH and CCD	Quane *et al.*[7]
Arg to Trp	552	MH	*McCarthy *et al.*
Arg to Cys	552	MH	*McCarthy *et al.*
Arg to Cys	614	MH	Gillard *et al.*[50]
Arg to Leu	614	MH	*McCarthy *et al.*
Arg to Cys	2162	MH	*McCarthy *et al.*
Arg to His	2162	MH	*McCarthy *et al.*
Val to Met	2167	MH	*McCarthy *et al.*
Gly to Arg	2433	MH	Philips *et al.*;[97] Keating *et al.*[98]
Arg to His	2435	MH and CCD	Zhang *et al.*[92]
Arg to Cys	2458	MH	*McCarthy *et al.*
Arg to His	2458	MH	*McCarthy *et al.*

*Personal communication.

CCD pedigrees. The currently available list of RyR1 mutations linked to MH and CCD show that they are clustered about two regions of the RyR1, namely in the amino terminal region where 9 substitutions have been identified, and the central region with 7 closely located mutations (Table 12.1). As yet the function of these domains in RyR1 Ca^{2+} channel activity are not precisely defined. More extensive reviews of the molecular genetic studies leading to the discovery of RyR1 mutations are available.[54]

In spite of the significant number of studies linking specific RyR1 mutations to MHS, this linkage appears incomplete in some families. Thus, it has been shown that altering the thresholds of the in vitro caffeine contracture test enhances the likelihood of detecting linkage of MHS to the RyR1.[55] Another report has demonstrated discordance between in vitro contracture testing and RyR1 haplotypes in which an individual carrying the Arg614Cys mutation tested normal.[56] Much of this disparity could arise from the difficulties inherent in performing the contracture test that has an acknowledged, albeit very low, error rate.[57]

A number of studies have now definitively excluded linkage of the RyR1 or the q13.1 region of chromosome 19 with MH[51,58] which has stimulated the search for causative mutations in other genes. The only such non-RyR1 mutation yet identified has been recently reported by Monnier *et al.*[59] who documented linkage of an Arg to Cys mutation at residue 1096 of the DHPRα_1 subunit with MHS. This is very intriguing because mutations in this protein have been linked to hypokalemic periodic paralysis (HPP).[60] Linkage of MH to a microsatellite marker near the DHPRα_2/δ gene on chromosome 7q has also been reported.[52] An MHS locus on chromosome 17q has been suggested,[61] with the possibility of linkage to the SCN4A that is proximal to the genes encoding the DHPRβ and γ subunits.[62] Finally, in genome-wide scans, MHS has been linked to markers on human chromosome 3[63] and 5[59] to where no candidate genes have been mapped to date.

FUNCTIONAL ALTERATIONS IN MHS CA^{2+} RELEASE AND EXCITATION-CONTRACTION COUPLING

There have been a number of reports characterizing the biochemical and physiological properties of muscle tissue obtained from MHS and normal individuals. To date the most informative studies are derived from porcine tissues, no doubt due to the limited amount of material available from human muscle biopsies, difficulties in performing membrane preparations and biochemical measurements on small amounts of material and the now well-recognized heterogeneity in human MH. The altered Ca^{2+} homeostasis in MHS muscle fibers that is responsible for the increased sensitivity of MHS muscles to halothane- and caffeine-induced contractures, MH episodes, and possibly an increased resting Ca^{2+} concentration, could result from a defect in any of the mechanisms regulating myoplasmic Ca^{2+} concentration. However, a number of studies have clearly demonstrated that there appears to be no abnormalities in any aspect of the contractile machinery of MHS muscle or SR Ca^{2+}-ATPase function.[6]

Ca^{2+} Dependence of MHS SR Ca^{2+} Release

Descriptions of the functional affects of the porcine RyR1 mutation on SR Ca^{2+} release depend on the tissue preparation and the analytic techniques utilized. To date most information has been obtained on the altered Ca^{2+} dependence of several MHS pig RyR1 activities including Ca^{2+} release and ryanodine binding to isolated SR vesicles, Ca^{2+} release from intact muscle fiber preparations, and single channel kinetics of the RyR1 channels.

Abnormalities in the MHS pig SR Ca^{2+} release mechanism were first reported in 1983.[64] The SR Ca^{2+} content at which Ca^{2+} release was initiated by Ca^{2+} or halothane (termed the intralumenal Ca^{2+} threshold for Ca^{2+}-induced Ca^{2+} release) was significantly lower for MHS pig SR than for normal pig SR .[64] However, the lumenal threshold for Ca^{2+}-induced Ca^{2+} release from MHS and normal human SR does not differ, indicating that an abnormality consistently associated with MHS pig SR was not observed in this human sample population.[65]

Many studies of the rate of Ca^{2+} release from isolated MHS and normal pig SR have been performed. Kim *et al.*[66] used a rapid mixing, stopped flow spectrophotometric technique to demonstrate that MHS SR released Ca^{2+} at a significantly greater rate than did normal SR over a range of Ca^{2+} concentrations (0.1–10 μM), in either the presence or absence of 0.1 mM halothane. The amount of Ca^{2+} released following addition of halothane at low ionized Ca^{2+} was also greater for MHS than normal SR. The rate constant and the Ca^{2+} dependence of Ca^{2+} release from MHS and normal pig SR vesicles which were passively loaded with 5 mM $^{45}Ca^{2+}$ to equal contents has also been measured by a rapid mixing chemical quench technique as well as a rapid filtration technique.[67,68] Rate constants for Ca^{2+} release from MHS and normal SR exhibit a bell-shaped dependence on Ca^{2+} that was maximal at 3–6 μM Ca^{2+}.[67,68] The rate constant for Ca^{2+}-induced Ca^{2+} release from MHS SR was two to three fold greater than for normal SR at any Ca^{2+} concentration over the entire range of activating Ca^{2+}.

The landmark study of Endo *et al.*[69] demonstrated that, in the presence of varying stimulatory Ca^{2+} concentrations, the rate of SR Ca^{2+} release in single saponin-permeabilized fibers prepared from an MHS patient was significantly greater than that in fibers prepared from a normal patient. The half-maximal Ca^{2+} concentration that stimulated SR Ca^{2+}-induced Ca^{2+} release was approximately 3.0 μM and 5.4 μM for MHS and normal fibers respectively. This work has now been extended to include a larger number of patients which confirmed an association between elevated Ca^{2+} release rate and anesthetic complications. Similar studies have been performed with MHS and normal porcine fibers, where the rate of Ca^{2+} release from the MHS fibers was approximately three times greater than from normal fibers at all stimulatory Ca^{2+} concentrations.[70]

Ca^{2+} Dependence of MHS SR [^{3}H]ryanodine Binding

Like SR Ca^{2+} release, the Ca^{2+} dependence of [^{3}H]ryanodine binding to MHS and normal pig SR in 0.1 M KCl (pH 7.0) was described by a bell-shaped curve (Fig. 12.1) that was maximal at 3–10 μM Ca^{2+}.[71] Although the Ca^{2+} concentration that half maximally stimulated

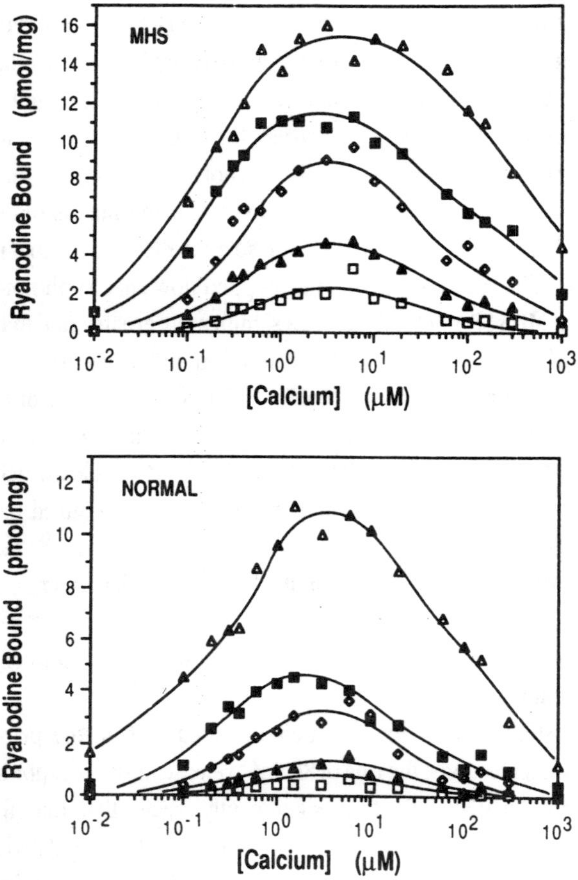

Fig. 12.1 Ca^{2+} dependence of [^{3}H]ryanodine binding to MHS and normal pig SR vesicles. [^{3}H]ryanodine binding to SR vesicles was determined in the presence of 10 nM (•), 30 nM (▲), 100 nM (◇), 300 nM (■) and 1000 nM (Δ) [^{3}H]ryanodine at various Ca^{2+} concentrations. MHS SR bound more [^{3}H]ryanodine at a given Ca^{2+} and [^{3}H]ryanodine concentration, and the Ca^{2+} dependence of the inhibition of ryanodine binding to MHS SR was significantly different from that of normal SR. Reproduced with permission of Mickelson *et al.*[71] with copyright permission.

[^{3}H]ryanodine binding to both MHS and normal SR did not differ, the Ca^{2+} concentration that half-maximally inhibited [^{3}H]ryanodine binding to MHS SR was significantly greater than that for normal SR. Furthermore, in the presence of maximally activating Ca^{2+} concentrations (6 μM) (Fig. 12.2), the apparent K_d of the MHS SR for [^{3}H]ryanodine was

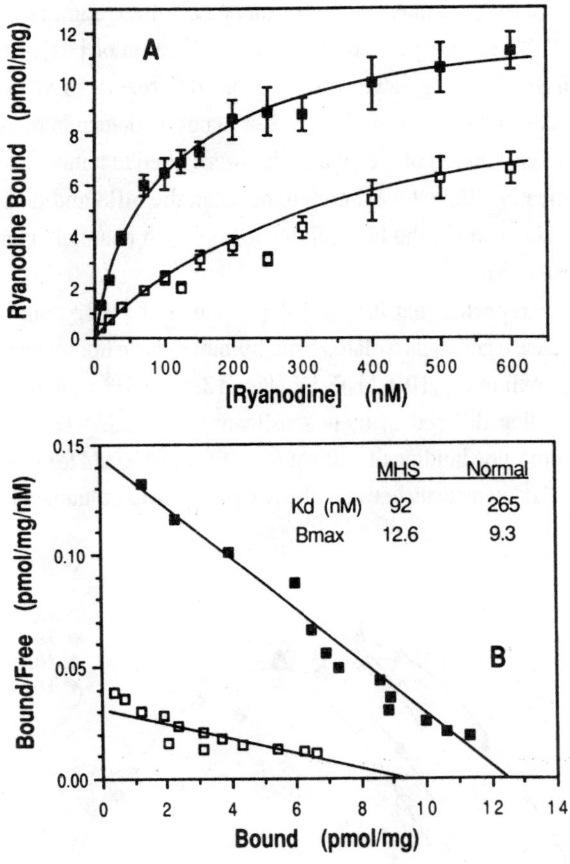

Fig. 12.2 [^{3}H]ryanodine binding to MHS and normal pig SR vesicles. Analysis of the binding data (A) and Scatchard plot (B) indicated that the B_{max} for [^{3}H]ryanodine binding to MHS and normal SR vesicles were not significantly different whereas the K_d values were significantly different. Reproduced with permission of Mickelson *et al.*[71] with copyright permission.

significantly lower than that of normal SR (92 vs 265 nM).[71,72] The affinity of the MHS SR RyR for [^{3}H]ryanodine in the presence of μM Ca^{2+} plus 10 mM ATP (K_d= 6.0 nM) was almost identical to that of normal SR (K_d= 8.5 nM), indicating little difference between MHS and normal RyRs when the RyR is maximally activated and has a high affinity for ryanodine.[73]

Hawkes *et al.*[74] were also able to identify various conditions in which the K_d for [^{3}H]ryanodine binding to the MHS SR was significantly lower than that of normal RyR.

However, in contrast to the studies of Mickelson *et al.*[71] these authors did not identify any difference in the Ca^{2+} sensitivity for stimulation or inhibition of [^{3}H]ryanodine binding to MHS and normal SR. This may have resulted from differences in salt concentrations or assay conditions used in these two studies.[67,75,76] Accurate determination of the kinetics of association and disassociation of [^{3}H]ryanodine were fitted to a model by these authors in which the difference in affinity for ryanodine between the MHS and normal RyR could be attributed to the rate at which the high affinity form (open channel) converted to the low affinity form (closed channel).

Valdivia *et al.*[77] reported that the Ca^{2+} dependence of [^{3}H]ryanodine binding to the solubilized Ca^{2+} release channels isolated from human muscle homogenates was maximally activated by approximately 10 μM Ca^{2+} (Fig. 12.3). MHS and normal human RyR [^{3}H]ryanodine binding differed in their sensitivity to activating Ca^{2+}, with a $Ca^{2+}{}_{0.5}$ for stimulating [^{3}H]ryanodine binding of 120 nM for MHS and 250 nM for normal preparations. This difference in Ca^{2+} sensitivity between the two groups was enhanced by 20 mM caffeine,

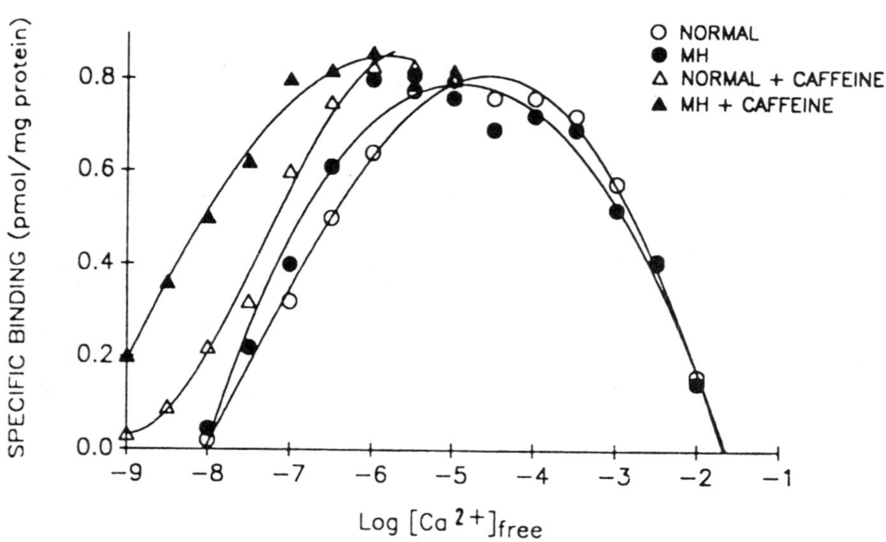

Fig. 12.3 Ca^{2+} dependence of [^{3}H]ryanodine binding to MHS and normal human CHAPS-solubilized RyRs in the presence and absence of caffeine. [^{3}H]ryanodine binding was determined in the absence or presence of 20 mM caffeine and normalized to the maximal [^{3}H]ryanodine bnding. The $C_{1/2}$ for Ca^{2+} stmulation of [^{3}H]ryanodine binding to MHS RyRs was sgnificantly less than that for normal RyRs in either the absence (120 nM versus 250 nM) or presence of 20 mM caffene (7 nM versus 40 nM). Reproduced from Valdivia *et al.*[77] (with copyright permission).

which lowered the Ca^{2+} requirement for activation of [^{3}H]ryanodine binding to 7 nM for MHS and 40 nM for normal channels. Valdivia *et al.*[77] reported that the K_d values for [^{3}H]ryanodine binding to the MHS and normal human Ca^{2+} channels were significantly different at an activating Ca^{2+} of 0.3 μM (4.1 nM versus 7.1 nM for MHS and normal respectively), while there was no difference between MHS and normal K_d values for [^{3}H]ryanodine binding at activating Ca^{2+} concentrations of either 30 μM, and 3 μM.

In most reported studies of human MHS ryanodine receptors, the ryanodine receptor genotype of the affected families has been unknown so it has not been possible to ascribe a specific functional alteration to a specific ryanodine receptor mutation. However, very recently, Richter *et al.*[78] have reported the [^{3}H]ryanodine binding properties of SR membranes isolated from five individuals carrying the Gly2434 → Arg2334 RyR1 mutation. This mutation resulted in an increased affinity of the receptor for ryanodine, and an enhanced sensitivity of the RyR1 to activating concentrations of Ca^{2+} as well as the exogenous and diagnostically used ligands caffeine and 4-chloro-*m*-cresol; the sensitivity to inhibiting concentrations of Ca^{2+} and calmodulin was reduced. In general, the effects of this human mutation were similar to the porcine Arg^{615} → Cys^{615} mutation.

MHS RyR Single Channel Activity

Confirmation that the defects in MHS SR Ca^{2+} release are directly due to the RyR1 mutation is provided by single channel recording and expression of the mutant RyR1 in cell cultures. The Cs^+ conductance of single RyR porcine SR Ca^{2+} release channels in native SR vesicles were recorded following fusion of SR vesicles into planar lipid bilayers.[68] The MHS and normal single native RyR channels were progressively activated in a similar fashion by cis Ca^{2+} with a half maximal stimulation at 1 μM Ca^{2+} and a maximal stimulation at 10 to 100 μM Ca^{2+} (Fig. 12.4). In contrast, at Ca^{2+} concentrations greater than 100 μM the Ca^{2+} concentration required to decrease the maximal P_o was significantly different for normal ($Ca^{2+}{}_{0.5}$ of 200 μM) and MHS channels ($Ca^{2+}{}_{0.5}$ of 2.5 mM). To confirm that this defect was intrinsic to the RyR, the single channel activities of purified MHS and normal porcine RyRs were examined.[76] At inactivating levels of cis-Ca^{2+} (> 0.1 mM), MHS channel opening again had a greater P_o than did normal channels and a higher cis $Ca^{2+}{}_{0.5}$ for inhibition of channel opening than did normal channels (235 vs. 79 μM for MHS and normal, respectively at pH 7.4).[76] At cis Ca^{2+} concentrations greater than 0.1 mM MHS channels also had longer mean open times (approximately two fold greater open time constants), and shorter mean closed times (approximately four fold lower closed time constants) than did normal channels.

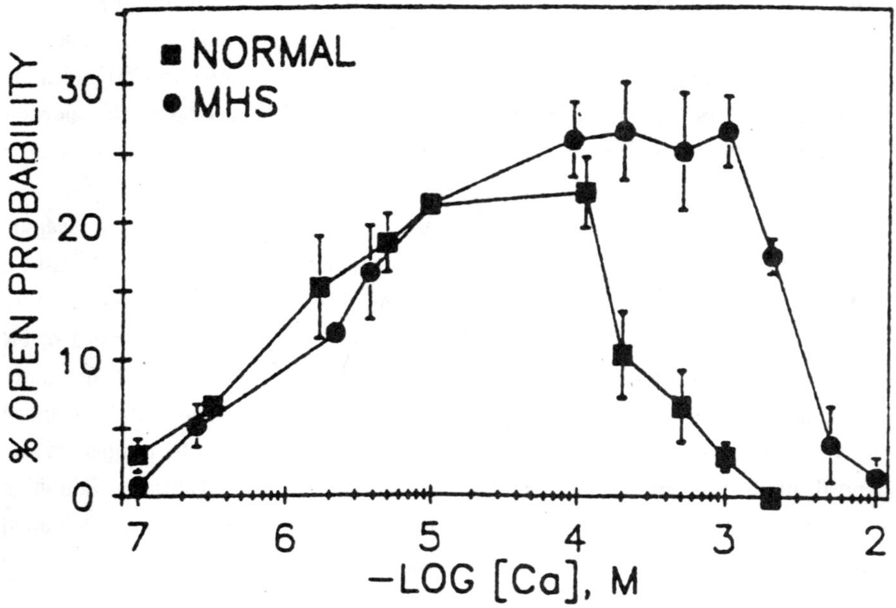

Fig. 12.4 Ca^{2+} dependence of MHS and normal pig Ca^{2+} release channel activities. Channel open probability (P_o) (normalized to the maximum channel P_o) is reported at various cis Ca^{2+} concentrations at pH 7.2 with 250 mM CsCl cis/50 mM CsCl trans at a holding potential of +20 mV. MHS and normal P_o were significantly different at Ca^{2+} concentrations greater than 100 μM (pCa 4). Reproduced from Fill *et al.*[77] (with copyright permission).

Thus, both ryanodine binding and single channel studies indicate that a defect in a low-affinity (inhibitory) Ca^{2+} binding site is responsible for the altered gating of MHS RyR channels.

In contrast to the single channel studies at pH 7.4, the P_o of MHS channels at pH 6.8 was significantly greater than that of normal channels at maximally activating Ca^{2+} concentrations (20% vs 5%, respectively);[79] at inactivating concentrations of cis Ca^{2+} (>0.01 mM Ca^{2+}) cis $Ca^{2+}_{0.5}$ for inhibition of channel opening for MHS channels was again significantly greater than that of normal channels (45 mM vs 15 mM respectively) indicating the pH independence of this RyR defect. A study by Nelson[80] of the Cs^+ conductance of MHS and normal native SR indicated that the mean open life time of the MHS channels was approximately twice that of normal channels. In addition, the open time constant of MHS channels was approximately twice that of normal channels, accounting for the difference in mean open times of these two preparations.

In studies of the single channel conductances of MHS and normal human RyR channels, Cs^+ conductances were shown to be very similar, and at a maximally activating Ca^{2+} concentration of 10 μM, the open and closed channel time constants and P_o, of MHS and normal channels were not significantly different.[81] The Ca^{2+} dependence of MHS and normal Ca^{2+} channel P_o also followed identical bell-shaped curves. In another study of single channel activity of MHS and normal human RyR channels, Nelson[82] arbitrarily subdivided the channel data recorded from MHS subjects at 6 μM Ca^{2+} into two groups of channels; those that increased in P_o in response to halothane (7/13), and those that did not respond to halothane (6/13); none of the normal channels were affected by halothane. The P_o of the normal channels and the halothane insensitive MHS channels were both significantly less (50–75%) than the P_o of the halothane-sensitive MHS channels. The increased (45%) current amplitude of the MHS halothane-sensitive channels reported by this author contrasts with reports indicating no significant difference in the single channel conductance of MHS and normal RyRs.[81]

MHS RyR1 EXPRESSION STUDIES

A number of studies have now been conducted using the MHS and normal RyR1 cDNAs expressed in cell cultures. When the normal (Arg615) and MHS (Cys615) rabbit RyR1 cDNAs were transfected into a mouse myoblast cell line and intracellular Ca^{2+} determined with the Ca^{2+} indicator indo-1, basal ionized Ca^{2+} concentrations of the Cys615 and Arg615 cell line were not significantly different.[83] Intracellular Ca^{2+} increased in the Cys615 transfected cells at lower caffeine concentrations ($C_{0.5}$ = 3.3 mM) than did the Arg615 transfected cells ($C_{0.5}$ = 7.2 mM). Furthermore, halothane increased Ca^{2+} in the Cys_{615} transfected cells but not in either the Arg_{615} transfected or untransfected cells.

The Arg615 Cys and Cys615 rabbit RyR1 cDNAs were also expressed in cos-7 cells and intracellular Ca^{2+} measurements were determined with the Ca^{2+} indicator Fura-2. Both Arg or Cys615 RyRs transfected cells appeared to have a similar resting Ca^{2+} concentration of approximately 80 nM. However, Cys615 transfected cells had a significantly lower 4-chloro-*m*-cresol threshold concentration than did the Arg615 transfected cells (56 vs 166 μM 4-chloro-*m*-cresol, respectively).

These studies indicate that cell lines expressing an Arg615Cys mutant RyR1 have a greater sensitivity to RyR-activating ligands than does the normal (Arg615) RyR1. This could be due to an increased Ca^{2+} channel activity in the presence of these ligands or an elevated resting cytosolic Ca^{2+}.

SURFACE MEMBRANE VOLTAGE DEPENDENCE OF MHS RYR1 ACTIVITY

Abnormalities associated with the transverse tubule-mediated SR Ca^{2+} release process have now been described using SR/t-tubule triad preparations.[84] Ca^{2+} release in these preparations can be initiated by rapidly changing the ionic composition of the medium from one high in K^+ to one high in Na^+, which depolarizes the transverse tubules. At a given level of t-tubule depolarization, the rate constant for Ca^{2+} release from MHS triads was always greater than that from normal triads.[84] When Ca^{2+} release was stimulated by polylysine (which appears to act directly on the Ca^{2+} release channel), the rate constant for Ca^{2+} release from MHS triads was also greater than for normal triads. Thus, the MHS SR Ca^{2+} release activity in isolated triads was greater than that in normal triads when the RyR is stimulated either directly by polylysine, or indirectly via t-tubule depolarization.

Single mechanically peeled fibers, in which the transverse-tubule SR coupling appears intact, have been utilized to examine the surface membrane control of Ca^{2+} release in MHS pig muscle.[85] SR Ca^{2+} release is achieved by changing the Cl^- composition of the bath from the resting value of 4 mM to 66 mM to depolarize the T-tubules.[86] Depolarization resulted in tension transients referable to SR Ca^{2+} release that consisted of a primary component triggered by the transverse tubule depolarization that was not inhibitable by procaine, and a secondary procaine-inhibitable Ca^{2+}-induced Ca^{2+} release component.[85] The transverse tubule mediated component of Ca^{2+} release was not different for MHS and normal fibers. However, the secondary component attributable to Ca^{2+}-induced Ca^{2+} release was significantly greater for MHS fibers. This caused the typical tension transient due to Cl^--induced depolarization of MHS fibers to be greater than that of normal fibers. This demonstrates that the MHS SR Ca^{2+} release abnormality can be expressed following t-tubule (and thus likely DHPR)-mediated initiation of Ca^{2+} release.

Gallant *et al.*[87] described the relationship between surface membrane potential changes generated by exposure to varying K^+ concentrations and the induced, transient contracture-force of MHS and normal pig intact muscle bundles (Fig. 12.5). MHS and normal muscles respond differently to raised K^+, in that MHS muscle bundles generated force at lower extracellular K^+ concentrations (i.e. less depolarization) than did normal muscles. Resting

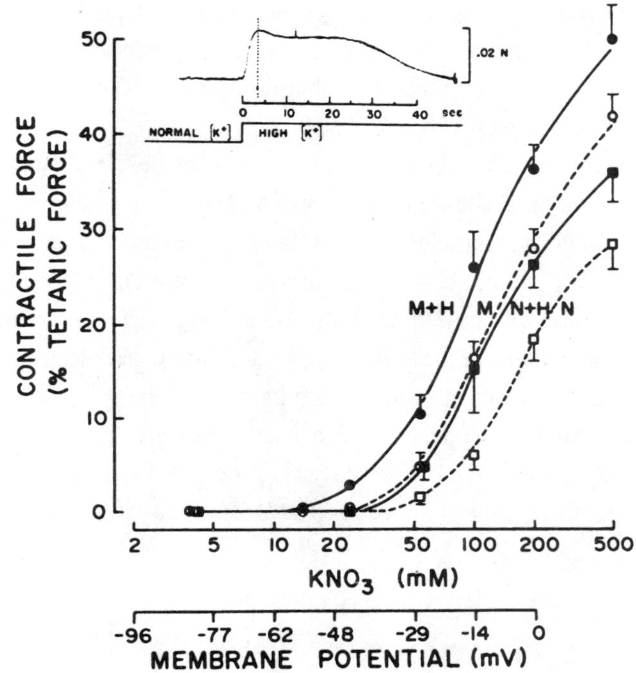

Fig. 12.5 Relationship between contracture force, external K^+ concentration and membrane potential of MHS and normal pig intact muscle fiber bundles. Intact muscle fiber bundles from the forelimb extensor muscle were exposed to various K^+ concentrations, in the presence or absence of halothane (H) and peak contracture force as a percentage of the maximum tetanus was measured. Inset shows record of contracture force produced in response to elevated K^+. MHS (M) muscles developed contractures at lower K^+ concentrations than normal (N) muscles in both the presence or absence of halothane. Reproduced from Gallant *et al.* with copyright permission.

hyperpolarization (preincubation in 2 mM K^+) reduced subsequent MHS muscle K^+-contracture force towards that of normal muscles, while resting depolarization (preincubation in 8 mM K^+) increased both MHS and normal cell contracture force.[88]

MHS MUSCLE CONTRACTILE ACTIVITY

Intact muscle bundles from MHS and normal pigs show no difference in their maximal tetanic force production (per cm^2 fiber cross-sectional area), however, MHS pig muscle bundles consistently demonstrate a significantly increased twitch force, and thus a greater twitch-to-tetanus ratio.[89] Most other measured mechanical properties of MHS and normal muscle twitches and tetani were similar, including the latent period from stimuli to onset of force production, and the half-relaxation times.

The contractile properties of intact MHS and normal pig muscles have been measured in situ during anesthesia.[90] The twitch tension of MHS muscles was approximately 30% greater than that of normal muscles, with a 15% longer contraction time. MHS muscles also exhibited a slightly greater (17%) mean half relaxation time than did normal muscles.[90] There was no difference between MHS and normal muscles in the electromechanical latency, tension latency, contraction time, half-relaxation time, or time to peak tension development, although a significantly greater peak twitch tension (approximately 79% greater than normal), and rate of peak tension development (approximately 40% greater than normal) for the MHS muscles was reported. The increased twitch force in situ in these studies correlates with the increased twitch force observed in isolated intact muscle bundle preparations in vitro.[88]

MUSCLE CA^{2+} REGULATION IN CCD

That both CCD patients as well as MH patients experience hyperthermic episodes during halothane anesthesia and have positive muscle biopsy contracture tests, suggests that CCD patients also have a defect in myoplasmic Ca^{2+} regulation. However, in CCD this can result in the disorganization of the contractile proteins, a proliferation of the SR and t-tubules, and the destruction of mitochondria in the central core.[91] Zhang *et al.*[92] have suggested that these alterations occur after the functional muscle fiber is formed and are the result of physiological adaptation to functional alterations in the RyR1 that have resulted in an elevation of Ca^{2+} within the muscle fiber. If myoplasmic Ca^{2+} is sufficiently elevated, this will now be actively accumulated by mitochondria that it turn may be destroyed in an attempt to protect muscle fiber from Ca^{2+}-induced necrosis.[93] Destruction of mitochondria would lead to a decreased rate of ATP synthesis leading to the muscle weakness and atrophy characteristic of CCD. Alternatively, as suggested by MacLennan and Phillips,[91] the high levels of Ca^{2+} in the central core region in CCD could cause localized contractures that could rupture t-tubule longitudinal SR connections leading to streaming and disruption of the characteristic skeletal muscle structure in the center of the myofibril. Whether Ca^{2+} is elevated in CCD muscles to levels that could activate the mitochondrial Ca^{2+} uptake mechanism remains to be determined. At present these explanations are only hypotheses. Studies using mutant RyR1 isoforms expressed in different mammalian cell types should provide important new insights into this issue.

FUNCTIONAL AFFECTS OF THE PORCINE MHS RyR1 MUTATION IN HETEROZYGOTES

The story of the expression of the RyR1 defect in pigs heterozygous for the RyR1 mutation provides many insights into the MH disorder. Pigs heterozygous for the Arg615Cys mutation do not develop MH episodes in response to halothane challenge although there are definite effects on O_2 consumption, arterial lactate levels, venous partial pressure of CO_2, and arterial and venous K^+ concentration which distinguished them form normals.[94,95] There is also muscle rigidity of heterozygote pigs following a halothane or halothane plus succinylcholine challenge. Intact muscle bundles from MHS pigs clearly showed that the caffeine sensitivity of contracture development in heterozygote muscles was intermediate to that of the two types of homozygotes.[96] SR isolated from pigs heterozygous for the MHS RyR1 allele demonstrated intermediate values for the rate of Ca^{2+} release and the affinity for [^{3}H]ryanodine.[43,74] Thus, heterozygotes are clearly expressing a RyR1 defect that is causing alterations at the cellular and whole animal levels.

That the RyR exists as a homotetramer of 565 kD subunits means that pigs heterozygous for the MHS RyR allele could have RyR populations consisting of both MHS and normal homotetramers, or heterotetrameric RyRs with properties distinct from those of the MHS or normal RyRs. To distinguish between these possibilities, single channel properties of purified MHS, normal, and heterozygous RyR preparations were examined in recording conditions chosen to maximize the differences between the normal and MHS RyR activities.[79] Channel P_o were 4.4 ± 0.6%, 21.8 ± 4.2%, and 30.0 ± 3.5% for normal, heterozygous, and MHS channels, respectively, with the range of heterozygous P_o values overlapping those for both the MHS and normal groups. Mean open and closed dwell times of heterozygous channels were also intermediate between those of MHS and normal channels. These data indicated that the heterozygous RyR distribution must include heterotetramers comprised of MHS and normal RyR subunits with distinct gating kinetics, and that more than one MHS subunit per channel tetramer is required to produce an altered channel function that is detectable at the single-channel level. That the heterozygous porcine RyR channels contain heterotetramers composed of mixtures of MHS and normal subunits suggests that RyR channels in MHS humans could have a similar heteromeric composition.

CONCLUSIONS REGARDING THE EFFECT OF THE MHS MUTATION ON RyR1 FUNCTION

It is clear that the MHS porcine RyR1 mutation is directly responsible for an alteration in the SR Ca^{2+} release mechanism, although interpretation as to the precise nature and expression of this defect differs in different laboratories. The increased rates of MHS SR Ca^{2+} release, the increased affinity of the Ca^{2+} channel for ryanodine, and the increased P_o of the single channel activity of the MHS RyR in vitro are all compatible with the hypothesis of an increased rate of MHS SR Ca^{2+} release upon stimulation in vivo. This increased channel activity may be due to a difference in the rate at which the channel converts from the open to closed states, leading to longer open times. Whether the sensitivity of the MHS SR RyR to stimulatory Ca^{2+} is increased is not clear, although most studies indicate that if it is different from normal, this difference is small.

While the single channel studies show a decreased inactivation of the MHS channel by inhibitory Ca^{2+}, the role that this low affinity Ca^{2+} inhibitory site may play in vivo remains to be defined. Inhibition of the RyR by millimolar Ca^{2+} in the restricted junctional space between the T-tubule and SR membranes may contribute to the closure of this channel during E-C coupling which could explain how a defect in this process results in an increased Ca^{2+} efflux from MHS SR in vivo.

That the porcine MHS Arg615Cys mutation does not affect the conductance or ion selectivity properties of the RyR indicates that this mutation is unlikely to be in a region of the channel that either forms or effects the ion conductance pathway. The alteration in the Ca^{2+} threshold for Ca^{2+}-induced Ca^{2+} release from MHS SR has been proposed to reflect a lumenal Ca^{2+} regulatory site defect, however, what role this Ca^{2+} threshold plays in the regulation of Ca^{2+} release has yet to be defined. The cytoplasmic foot has been considered to play a role in the transmission of a signal from the t tubule DHPR to the SR where it effects the opening of the RyR channel. Thus a mutation in this region of the RyR could conceivably affect the regulation of Ca^{2+} channel activity, although how it could alter a putative lumenal Ca^{2+} binding site remains to be defined. The N-terminal region does not appear to contain sites to which known RyR regulatory ligands such as calmodulin, Ca^{2+} and ruthenium red bind so it is puzzling why the Arg615Cys MHS mutation affects the regulation of RyR activity by so many different ligands.

With the now recognized genetic heterogeneity in human MH it is easy to understand how conflicting results have been obtained when searching for defects in RyR1 activity in human MHS sample populations. In hindsight it is probably not valid to group all MHS human subjects in a single group for statistical analysis. That statistically significant differences in RyR activity between the human MHS and normal groups were obtained in many of these studies, given the likely heterogeneity within the MHS group, leads one to question the significance of many of the differences reported. However, the data of Richter *et al.*[78] does indicate that in those MHS human families carrying RyR1 mutations, the functional alterations in human RyR1 function are similar to those reported for the pig Arg615 Cys 615 mutation. Calsequestrin, FKBP-12, and triadin are now no doubt being examined as possible candidate genes for both MH and CCD in families where there is no linkage of these syndromes to the RyR1, as these proteins have been proposed to play a role in regulating SR Ca^{2+} release.

The major result of the [^{3}H]ryanodine binding studies is that MHS human RyRs may have an altered Ca^{2+} dependence of [^{3}H]ryanodine binding, and an increased affinity for [^{3}H]ryanodine. However, there does not appear to be any difference in the Ca^{2+} dependence of activation of MHS and normal human single RyR channels. There is also some controversy as to whether the MHS and normal human channels differ in single channel conductance or P_o. It seems premature to conclude that a population of human Ca^{2+} release channels has an altered interaction with halothane, although if this conclusion were confirmed in future studies it would be a major advance in our understanding of MH. The problem with the single channel studies to date is that there are too few channels on which to base a data structure which subclassifies MHS human channels into two groups, given the large channel to channel variabilities that are encountered with this technique.

Future studies examining the possibility of altered RyR function in MHS humans would benefit greatly by the use of muscle samples from patients known to be linked to chromosome 19q and preferably with known RyR1 mutations.[78] Ideally, such data could be as informative as that for the MHS porcine RyR. However, such studies will be challenging in that the ability to statistically distinguish the multiple different channel types obtained from the assembly of MHS and normal subunits in heterozygous humans is in question.[79] The other possible route for examining the effects of human RyR1 mutations on the function of this protein is through site specific mutagenesis and expression studies, where the sensitivity of intracellular Ca^{2+} regulation to various pharmacological agents can be determined. Although this would require much time and effort to examine even a few mutations, such information would be valuable in defining structure-function relationships for various domains within the Ca^{2+} release channel protein.

ACKNOWLEDGEMENTS

We thank the many collaborators and laboratory personnel who assisted us in the studies in our laboratory that have been supported over a period of several years by the National

Institutes of Health, the Muscular Dystrophy Association, and the American Heart Association Minnesota Affiliate.

REFERENCES

1. George AL, Crackover MA, Abdalla JA *et al.* Molecular basis of Thomsen's disease (autosomal dominant myotonia congenita). *Nature Genetics* 1993;**3**:305–310.
2. Heine R, George AL, Pika U *et al.* Proof of a non-functional muscle chloride channel in recessive myotonia congenita (Becker) by detection of a 4 base pair deletion. *Hum Mol Gen* 1994;**3**: 1123–1128.
3. Hoffman EP, Lehmann-Horn F, Rüdel R. Overexcited or Inactive: Ion channels in muscle disease. *Cell* 1995;**80**:681–686.
4. Rojas CV, Wang J, Schwartz LS *et al.* A Met-to-Val mutation in the skeletal muscle Na^+ channel alpha-subunit in hyperkalemic periodic paralysis. *Nature* 1991;**354**:387–389.
5. Fujii J, Otsu K, Zorzato F, *et al.* Identification of a mutation in porcine ryanodine receptor associated with malignant hyperthermia. *Science* 1991;**253**:448–451.
6. Mickelson JR, Louis CF. Malignant hyperthermia: excitation-contraction coupling, Ca^{2+} release channel, and cell Ca^{2+} regulation defects. *Physiol Rev* 1996;**76**:537–591.
7. Quane KA, Healy JMS, Keating KE, *et al.* Mutations in the ryanodine receptor gene in central core disease and malignant hyperthermia. *Nature Genetics* 1993;**5**:51–55.
8. Denborough MA, Lovell RRH. Anesthetic deaths in a family. *Lancet* 1960;**2**:45.
9. Britt BA, Kalow W. Malignant hyperthermia: a statistical review. *Can Anesth Soc* J1970;**17**: 293–315.
10. Ording H. Incidence of malignant hyperthermia in Denmark. *Anesthesia Analgesia* 1985;**64**: 700–704.
11. Hall LW, Woolf M, Bradley JW, Jolly DW. Unusual reaction to suxamethonium chloride. *B M J* 1966;**2**:1305.
12. O'Brien PJ. *Canine malignant hyperthermia/canine stress syndrome.* In: Ohnishi ST, Ohnishi T, eds. Malignant hyperthermia, a genetic membrane disease. Boca Raton: CRC Press, **1993**: 105–116.
13. Nelson TE. Malignant hyperthermia in dogs. *J Am Vet Med Assoc* 1991;**198**:989–994.
14. Webb AJ, Simpson SP, Southwood OI. Porcine stress syndrome: recent advances. Royal Agr. Soc. England Animal Health and Productivity Symposium, 1985:143–464.
15. Gronert GA. *Malignant hyperthermia.* In: Engel AG, Armstrong CF, eds. Myology. New York: McGraw Hill, **1994**:1661–1668.
16. Gronert GA, Ahern CP, Milde JH. Treatment of porcine malignant hyperthermia: lactate gradient from muscle to blood. *Can Anesth Soc J* 1986;**33**: 729–736.
17. Britt BA. Dantrolene - an update. In: Britt BA, ed. Malignant hyperthermia. Boston: Martinus Nijhoff, 1987:325–368.
18. Harrison CG. Porcine malignant hyperthermia. *British International Anesthesia Clinics* 1979;**17**:25–61.

19. Nelson TE, Sweo T. Ca^{2+} uptake and Ca^{2+} release by skeletal muscle sarcoplasmic reticulum: differing sensitivity to inhalational anesthetics. *Anesthesiology* 1988;**69**:571–577.
20. Louis CF, Zualkernan K, Roghair T, et al. The effects of volatile anesthetics on calcium regulation by malignant hyperthermia-susceptible sarcoplasmic reticulum. *Anesthesiology* 1992:114–125.
21. Shy GM, Magee KR. A new congenital nonprogressive myopathy. *Brain* 1956;**79**:610–619.
22. Shuaib A, Paasuke RT, Brownell KW. Central core disease. Clinical features in 13 patients. *Medicine* 1987;**66**:389–396.
23. Mulley JC, Kozman HM, Phillips HA, *et al.* Refined genetic localization for central core disease. *American Journal of Human Genetics* 1993;**52**:398–405.
24. Kausch K, Lehmann-Horn F, Janka M, *et al.* Evidence for the linkage of the central core disease locus to the proximal long arm of human chromosome 19. *Genomics* 1991;**14**:765–769.
25. Haan EA, Freemantle CJ, McClure JA, *et al.* Assignement of the gene for central core disease to chromosome 19. *Human Genetics* 1990;**86**:187–190.
26. Schwemmle S, Wolff K, Palmucci LM, *et al.* Multipoint mapping of the central core disease locus. *Genomics* 1993;**17**:205–207.
27. Gronert GA. Malignant hyperthermia. *Anesthesiology* 1980;**53**:395–423.
28. Lopez JR, Alamo LA, Jones DE, *et al.* $[Ca^{2+}]_i$ in muscles of malignant hyperthermia susceptible pigs determined in vivo with Ca^{2+} selective microelectrodes. *Muscle and nerve* 1986;**9**:85–86.
29. Lopez JR, Alamo L, Caputo C, *et al.* Intracellular ionized calcium concentration in muscles from humans with malignant hyperthermia. *Muscle Nerve* 1985;**8**:355–358.
30. Lopez JR, Allen PD, Alamo L, *et al.* Myoplasmic free $[Ca^{2+}]$ during a malignant hyperthermia episode in swine. *Muscle and Nerve* 1988;**11**:82–88.
31. Lopez JR, Medina P, Alamo L. Dantrolene sodium is able to reduce the resting ionic $[Ca^{2+}]_i$ in muscle from humans with malignant hyperthermia. *Muscle and Nerve* 1987;**10**:77–79.
32. Lopez JR, Gerardi A, Lopez MJ, *et al.* Effects of dantrolene on myoplasmic free $[Ca^{2+}]$ measured in vivo in patients susceptible to malignant hyperthermia. *Anesthesiology* 1992;**76**:711–719.
33. Iaizzo PA, Klein W, Lehmann-Horn DF. Fura-2 detected myoplasmic calcium and its correlation with contracture force in skeletal muscle from normal and malignant hyperthermia susceptible pigs. *Pflugers Archives of Physiology* 1988;**411**:648–653.
34. Iaizzo PA, Seewald M, Oakes SG, *et al.* The use of Fura-2 to estimate myoplasmic $[Ca^{2+}]$ in human skeletal muscle. *Cell Calcium* 1989;**10**:151–158.
35. Ellis FR, Harriman DGF. A new screening test for susceptibility to malignant hyperpyrexia. *British Journal of Anesthesiology* 1973;**45**:638.
36. Group TEMH. A protocol for the investigation of malignant hyperpyrexia. *British Journal of Anesthesiology* 1984;**56**:1267–1269.
37. Larach MG. Standardization of the caffeine halothane muscle contracture test. *Anesthesia Analgesia* 1989;**69**:511–515.
38. Archibald AL, Imlah P. The halothane sensitivity locus and its linkage relationships. *Animal Blood Group Biochemistry* 1985;**16**:253–263.
39. Davies W, Harbitz I, Fries R, *et al.* Porcine malignant hyperthermia carrier detection and chromosomal assignment using a linked probe. *Animal Genetics* 1988;**19**:203–212.

40. Harbitz I, Chowdary B, Thomsen PD, *et al.* Assignment of the porcine calcium release channel gene, a candidate for the malignant hyperthermia locus, to the 6p11q21 segment of chromosome 6. *Genomics* 1990:243–248.
41. Chowdary BP, Thomsen PD, Harbitz I, *et al.* Precise localization of the genes for glucose phosphate isomerase (GPI), calcium release channel (CRC), hormone-sensitive lipase (LIPE) and growth hormone (GH) in pigs using nonradioactive in situ hybridization. *Cytogenetics and Cell Genetics* 1994;**67**:211–214.
42. Otsu K, Khanna VK, Archibald AL, *et al.* Cosegregation of porcine malignant hyperthermia and a probable causal mutation in the skeletal muscle ryanodine receptor gene in backcross families. *Genomics* 1991;**11**:744–750.
43. Mickelson JR, Gallant EM, Rempel WE, *et al.* Halothane-sensitivity gene and sarcoplasmic reticulum function in malignant hyperthermia. *American Journal of Physiology* 1989;**257**:C787–C794.
44. MacKenzie AE, Korneluk RG, Zorzato F, *et al.* The human ryanodine receptor gene: its mapping to 19q13.1, placement in a chromosome 19 linkage group, and exclusion as the gene causing myotonic dystrophy. *American Journal of Human Genetics* 1990;**46**:1082–1089.
45. Bender K, Senff H, Wienker TF, *et al.* A linkage study of malignant hyperthermia (MH). *Clinical Genetics* 1990;**37**:221–225.
46. McCarthy TV, Healy JMS, Heffron JJA, *et al.* Localization of the malignant hyperthermia susceptibility locus to human chromosome 19q12-13.2. *Nature (London)* 1990;**343**: 5662–564.
47. MacLennan DH, Duff C, Zorzato F, *et al.* Ryanodine receptor gene is a candidate for predisposition to malignant hyperthermia. *Nature (London)* 1990;**343**:559–561.
48. Healy JMS, Heffron JJA, Lehane M, *et al.* Diagnosis of susceptibility to malignant hyperthermia with flanking DNA markers. *British Medical Journal* 1991;**303**:1225–1228.
49. Hogan K, Couch F, Powers PA, *et al.* A Cysteine-for-Arginine substitution (R614C) in the human skeletal muscle calcium release channel cosegregates with malignant hyperthermia. *Anesthesia Analgesia* 1992;**75**:441–448.
50. Gillard EF, Otsu K, Fujii J, *et al.* A substitution of cysteine for arginine 614 in the ryanodine receptor is potentially causative of malignant hyperthermia. *Genomics* 1991;**11**:751–755.
51. Levitt RC, Nouri N, Jedlicka AE, *et al.* Evidence for genetic heterogeneity in malignant hyperthermia susceptibility. *Genomics* 1991;*11*:543–547.
52. Iles DE, Segers B, Heytens L, *et al.* High-Resolution physical mapping of four microsatellite repeat markers near the RyR1 locus on chromosome 19q13.1 and apparent exclusion of the MHS locus from this region in two malignant hyperthermia susceptible families. *Genomics* 1992;**14**:749–754.
53. Ball SP, Dorkins HR, Ellis FR, *et al.* Genetic linkage analysis of chromosome 19 markers in malignant hyperthermia. *British Journal of Anesthesiology* 1993;**70**:70–75.
54. MacLennan DH, Phillips MS. The role of the skelal muscle ryanodine receptor (ryr1) gene in malignant hyperthermia and central core disease. *Society of General Physiologists Series* 1995;**50**:89–100.

55. MacKenzie AE, Allen G, Lahey D, *et al.* A comparison of the caffeine halothane muscle contracture test with the molecular genetic diagnosis of malignant hyperthermia. *Anesthesiology* 1991;**73**: 4–8.
56. Deufel T, Sudbrak R, Feist T, *et al.* Discordance in a malignant hyperthermia pedigree between in vitro contracture test phenotypes and haplotypes for the MHS1 region on chromosome 19q12-13.2 comprising the C1840T transition in the RyR1 gene. *American Journal of Human Genetics* 1995;**56**:1334—1342.
57. MacLennan D. Discordance between phenotype and genotype in malignant hyperthermia. *Current Opinions in Neurology* 1995;**8**:397–401.
58. Deufel T, Golla A, Iles D, *et al.* Evidence for genetic heterogeneity of malignant hyperthermia susceptibility. *American Journal of Human Genetics* 1992;**50**:1151–1161.
59. Monnier N, Procaccio V, Stieglitz P,et al. Malignant hyperthermia susceptibility is associated with a mutation of the alpha1 subunit of the human dihydropyridine-sensitive L-type voltage sependent calcium-channel receptor in skeletal muscle. *American Journal of Human Genetics* 1997;**60**:1316–1325.
60. Jurkat-Rott K, Lehmann-Horn F, Elbaz A, *et al.* A calcium channel mutation causing hypokalemic periodic paralysis. *Human Molecular Genetics* 1994;**3**:1415–1419.
61. Levitt RC, Oilckers A, Meyers S, *et al.* Evidence for the localization of a malignant hyperthermia susceptibility locus (MHS2) to human chromosome 17q. *Genomics* 1992;**14**:562–566.
62. Olckers A, Meyers DA, Meyers S, *et al.* Adult muscle sodium channel -subunit is a gene candidate for malignant hyperthermia susceptibility. *Genomics* 1992;**14**:829–831.
63. Sudbrak R, Procaccio V, Klausnitzer M, *et al.* Mapping of a further malignant hyperthermia susceptibility locus to chromosome 3q13.1. *American Journal of Human Genetics.* 1995;**56**:684–691.
64. Nelson TE. Abnormality in calcium release from skeletal sarcoplasmic reticulum of pigs susceptible to malignant hyperthermia. *Journal of Clinical Investigation* 1983;**72**:862–870.
65. Fletcher JE, Mayberger S, Tripolitis L, *et al.* Fatty acids markedly lower the threshold for halothane-induced calcium release from the terminal cisternae in human and porcine normal and malignant hyperthermia susceptible skeletal muscle. *Life Sciences* 1991;**49**:1651–1657.
66. Kim DH, Sreter FA, Ohnishi ST, *et al.* Kinetic studies of Ca^{2+} release from sarcoplasmic reticulum of normal and malignant hyperthermia susceptible pig muscles. *Biochimica et Bipohysica Acta* 1984;**775**:320–327.
67. Carrier L, Villa M, Dupont Y. Abnormal rapid Ca^{2+} release from sarcoplasmic reticulum of malignant hyperthermia susceptible pigs. *Biochimica et Biophysica Acta* 1991;**1064**: 175–183.
68. Fill MR, Coronado R, Mickelson JR, *et al.* Abnormal ryanodine receptor channels in malignant hyperthermia. *Biophysical Journal* 1990;**50**:471–475.
69. Endo M, Yagi S, Ishizuka T, *et al.* Changes in the Ca-induced Ca release mechanism in the sarcoplasmic reticulum of the muscle from a patient with malignant hyperthermia. *Biomedical Research* 1983;**4**:83–92.

70. Ohta T, Endo M, Nakano T, *et al.* Ca-induced Ce release in malignant hyperthermia-susceptoible pig skeletal muscle. *American Journal of Physiology* 1989;**256**:C358–C367.
71. Mickelson JR, Gallant EM, Litterer LA, *et al.* Abnormal sarcoplasmic reticulum ryanodine receptor in malignant hyperthermia. *Journal of Biological Chemistry* 1988;**263**:9310–9315.
72. Herrmann-Frank A, Richter M, Lehmann-Horn F. 4-Chloro-m-cresol: a specific tool to distinguish between malignant hyperthermia-susceptible and normal muscle. *Biochemical Pharmacology.* 1996;**52**:149–155.
73. Mickelson JR, Litterer LA, Jacobson BA, *et al.* Stimulation and inhibition of [^{3}H]ryanodine binding to sarcoplasmic reticulum from malignant hyperthermia susceptible pigs. *Archives of Biochemistry and Biophysics* 1990;**278**:251–257.
74. Hawkes MJ, Nelson TE, Hamilton SL. [^{3}H]Ryanodine as a probe of changes in the functional state of the Ca^{2+}-release channel in malignant hyperthermia. *Journal of Biological Chemistry* 1992;**267**:6702–6709.
75. Vita GM, Fletcher JE, Tripolitis L, *et al.* Altered [^{3}H]ryanodine binding is not associated with malignant hyperthermia susceptibility in terminal cisternae preparations from swine. *Biochemistry International* 1991;**23**:563–570.
76. Shomer NH, Louis CF, Fill M, *et al.* Reconstitution of abnormalities in the malignant hyperthermia-susceptible pig ryanodine receptor. *American Journal of Physiology* 1993;**264**:C125–C135.
77. Valdivia HH, Hogan K, Coronado R. Altered binding site for Ca^{2+} in the ryanodine receptor of human malignant hyperthermia. *American Journal of Physiology* 1991;**261**:C237–C245.
78. Richter M, Schleithoff L, Deufel T, Lehmann-Horn F, Hermann-Frank A. Functional characterization of a distinct ryanodine recepotor mutation in human malignant hyperthermia-susceptible muscle. *Journal of Biological Chemistry* 1997;**272**:5256–5260.
79. Shomer NH, Mickelson JR, Louis CF. Ca^{2+} release channels of pigs heterozygous for malignant hyperthermia. *Muscle and Nerve* 1995;**18**:1167–1176.
80. Nelson TE, Lin M. Abnormal function of porcine malignant hyperthermia calcium release channel in the absence and presence of halothane. *Cell Physiology and Biochemistry* 1995;**5**:12–22.
81. Fill M, Stefani E, Nelson TE. Abnormal human sarcoplasmic reticulum Ca^{2+} release channels in malignant hyperthermic skeletal muscle. *Biophysical Journal* 1991;**59**:1085–1090.
82. Nelson TE. Halothane effects on human malignant hyperthermia skeletal muscle single calcium-release channels in planar lipid bilayers. *Anesthesiology* 1992;**76**:588–595.
83. Otsu K, Willard HF, Khanna VK, *et al.* Molecular cloning of cDNA encoding the Ca^{2+} release channel (ryanodine receptor) of rabbit cardiac muscle sarcoplasmic reticulum. *Journal of Biological Chemistry* 1990;**265**:13472–13483.
84. El-Hayak R, Yano M, Antoniu B, *et al.* Altered E-C coupling in triads isolated from malignant hyperthermia-susceptible porcine muscle. *American Journal of Physiology* 1995;**268**: C1381–C1386.
85. Donaldson SK, Gallant EM, Huetteman DA. Skeletal muscle excitation-contraction coupling I: transverse tubule control of peeled fiber Ca^{2+}-induced Ca^{2+} release in normal and malignant hyperthermic muscles. *Pflugers Archives of Physiology* 1989;**414**:15–23.
86. Donaldson SK. Peeled mammalian skeletal muscle fibers: possible stimulation of Ca^{2+} release via a TT-SR mechanism. *Journal of General Physiology* 1985;**86**:501–525.

87. Gallant EM, Gronert GA, Taylor SR. Cellular membrane potential and contracture threshold in mammalian skeletal muscle susceptible to malignant hyperthermia. *Neuroscience Letters* 1982;**28**:181–186.
88. Gallant EM, Donaldson SK. Skeletal muscle excitation-contraction coupling II. Plasmalemma voltage control of intact bundle contractile properties in normal and malignant hyperthermic muscles. *Pflugers Archives of Physiology* 1989;**414**:24–30.
89. Gallant EM, Fletcher TF, Goettl VM, *et al.* Porcine malignant hyperthermia: cell injury enhances halothane sensitivity of biopsies. *Muscle and Nerve* 1986;**9**:174–184.
90. Campion DR, Eikenbloom G, Cassens RG. Isometric contractile properties of skeletal muscle from stress susceptible and stress resistant pigs. *Journal of Animal Science* 1974;**39**:68–72.
91. MacLennan DH, Phillips M. The role of the skeletal muscle ryanodine receptor (RyR1) gene in malignant hyperthermia and central core disease. *Society of General Physiologists Series* 1995;**50**:89–100.
92. Zhang Y, Chen HS, Khanna VK, *et al.* A mutation in the human ryanodine receptor gene associated with central core disease. *Nature Genetics* 1993;**5**:46–50.
93. Wrogemann K, Pena SD. Mitochondrial calcium overload: A general mechanism for cell-necrosis in muscle diseases. *The Lancet* 1976;**1**:672–674.
94. Nelson TE, Flewellen EH, Gloyna DF. Spectrum of susceptibility of malignant hyperthermia-diagnostic dilemma. *Anesthesia Analgesia* 1983;**62**:545–552.
95. Gallant EM, Mickelson JR, Roggow BD, *et al.* Halothane-sensitivity gene and muscle contractile properties in malignant hyperthermia. *American Journal of Physiology* 1989;**257**:C781–786.
96. Gallant EM, Lentz LR. Excitation-contraction coupling in pigs heterozygous for malignant hyperthermia. *American Journal of Physiology* 1992;**262**:C422—C426.
97. Phillips MS, Khanna VK, Dellon S, *et al.* The substitution of Arg for Gly2433 in the human skeletal muscle ryanodine receptor is associated with malignant hyperthermia. *Human Molecular Genetics* 1993;**3**:2181–2186.
98. Keating KE, Quane KA, Manning BM, *et al.* Detection of a novel RyR1 mutation in four malignant hyperthermia pedigrees. *Human Molecular Genetics* 1994;**3**:1855–1858.

Chapter 13

Ryanodine Receptor Adaptation

Patricio Vélez, Rafael Mejía-Alvarez and Michael Fill

It is known that during excitation-contraction coupling (E-C coupling) in cardiac myocytes, the intracellular Ca^{2+} release from the sarcoplasmic reticulum (SR) is directly triggered by Ca^{2+} influx through L-type Ca^{2+} channels in the transverse tubular membranes. This process is called Ca^{2+}-induced Ca^{2+} release (CICR). Intuitively, CICR may appear as a self regenerating process ended only by depletion of the SR Ca^{2+} stores. However, there are many reports indicating that CICR is accurately graded and precisely controlled by the trigger Ca^{2+} signal. This implies that a negative feedback control must exist to counter the inherent positive feedback of CICR. The nature of such a negative feedback mechanism is not yet clear.

Significant advancements in our understanding of CICR in heart have occurred over the last decade. For example, the SR Ca^{2+} release channel protein was isolated and identified as the ryanodine receptor (RyR).[1,2] Further, the structure of the RyR channel[3,4] and its morphological features in the cell are better defined.[5] The biophysical profile of RyR has also been characterized at the single-channel level by incorporating the RyR protein into artificial planar lipid bilayers.[2,6,7,8,9] The RYR encoding gene has even been cloned from several tissues. This body of information has provided a detailed functional picture of cardiac E-C coupling, in which RYR plays a central role mediating CICR. Nevertheless, our increased

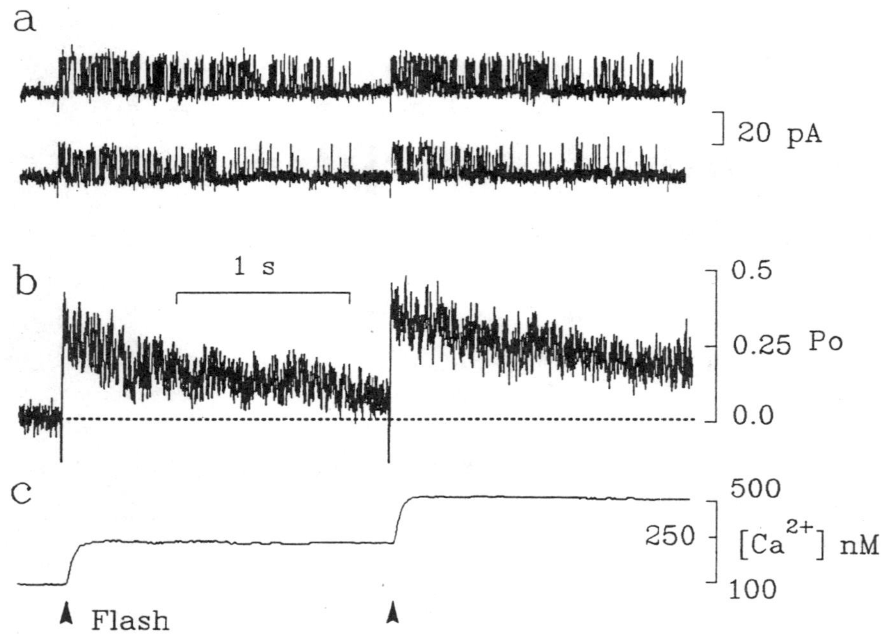

Fig. 13.1 Activation of cardiac RyR channel by two incremental increases in [Ca^{2+}]. Two UV flashes were applied 2 s apart (at arrowheads). (**a**) Single-channel records. Open events are shown as upward deflections. (**b**) Ensemble currents generated from 92 data sweeps collected from a single channel. (**c**) Time course of the change in [Ca^{2+}]. [Reprinted with permission from Györke and Fill, *Science*, 260: 807–809, 1993. Copyright 1993 American Association for the Advancement of Science].

knowledge of RyR structure and function has not yet clearly identified the nature of the mechanisms responsible for the negative feedback which regulate channel operation in the living cell.

In this regard, Fabiato (1985) originally proposed that Ca^{2+}-dependent inactivation may be the mechanism by which the self-regenerating nature of CICR is countered in heart. Since cardiac RyR channels reconstituted in planar lipid bilayers are consistently active in the presence of relatively high free Ca^{2+} concentrations, there is no evidence of steady-state Ca^{2+}-dependent inactivation at the single-channel level. However, the steady-state properties of the RyR channel may not reflect how the channel responds to fast Ca^{2+} transients in the cell. Györke and Fill (1993) provided the first evidence at the single-channel level that a negative feedback mechanism regulates the operation of RyR. They proposed that single cardiac RyR channels adapt, instead of inactivate, in response to a fast, sustained Ca^{2+}

stimulus. This review summarizes a growing body of results from several laboratories which address the RyR adaptation hypothesis and discusses the potential role of RyR adaptation the regulation of CICR in heart.

SINGLE RyR CHANNEL ADAPTATION

Györke and Fill[11] observed that single cardiac RyR channel activity peaks and then spontaneously decays in response to a fast elevation in resting [Ca^{2+}] when applied by photolysis of caged-Ca^{2+} (Fig. 13.1). In the historical context of Fabiato,[10] the spontaneous decay could be interpreted as evidence of Ca^{2+}-dependent inactivation since channel activity spontaneously decreased in the presence of stimulating Ca^{2+} concentrations. Traditionally, channel inactivation has been defined as the transition to an absorbing/non-conducting state that follows an all-or-none behavior at the single-channel level (i.e. individual channels are either inactivated or not). Therefore, the degree of inactivation of a macroscopic current simply reflects the number of inactivated channels in a large population of channels. Within this framework, if single RyR channels do indeed inactivate they will be refractory and thus insensitive to further stimulation. Györke and Fill[11] demonstrated that apparently "inactivated" single RyR channels were not refractory and activated in response to a subsequent incremental Ca^{2+} stimulus, indicating that the non-conducting state was not absorbing. Therefore, the spontaneous decay in single RyR channel activity was not mediated by a conventional inactivation mechanism. To explain their resul the RyR adaptation hypothesis using a variety of experimental and theoretical approaches. Information gleaned from these studies is briefly detailed below.

FLASH PHOTOLYSIS STUDIES

Fast calibrated free [Ca^{2+}] stimuli can be applied to single RyR channels in planar lipid bilayers using laser flash photolysis of caged-Ca^{2+} compounds. This methodology eliminates diffusion delays and allows the free [Ca^{2+}] in the microenvironment of the channel to be changed in less than 100 ms. This is very close to the speed of free [Ca^{2+}] changes that are thought to activate single RyR channels in vivo.[12] Using the flash photolysis methodology, single cardiac and skeletal RyR channels activate with a time constant of approximately 1 ms.[11,13,14,15] In these studies, single RyR channel activity peaked and then spontaneously decayed in response to a fast sustained elevation in the free [Ca^{2+}]. The spontaneous decay was interpreted as RyR adaptation.[11,13,14,15] In cardiac RyR channels, the time constant of adaptation was about 1 s.[11,14] In skeletal RyR channels, adaptation was about 4 times slower.[13]

Valdivia *et al.* reported that the rate of cardiac RyR adaptation was significantly faster (~10 fold) in the presence of physiological levels of Mg^{2+} or after the cardiac RyR channel was phosphorylated by a cAMP-dependent protein kinase.[14]

Györke and Fill showed that Ca^{2+} sensitivity of single cardiac RyR was different immediately after a fast elevation in free [Ca^{2+}] and proposed that shifting the RyR Ca^{2+} sensitivity may be an important factor in explaining the adaptation phenomenon.[11] The shift in RyR Ca^{2+} sensitivity was limited by the channel's steady-state Ca^{2+} dependence. This means that adaptation did not reduce single-channel open probability (P_o) below the value predicted by the channel's steady-state Ca^{2+} dependence. Regardless of mechanism the implication of these results is that RyR adaptation appears to operate over a fairly narrow free [Ca^{2+}] range (~0.1 to ~5 μM).

Valdivia *et al.* showed that single cardiac RyR channel activity (P_o) was proportional to the free [Ca^{2+}] if the rate of Ca^{2+} application is adequately slow (~ 8 s).[14] When Ca^{2+} was applied rapidly (~0.1 ms), P_o peaked at values well above those predicted by the steady-state Ca^{2+} dependence of the channel and then adapted to lower values. These data indicate that single RyR channel activity depends on both the absolute [Ca^{2+}] and on the rate of the free [Ca^{2+}] change. A similar conclusion was reached by Fabiato in skinned cardiac myocytes.[10] The interesting ability of the RyR channel protein to sense "rate of change" suggests that RyR channel acts as a [Ca^{2+}] differentiator. This property may explain the capacity of the RyR channel to repeatedly and transiently respond to successive incremental [Ca^{2+}] stimuli.

Flash photolysis studies have contributed significant insights into Ca^{2+} regulation of RyR channels. This methodology, however, introduces a few undesired experimental variables. First, the flash is a brief (10 ns) but intense light pulse which may disturb the bilayer by itself (e.g. heating, altering lipids, etc.). Flash-dependent bilayer disturbances do not appear to be significant because long trains of high frequency flashes in the absence of caged compounds do not alter RyR channel function. Second, flash photolysis of caged-Ca^{2+} can generate free radicals which may damage the channel or alter its function. However, results from photolysis studies in the presence[11] and absence[14] of free radical scavengers are amazingly similar. Third, the photochemistry of the caged-Ca^{2+} compounds can complicate the time course of the free Ca^{2+} waveform. Laser flash photolysis of DM-nitrophen or NP-EGTA generates a very fast (~100 μs) overshoot at the leading edge of the free Ca^{2+} waveform.[16,17] The impact of this fast overshoot on the interpretation of the flash photolysis results has been a concern.[18] This concern is discussed further in the "RyR Adaptation Challenges" section below.

MECHANICAL SOLUTIONS CHANGE STUDIES

Sitsapesan *et al.* were the first to mechanically apply rapid free [Ca^{2+}] changes to RyR channels in planar bilayers.[19,20] A solenoid valve was used to rapidly move the interface between two solution streams from a theta tube across a planar bilayer on the tip of a glass pipette. Liquid junction potential measurements indicated that solutions exchanged in less than 10 ms. Sitsapesan *et al.* showed that Ca^{2+}-dependent activation and deactivation of RyR are fast, with time constants probably less than the solution exchange time (~10 ms).[19] The P_o of RyR channels immediately following the [Ca^{2+}] step was similar to steady-state values. They also reported that time-dependent "inactivation" occurred at +40 mV but not at -40 mV. Interestingly, they suggested that RyR activation was independent of the rate of the free [Ca^{2+}] stimulus. This latter suggestion is contrary to the classical results of Fabiato[10] and the flash photolysis studies described above. A possible explanation for this apparent discrepancy is the presence of unstirred layers, an inherent drawback of any mechanical solution exchange methodology. The unavoidable existence of an unstirred layer near the surface of the bilayer may imply that the solution exchange rate in the microenvironment of the RyR channel could be significantly underestimated by liquid junction potential measurements.[21]

Laver and Curtis also used a mechanical flow method to apply rapid free [Ca^{2+}] changes to RyR channels in planar lipid bilayers.[22] A pipette positioned near the bilayer was used to "puff" solution at the bilayer in order to exchange solutions in the microenvironment of the channel in ~20 ms. These authors showed that RyR channels activate and deactivate rapidly in response to a rapid increase or decrease in free [Ca^{2+}]. The initial P_o reached after a rapid [Ca^{2+}] increase was higher than at steady-state. The P_o then spontaneously decreased in most RyR channels (70%) even though the free [Ca^{2+}] was constant. As Laver and Curtis pointed out, the physical constraints of rapidly exchanging solutions near a large bilayer surface led to an ~25 ms uncertainty in synchronizing their single-channel records.[21,22] This limitation made it impossible to define fast phenomena ($\tau < 25$ ms, like Ca^{2+}-dependent activation and deactivation) by ensemble averaging. However, Laver and Curtis did report that the time constant of the spontaneous decay ranged from 0.5 to 15 s.[22]

Impressive results by Schiefer *et al.* were generated using a very fast piezo operated liquid-filament switch which rapidly changed solutions near a lipid bilayer formed in the tip of a small (1 μm) patch pipette.[23] The relatively small area of the bilayer minimized complications due to unstirred layers but introduced another experimental complication. The small bilayer area made the probability of incorporating a RyR channel very low ($< 10^{-4}$) compared to the other studies discussed in this review. Thus, the Schiefer *et al.*

data are based on repeated solution changes carried out on a low number of successful RyR channel incorporations.[23] Nevertheless, they showed that rapid free [Ca^{2+}] changes activated and deactivated RyR channels on a millisecond time scale. In addition, this mechanical solution change study is unique because it shows that at physiological [Ca^{2+}]'s (< 10 μM) the channel exhibits an apparent "inactivation" without becoming refractory.

The RyR adaptation phenomenon occurs over a relatively narrow [Ca^{2+}] range (~0.1 to ~5 mM).[11] Schiefer *et al.* (1995) reported that the "inactivation" time constant of RyR channels was about 1 s when the free [Ca^{2+}] was raised to 3 μM.[23] This is consistent with the time constants of the spontaneous decay observed in flash photolysis studies when the free [Ca^{2+}] was raised to 1 mM (i.e. $\tau \sim 1.3$ s).[11,14] Schiefer *et al.* also reported that refractory behavior was not observed when [Ca^{2+}] was stepped to concentrations less than 10 μM.[23] These observations are consistent with the data of Györke and Fill in that single RyR channels that apparently "inactivate" show no signs of being refractory.[11] Evidence for refractory behavior was only apparent at very high [Ca^{2+}]'s (1 mM) and thus are not likely to be related to the RyR adaptation phenomenon.[23]

The results of the flash photolysis[11,14] and mechanical solution change studies[19,22,23] are in fact remarkably consistent. All these studies reported that single RyR channels activated and deactivated on a millisecond time scale in response to fast [Ca^{2+}] changes. Although the extent, rate and probability of observing the decay varied between studies, all the reports agree that RyR channel activity peaked and then spontaneously decayed when the free [Ca^{2+}] was rapidly elevated. The RyR adaptation hypothesis is based on flash photolysis results that showed that the activity of single RyR channels spontaneously decayed in response to a fast sustained submicromolar [Ca^{2+}] stimuli and that the mechanism which mediated the decay did not make the channel refractory.[11] One mechanical solution change study reported that no refractory behavior was observed when RyR channels were challenged with micromolar [Ca^{2+}] stimuli (3 and 10 μM).[23] Thus, the experimental results are quite consistent despite significant differences in interpretation of the data.

RyR ADAPTATION CHALLENGES

The physiological role of RyR adaptation is not clearly established, since it may be too slow to explain beat to beat regulation of cardiac CICR. The time constant for adaptation reported by Györke and Fill was 1.3 s.[11] To regenerate CICR at the single-channel level, feedback regulation would have to occur at least 10 times faster in cells. However, it should be noted that even a relatively slow negative feedback mechanism may introduce enough tendency toward closure to make stochastic attrition in clusters of RyRs an important factor for termination of the Ca^{2+} release process in cells. Interestingly, Valdivia *et al.* showed that RyR adaptation is faster (~10 fold) in the presence of physiological $[Mg^{2+}]$'s or after PKA-dependent phosphorylation.[14] Additionally, Yasui *et al.* have shown that Ca^{2+} release in intact cardiac myocytes is controlled by an adaptation-like mechanism that occurs at physiologically relevant rates.[24] Thus, RyR adaptation may be significantly faster when the channel is operating in its physiological environment. If so, it would be a likely candidate for the negative feedback mechanism responsible of controlling CICR in heart.

A second challenge of RyR adaptation is the discrepancy of results that different laboratories have reported. The controversy probably arises from the variation in Ca^{2+} application rates associated with the different methodologies. Considering that RyR channels are highly sensitive to the speed of the Ca^{2+} stimulus,[10,25] we believe that the experimental results are amazingly consistent and the controversy essentially arises from differences in data interpretation. Thus, the experimental foundation for the adaptation hypothesis is sound.

A third concern about RyR adaptation was that the slow spontaneous decay in channel activity originally observed by Györke and Fill may be due to simple deactivation of RyR channels following the fast Ca^{2+} overshoot generated by flash photolysis of DM-nitrophen (caged-Ca^{2+}).[11] This legitimate concern was based on the possibility that Ca^{2+} bound to the activation site during the overshoot may slowly "fall off" the activation site leading to a spontaneous decay in RyR channel activity (i.e. slow Ca^{2+} deactivation).[18] It is now clear that RyR deactivation after a rapid decrease in free $[Ca^{2+}]$ occurs within few milliseconds.[19,22,23,26] Furthermore, two groups have reported that the Ca^{2+} overshoots are actually much faster than previously thought.[16,17] Thus, the temporal mismatch between the Ca^{2+} overshoot and the rate of RyR adaptation is even larger than originally foreseen. In any event, if adaptation were due to RyR deactivation, then the rate of RyR adaptation would be keenly sensitive to the kinetic features of the Ca^{2+} overshoot. Györke and Fill[11] and Valdivia *et al.*[14] used different caged-Ca^{2+} compounds and consequently applied Ca^{2+} overshoots with dramatically different kinetics[16] to single RyR channels. Since the rate of RyR adaptation observed by these two groups was identical, it appears that RyR adaptation is not highly dependent on the kinetics of the Ca^{2+} overshoot. Thus, the Ca^{2+} overshoot concern has been experimentally examined and appears to be unwarranted.[18]

RyR ADAPTATION MODELS

Although it is clear that additional experimental data is required to definitively model RyR adaptation, several groups have proposed interesting models based on the existing body of information.[27,28,29,30] The hope is that a better understanding of the RyR adaptation phenomenon can be gained by establishing clear theoretical modeling constraints which may help to guide future experimentation.

Stern has examined the thermodynamic limitations of kinetic mechanisms which describe adaptive RyR channel behavior.[31] In his model, Stern convincingly argues that adaptation would require several Ca^{2+} binding sites, if it is to occur in the absence of exogenous sources of free energy.[31] There is evidence that the four RyR monomers that form the channel superstructure may each have at least three Ca^{2+} binding sites.[32] Thus, Ca^{2+} binding may provide sufficient free energy to meet the thermodynamic limitations articulated by Stern.[31] However, there is no direct experimental evidence that more than one Ca^{2+} binding site is involved in the activation and subsequent adaptation of the RyR channel. It is possible that an exogenous source of free energy contributes to the RyR adaptation phenomenon. One potential exogenous source of free energy is the interaction of the RyR protein with closely associated regulatory proteins. In preliminary work, Vélez et al. showed that RyR adaptation is absent (or very slow) when the RyR channel protein is isolated from other SR proteins by a CHAPS solubilization procedure.[15] This could suggest that the action of closely associated regulatory proteins could be important in the RyR adaptation process. Another potentially large source of free energy is the flux of ions down their electrochemical gradient. Several transport proteins (exchangers/carriers) routinely tap this source of energy to drive the "uphill" movement of solutes. Although it would be unusual for an ion channel to be coupled to the electrochemical gradient, this situation has been observed in some chloride channels.[33] If RyR adaptation were coupled to ion permeation, then the RyR adaptation would be quite sensitive to direction and magnitude of current flow. Sitsapesan *et al.* reported that "inactivation" of cardiac RyR channel activity was voltage dependent (i.e. occurred at +40 but not -40 mV).[19] However, definitive demonstration that permeation and RyR gating are indeed coupled would require substantially more experimental information. In any event, defining the thermodynamic requirements of RyR adaptation and then determining how those requirements are met is a logical first step in defining the mechanism which governs RyR adaptation. At this point, our best estimation is that the thermodynamic requirements of RyR adaptation will most likely be met by a combination of several free energy sources (i.e. Ca^{2+} binding sites and interaction with regulatory proteins).

Cheng *et al.* proposed a RyR adaptation model in which each RyR monomer has at least two Ca^{2+} binding sites, Ca^{2+} binding to one type of site (O) opens the channel while Ca^{2+} binding to the other type of site (A) causes channel closure. In this model, the intact RyR channel tetramer has eight Ca^{2+} binding sites (i.e. 4 "O" and 4 "A" sites).[27] The model predicts adaptive channel behavior based on, (1) the probability that Ca^{2+} binding sites are occupied and, (2) the affinities of the two types of Ca^{2+} binding sites. This model assumes that the "O" sites are cooperative (n=2), have lower affinity and faster kinetics than the "A" sites. The finite number of Ca^{2+} binding sites limits (to 3–5) the number of times a single channel can "adapt". Thus, the Cheng *et al.* model predicts that RyR adaptation is governed by multiple cooperative Ca^{2+} binding sites and can occur only a limited number of times.[27] There is evidence that the RyR channel has multiple Ca^{2+} binding sites[32] and that the nonstationary activation of RyR channels may be cooperative.[11] However, it is not yet determined if there is a limit to how many times a single RyR can adapt.

Sachs *et al.* have proposed an alternative model of RyR adaptation based on a relatively simple four-state Markovian scheme in thermal equilibrium.[28] This model based largely on a scheme previously proposed to explain bacterial chemoreception,[34] assumes that Ca^{2+} binding to one site alters the probability that the channel fluctuates between one set of open/closed states (i.e. gating mode) or another. The predictions of this model are that the channel gating behavior will be modal (i.e. modes generated by at least two sets of open/closed states), and that the transitions to one open and one closed state will be Ca^{2+} dependent. Under steady-state conditions, the closed times of spontaneously gating RyR channels do not appear to be Ca^{2+}-dependent.[30,35] However, it should be noted that the closed time kinetic data collected at low steady-state $[Ca^{2+}]$'s were based on a relatively small pools of events. In support of the Sachs *et al.* model,[28] two groups have shown that single cardiac RyR channels gating is indeed modal under steady-state experimental conditions.[30,36] However, it is not yet established if modal gating behavior also occurs under nonstationary conditions (i.e. when the channel is responding to a fast Ca^{2+} stimulus). Another important, yet not readily evident limitation of the Sachs *et al.* model is that it predicts that small changes in P_o would be induced by relatively small Ca^{2+} steps.[28] Györke and Fill showed that relatively small Ca^{2+} stimuli induce rather large changes in P_o.[11] This model limitation, however, could be overcome if the single Ca^{2+} binding site were replaced by multiple binding sites which operate in a cooperative manner.[27,31]

Zahradníková and Zahradník also proposed a RyR adaptation model based on a Markovian scheme composed by two open, four closed states and one Ca^{2+}-dependent transition.[30] Like the Sachs *et al.* model,[28] a fundamental prediction of the model is that RyR channel gating is modal. However, the way Ca^{2+} alters open and closes times is different in the Zahradníková and Zahradník[30] approach. Thus, the predictions of the Zahradníková and Zahradník and Sachs et al. models could be experimentally tested by, (1) clearly defining if single RyR channels do indeed gate in a modal fashion under nonstationary conditions and, (2) determining how many of open/closed states are indeed Ca^{2+}-dependent.[28,30]

Tang and Othmer proposed a relatively simple four-state Markovian model to describe RyR adaptation.[29] These authors, however, only apply their model to predict channel behavior in response to relatively large [Ca^{2+}] stimuli. Unfortunately, it predicts that P_o falls to near zero at steady-state [Ca^{2+}]'s $\leq 10\ \mu M$. If small Ca^{2+} stimuli similar to those applied by Györke and Fill[11] are applied to drive the model, the predictions are: (a) high resting P_o's at very low [Ca^{2+}]'s, and (b) almost no adaptive behavior. Since these predictions are clearly contrary to what has been observed experimentally, the limitations of this model should be considered carefully before it is used to describe the RyR adaptation process.

We believe that the substantial free energy requirements of adaptation are supplied by multiple Ca^{2+} binding sites and interaction with closely associated regulatory proteins.[31] Our "working" hypothesis is that RyR adaptation will eventually be described by a Markovian scheme which predicts modal gating of the RyR channel under stationary and nonstationary conditions. In this scheme, Ca^{2+} binding governs the transition between a low P_o mode and a high P_o mode. At low steady-state [Ca^{2+}]'s, channel gating behavior is dominated by the low P_o mode. A rapid elevation in [Ca^{2+}], promptly shifts the channel into the high P_o mode. Relaxation to a new equilibrium between the high and low P_o modes is the process responsible for the observed adaptation phenomenon. A second rapid elevation in [Ca^{2+}], disrupts the equilibrium again causing another transient increase in P_o. This model is most similar to that proposed by Zahradníková and Zahradník[30] and Sachs *et al.* [28] although we believe that multiple Ca^{2+} binding sites regulate the channel. The one thing that is certain about RyR adaptation modeling is that definitive models can not be developed until more experimental data become available.

RyR ADAPTATION IN A PHYSIOLOGICAL CONTEXT

To define RyR function at the single-channel level, the RyR protein must be literally ripped out of its cellular surroundings and placed into the artificial environment of the planar lipid bilayer. As a consequence, functional attributes of single RyR channels defined in bilayers may or may not reflect RyR function in the cell. Thus, it is important to establish how RyR adaptation as defined in the bilayer may be manifested in the cell.

There is a substantial body of evidence that the RyR channel is the primary SR Ca^{2+} release pathway in cardiac cells. It has also been reported that Ca^{2+} currents through individual L-type Ca^{2+} channels may directly regulate one or two RyR channels. In voltage-clamped ventricular myocytes, depolarization activates the L-type Ca^{2+} channels and the resulting inward Ca^{2+} current activates the RyR channel initiating the SR Ca^{2+} release process.

The globally measured L-type Ca^{2+} current and SR Ca^{2+} efflux both "inactivate" during a maintained depolarization. Repolarization initiates a large inward Ca^{2+} tail current which re-activates the SR Ca^{2+} release process.[25] The re-activation of SR Ca^{2+} release by the tail current is a classical whole-cell observation which has been very difficult to explain. The problem is that the tail current by definition is mediated by open L-type Ca^{2+} channels. The RyR channels near open L-type Ca^{2+} channels should be "inactivated" and thus insensitive to the tail current. There are two possible explanations for tail current re-activation of SR Ca^{2+} release. The tail current may recruit an entirely new set of RyR channels or it may actually re-activate the same set of RyR channels. Recent observations suggest that cardiac RyR channels operate as clusters and that all RyR's in a cluster open when the cluster is activated.[37,38,39,40] This implies that there may not be new RyR channels near an open L-type Ca^{2+} channel for the tail current to recruit. On the other hand, recent single RyR channel data suggest that individual RyR adapt, instead of inactivate, allowing them to repeatedly and transiently respond to multiple Ca^{2+} channel stimuli.[11] This could indicate that RyR channels near open L-type Ca^{2+} channels may actually be re-activated by the tail current. Thus, tail current re-activation of SR Ca^{2+} release may be an obvious whole-cell manifestation of RyR adaptation.

In a now classical work, Fabiato demonstrated that SR Ca^{2+} release was dependent on both the amplitude and speed of the trigger Ca^{2+} stimulus.[10] Fabiato suggested that SR Ca^{2+} release may be regulated by a Ca^{2+}-dependent inactivation mechanism.[10] He suggested that the SR Ca^{2+} release channels may have two Ca^{2+} binding sites, a fast low affinity Ca^{2+} activation site and a slow high affinity Ca^{2+} inactivation site. Fast Ca^{2+} stimuli would transiently activate the channel until the inactivation process predominates. Slow Ca^{2+} stimuli would simply inactivate channels leaving them in a refractory state. A decade of intense effort by several laboratories, however, has failed to clearly demonstrate that Ca^{2+}-dependent inactivation occurs in intact myocytes.[25] At the single-channel level, it is clear that relatively high $[Ca^{2+}]$ do not "inactivate" cardiac RyR channels.[35,41,42] Interestingly, RyR adaptation would allow single RyR channels to "sense" the rate of free $[Ca^{2+}]$ change rendering the SR Ca^{2+} release process dependent on both the amplitude and speed of the trigger Ca^{2+} stimulus. This would be consistent with the original observations of Fabiato.[10] Additionally, Stern has theoretically shown that a rate-sensitive (i.e. adapting) RyR channel could explain the paradoxical stability of CICR in heart.[31]

In isolated cardiac myocytes, spontaneous openings of RyR channels are thought to generate small localized release events known as Ca^{2+} sparks.[43] The Ca^{2+} spark is thought to represent the elementary Ca^{2+} release unit that underlies global Ca^{2+} signaling events in heart. It is thought that Ca^{2+} sparks originate from clusters of RyR channels[37,38,39,40] and that RyR clusters may operate in a concerted manner.[44] It has also been argued that activation of one RyR channel in a cluster will activate all channels within that cluster.[40] Several laboratories have shown that Ca^{2+} sparks of nearly equal amplitude re-occur at the same site only a few milliseconds apart.[39,40,45] These latter data suggest that individual SR Ca^{2+} release sites do not become refractory and this could be easily explained by RyR adaptation.

It is interesting that RyR adaptation would reconcile a rather large body of apparently controversial results from intact cells, skinned fibers and single channels. Although this makes RyR adaptation an attractive candidate as the negative feedback mechanism which stabilizes CICR, it must be noted that the evidence that RyR adaptation exists in cells[24,46] is at best consequential at this point. Additional experimentation is clearly required to definitively establish a role (if any) for RyR adaptation in heart.

REFERENCES

1. Lai, F.A., Anderson, K., Rousseau, E., Liu, Q.Y., and Meissner, G. Evidence for a Ca^{2+} channel within the ryanodine receptor complex from cardiac sarcoplasmic reticulum. *Biochem. Biophys. Res. Comm.* 1988; **151(1)**:441–449.
2. Lai, F.A., Erickson, H.P., Rousseau, E., Liu, Q.Y., and Meissner, G. Purification and reconstitution of the calcium release channel from skeletal muscle. *Nature* 1988; **331**:315–319.
3. Wagenknecht T, Grassucci R, Frank, J., Saito, A., Inui, M., and Fleischer, S. Three-dimensional architecture of the calcium channel/foot structure of sarcoplasmic reticulum. *Nature* 1989; **338**:167–170.
4. Orlova, E.V., . Serysheva II. van Heel, M., Hamilton, S.L., and Chiu, W. Two structural configurations of the skeletal muscle calcium release channel. *Nature Structural Biology* 1996; **3**:547–552.
5. Block, B.A., Imagawa, T., Campbell, K.P. and Franzini-Armstrong, C. Structural evidence for direct interaction between the molecular components of the tranverse tubule/sarcoplasmic reticulum junction in skeletal muscle. *J. Cell Biol.* 1988; **107(6)**:2587–2600.
6. Smith, J.S., Imagawa, T., Ma, J., Fill, M., Campbell, K.P., and Coronado, R. Purified ryanodine receptor from rabbit skeletal muscle is the calcium-release channel of sarcoplasmic reticulum. *J. Gen. Physiol.* 1988; **92**:1–26.
7. Bull, R., Marengo, J.J., Suárez-Isla, B.A., Donoso, P., Sutjo, J.L., and Hidalgo, C. Activation of calcium channels in sarcoplasmic reticulum from frog muscle by nanomolar concentrations of ryanodine. *Biophys. J.* 1989; **56**:749–756.
8. Holmberg, S.R. and Williams, A.J. Single channel recordings from human cardiac sarcoplasmic reticulum. *Circulation Res.* 1989; **65**:1445–1449.
9. Murayama, T. and Ogawa, Y. Purification and characterization of two ryanodine-binding protein isoforms from sarcoplasmic reticulum of bullfrog skeletal muscle. *J. Biochem.* 1992; **112**: 514–522.

10. Fabiato A. Time and calcium dependence of activation and inactivation of calcium-induced release of calcium from the sarcoplasmic reticulum of a skinned canine cardiac Purkinje cell. *J. Gen. Physiol.* 1985; **85**:247–289.
11. Györke, S., Fill, M. Ryanodine receptor adaptation: control mechanism of Ca^{2+}-induced Ca^{2+} release in heart [see comments]. *Science* 1993; **260**:807–809.
12. Stern, M.D. Theory of excitation-contraction coupling in cardiac muscle. *Biophys. J.* 1992; **63(2)**:497–517.
13. Györke S, Vélez P, Suárez-Isla, B. A., and Fill, M. Activation of single cardiac and skeletal ryanodine receptor channels by flash photolysis of caged Ca^{2+} [see comments]. *Biophys. J.* 1994; **66**: 1879–1886.
14. Valdivia, H.H., Kaplan, J.H., Ellis-Davies, G.C., and Lederer, W.J. Rapid adaptation of cardiac ryanodine receptors: modulation by Mg^{2+} and phosphorylation. *Science* 1995; **267**:1997–2000.
15. Vélez, P., Li, X., Tsushima, R., Wasserstrom, A., and Fill, M. Adaptation of single cardiac RyR channels may involve a closely associated regulatory protein. *Biophys. J.* 1995; **68(2)**:A375.
16. Ellies-Davies, G.C.R., Kaplan, J.H., and Barsotti, R.J. Laser photolysis of caged calcium: rates of calcium release by Nitrophenyl-EGTA and DM-Nitrophen. *Biophys. J.* 1996; **70**:1006–1016.
17. Escobar, A.L., Cifuentes, F., and Vergara, J.L. Detection of Ca^{2+} -transients elicited by flash photolysis of DM-nitrophen with a fast calcium indicator. *FEBS Letters* 1995; **364**:335–338.
18. Lamb, G.D. and Stephenson, D.G. Activation of ryanodine receptors by flash photolysis of caged Ca^{2+}. *Biophys. J.* 1995; **68**:946–948.
19. Sitsapesan, R., Montgomery, R.A.P., Williams, A.J. New insights into the gating mechanisms of cardiac ryanodine receptors revealed by rapid changes in ligand concentration. *Circ. Res.* 1995a; 77:765–772.
20. Sitsapesan, R., Montgomery, R.A.P., and Williams, A.J. A novel method for incorporation of ion channels into a planar phospholipid bilayer which allows solution changes on a millisecond timescale. *Pflügers Arch.* 1995b; **430**:584–589.
21. Laver, D,R, and Curtis, B.A. Surface potentials measure ion concentrations near lipid bilayers during rapid solution changes. *Biophys. J.* 1996a; **71**:722–731.
22. Laver, D.R. and Curtis, B.A. Response of ryanodine receptor channels to Ca^{2+} steps produced by rapid solution exchange. *Biophys. J.* 1996; **71**:732–741.
23. Schiefer, A., Meissner, G., and Isenberg, G. Ca^{2+} activation and Ca^{2+} inactivation of canine reconstituted cardiac sarcoplasmic reticulum Ca^{2+}-release channels. *J. Physiol.* 1995; **489**: 337–348.
24. Yasui, K., Palade, P., and Györke, S. Negative control mechanism with features of adaptation controls Ca^{2+} release in cardiac myocytes [see comments]. *Biophys. J.* 1994; **67**:457–460.
25. Bers, D.M. *Excitation-contraction coupling and cardiac contractile force.* In: Developments in cardiovascular medicine. Vol. 122. 1991. Kluwer Academic Publishers.
26. Vélez, P., Györke, S., Escobar, A., and Fill, M. Adaptation of single cardiac ryanodine receptor channels. *Biophys. J.* 1997; **72**:691–697.
27. Cheng, H., Fill, M., Valdivia, H. and Lederer, W.J. Models of Ca^{2+} release channel adaptation. *Science* 1995; **267**:2009–2010.

28. Sachs, F., Qin, F., and Palade, P. Models of Ca^{2+} release channel adaptation. *Science* 1995; **267**:2010–2011.
29. Tang, Y., Othmer, H.G. A model of calcium dynamics in cardiac myocytes based on the kinetics of ryanodine-sensitive calcium channels. *Biophys. J.* 1994; **67**:2223–2235.
30. Zahradníková, A., and Zahradník, I. Description of modal gating of the cardiac calcium release channel in planar lipid membranes. *Biophys. J.* 1995; **69**:1780–1788.
31. Stern, M.D. "Adaptive" behavior of ligand-gated ion channels: Constraints by thermodynamics. *Biophys. J.* 1996; 2100-2109.
32. Chen, S.R.W. and MacLennan, D.H. Identification of calmodulin-, Ca^{2+}-, and ruthenium red-binding domains in the Ca^{2+} release channel (ryanodine receptor) of rabbit skeletal muscle sarcoplasmic reticulum. *J. Biol. Chem.* 1994; **269(36)**:22698–22704.
33. Richard, E.A. and Miller, C. Steady-state coupling of ion-channel conformations to a transmembrane ion gradient. *Science* 1990; **247**:1208–1210.
34. Segel, L.A., Goldbeter, A., Devreotes, P.N., and Knox, B.E. A mechanism for exact sensory adaptation based on receptor modification. *J. Theor. Biol.* 1986; **120**:151.
35. Chu, A., Fill, M., Stefani, E., and Entman, M.L. Cytoplasmic Ca^{2+} does not inhibit the cardiac muscle sarcoplasmic reticulum ryanodine receptor Ca^{2+} channel, although Ca^{2+}-induced Ca^{2+} inactivation of Ca^{2+} release is observed in native vesicles. *J. Membrane Biol.* 1993; **135(1)**: 49–59.
36. Armisen R., Sierralta J., Vélez P., Naranjo, D., and Suárez-Isla, B.A. Modal gating in neuronal and skeletal muscle ryanodine-sensitive Ca^{2+} release channels. *Am. J. Physiol.* 1996; **271(1 Pt 1)**:C144–53.
37. Wier, W.G., Egan, T.M., López-López, J.R., Balke, C.W. Local control of excitation-contraction coupling in rat heart cells. *J. Physiol.* 1994; **474(3)**:463–471.
38. Lipp, P. and Niggli, E. Submicroscopic calcium signals as fundamental events of excitation–contraction coupling in guinea-pig cardiac myocytes. *J. Physiol.* 1996; **492(1)**:31–38.
39. Parker, I., Zang, W.J., and Wier, W.G. Ca^{2+} sparks involving multiple Ca^{2+} release sites along Z-lines in rat heart cells. *J. Physiol.* 1996; **497(Pt 1)**:31–38.
40. Blatter L.A., Hüser J., and Ríos, E. Sarcoplasmic reticulum Ca^{2+} release flux underlying Ca^{2+} sparks in cardiac muscle. *Proc. Natl. Acad. Sci.; USA* 1997; **94**:4176–4181.
41. Rousseau, E. and Meissner, G. Single cardiac sarcoplasmic reticulum Ca^{2+}-release channel: Activation by carreing. *Am. J. Physiol.* 1989; **256**:H328–H333.
42. Laver, D.R., Roden, L.D., Ahern, G.P. Eager, K.R., Junankar, P.R., and Dulhunty, A.F. Cytoplasmic Ca^{2+} inhibits the ryanodine receptor from cardiac muscle. *J. Membrane Biol.* 1995; **147(1)**: 7–22.
43. Cheng, H., Lederer, W.J., and Cannell, M.B. Calcium sparks: Elementary events underlying excitation-contraction coupling in heart muscle. *Science* 1993; **262**:740–744.
44. Protasi, F. and Franzini-Armstrong, C. Significance of the distance between calcium release units in cardiac and skeletal muscle. *Biophys. J.* 1996; **70**:A245.

45. Sommer, J.R., Bossen, E., Dalen, H., Dolber, P., High, T. Jewett, P. Johnson, E.A., Junker, J., Leonard, S., Nassar, R. *et al.* To excite a heart: a bird's view. *Acta Phisiol. Scand.* 1991; **142(Suppl.599)**: 5–21.
46. Györke, I. and Györke, S. Adaptive control of intracellular Ca^{2+} release in C2C12 mouse myotubes. *Pflügers Arch.* 1996. **431(6)**:838–843.

Chapter 14

Adaptation, Inactivation and Inhibition in Ryanodine Receptors

G.D. Lamb and D.R. Laver

Many different processes have been reported to cause a decrease in the activity of ryanodine receptor (RyR)/Ca^{2+} release channels. These processes, which might be generally referred to as 'negative control mechanisms',[1] have many different names, and in many cases the same name is used for what are clearly different processes. Furthermore, some of these processes are very complex, and in fact may not be separate distinguishable processes but instead a composite of others. The confusion in terminology is not merely a semantic problem that perplexes newcomers to the field, but instead continues to cause major misinterpretations and inappropriate conclusions throughout the current literature. Here, we describe the major properties of the following processes, and try to distinguish between them where possible:

(a) 'Adaptation' in isolated RyRs

(b) Inactivation in isolated RyRs

(c) Ca^{2+}/Mg^{2+} inhibition in isolated RyRs

(d) Inhibition of RyRs by reduction in SR luminal [Ca^{2+}]

(e) Ca^{2+} inactivation and Ca^{2+} sparks in cardiac and skeletal muscle

(f) 'Adaptation' in cardiac and skeletal muscle

(g) Repolarization-induced stop of Ca^{2+} release (RISC) in skeletal muscle

ADAPTATION IN ISOLATED RyRs

Györke & Fill[2] described how individual cardiac RyRs in lipid bilayers show a very distinctive response when stimulated by a rapid rise in [Ca^{2+}] generated by flash photolysis of DM-nitrophen, a caged-Ca^{2+} compound. It was reported that stimulating a RyR with a 'step' rise in [Ca^{2+}] from 0.1 μM to 0.2 μM resulted in the average channel open probability (P_o) increasing very rapidly (time-constant <2 ms) from near zero to a high level and then declining slowly over several seconds to a low level again, even though the [Ca^{2+}] was still raised. Thus, the channel activity declined even though the supposed stimulus was still present. Importantly, if the RyR was then stimulated by a larger [Ca^{2+}] 'step', P_o once again increased rapidly and then subsequently declined slowly. Because the RyR apparently ceased responding to the first 'step' increase in [Ca^{2+}] and yet could still respond vigorously to a second larger 'step', this type of response was termed 'adaptation',[2] in order to distinguish it from inactivation or desensitization.

However, there is now general agreement that the flash photolysis procedure used to generate the [Ca^{2+}] 'step' in such experiments[2,3] actually causes the [Ca^{2+}] to transiently rise to a very high level before settling down to the final steady 'step' level (see Refs. 4–10). This 'Ca^{2+} spike' at the leading edge of the step is very brief (~1 ms), but probably reaches 60 to 300 μM, depending on the caged compound and exact conditions used. Thus, the [Ca^{2+}] during the spike is up to several hundred times the final level, and as most of the RyR activation occurs during the time-course of the Ca^{2+}-spike, it may be concluded, contrary to recent claims,[9] that the Ca^{2+} spike profoundly influences the activation of the RyR (also see Lamb[10]).

In fact, there is now direct experimental evidence that even smaller Ca^{2+} spikes (reaching ~17 μM) do activate the RyR.[11] Furthermore, such stimuli only caused brief (millisecond) activation of the RyR, even though the final (ie. 'step') level of Ca^{2+} was considerably higher than that found previously to cause prolonged activation and consequent adaptation.[2] This indicates that the prolonged activation is not dependent on the final 'step' Ca^{2+} level, but instead arises if the preceding Ca^{2+} spike stimulus is sufficiently large or prolonged.[10]

Moreover, it is clear from the data of Györke & Fill[2,12] that the Ca^{2+} spike is almost certainly driving RyR activation because it would otherwise (a) indicate that small Ca^{2+} 'steps' activate the RyR faster than large Ca^{2+} 'steps'[7,10] and (b) imply a Ca^{2+} activation rate constant (2×10^9 M^{-1} s^{-1}, for the 'step' to 0.2 μM Ca^{2+}) that is ten times higher than that obtained with true step changes in [Ca^{2+}] (2×10^8 M^{-1} s^{-1}, see Schiefer *et al.*[13]), and which considerably exceeds the maximum rate possible with diffusion limitations (~3×10^8 M^{-1} s^{-1}).

It is also important to note that such a large, brief Ca^{2+} spike could well affect a RyR in a very different way than would a small step rise in [Ca^{2+}]. For example, given the true activation rate constant of 2×10^8 M^{-1} s^{-1} (see Ref. 13), the Ca^{2+} spike associated with a 'step' from 0.1 to 0.2 μM would have a high probability of driving Ca^{2+} simultaneously onto several independent activation sites on a RyR, but similar occupancy would very rarely occur in the presence of a low steady [Ca^{2+}]. In this regard, we note that when the buffering conditions of the flash photolysis experiments are altered (e.g. case of 'step' rise from 0.05 to 0.067 μM Ca^{2+} in Ref. 12 - see Refs. 7 and 10) such that the Ca^{2+} spike becomes much briefer, and consequently has a much lower probability of driving Ca^{2+} onto any activation site, the stimulus no longer activates the RyR. In an alternative scenario, if Ca^{2+} binding to one activation site were to lower the affinity of other activation sites on a RyR tetramer, a large Ca^{2+} spike might still drive Ca^{2+} onto those sites, whereas such simultaneous occupation would occur only very rarely with smaller, briefer spikes or with low steady levels of Ca^{2+}. Thus, it is clear that the occurrence of a Ca^{2+} spike could possibly have a major effect on the particular activation behavior of the RyR.

Further evidence that it is the Ca^{2+} spike that is underlying the vigorous response to small 'step' increases in Ca^{2+} in the photolysis experiments comes from the fact that such an apparent increase in Ca^{2+} sensitivity is not seen when the [Ca^{2+}] is raised directly by solution change.[13,14A] Moreover, in the cases where a subsequent slow decline in RyR activity was observed in the fast solution change experiments, it only occurred at higher [Ca^{2+}] than reported in the flash photolysis experiments, and appeared to be due to an inactivation process rather than to 'adaptation' (see Section B). Thus, there is considerable evidence against the conclusion that individual RyRs 'adapt' to small increases in [Ca^{2+}]. Instead, it seems that if the Ca^{2+} spike produced by the flash photolysis is sufficiently large and prolonged, it causes the RyR to enter a prolonged state with a high P_o lasting several seconds or so. At first this seems in conflict with the reported ability of RyRs to close rapidly when the [Ca^{2+}] is suddenly reduced.[9,13–15] However, these experiments did not show whether the RyR is able to close rapidly after it has been activated by a very large, brief Ca^{2+} spike stimulus, which may be strongly influenced by the modal gating behavior of the RyRs (see below). Clearly, the transition from transient to prolonged RyR activity found when increasing the size of the Ca^{2+} spike[11] suggests that the precise stimulus is actually quite crucial to the channel behavior.

Modal Gating Behavior

The prolonged activity induced by a Ca^{2+} spike can be easily understood in terms of the modal gating observed in RyRs at steady $[Ca^{2+}]$.[16,17] The RyR shows both a low (L) and a high (H) mode of activity. Importantly, in both cardiac and skeletal RyRs, the time taken to drop out of H-mode (time-constant about 1 s and 3 to 5 s, respectively) is very similar to the time for the RyRs to 'adapt' after a flash.[2,17,18] Armisen *et al.*[17] have proposed that the 'adaptation' behavior observed in photolysis experiments is genuine and results from the fast small 'step' increase in Ca^{2+} (e.g. from 0.1 to 0.2 μM) inducing the H-mode behavior. However, in view of the discussion above, it must be the fast Ca^{2+} spike preceding the 'step', rather than the 'step' itself, that is inducing the transition to H-mode. If the Ca^{2+} spike is large enough to cause the RyR to enter H-mode, perhaps because it causes the simultaneous occupation of several activation sites (see above), P_o will remain high for a second or more even though the Ca^{2+} stimulus has largely disappeared, and when P_o has eventually declined it can be increased again by stimulating the RyR with another Ca^{2+} spike. If this is correct, the decline in RyR activity is really due to *deactivation* (i.e. closure following withdrawal of the stimulus) rather than to 'adaptation' or inactivation. It is also significant that the rate of 'adaptation' is the apparently the same for all $[Ca^{2+}]$ 'steps',[2] which is not consistent with the phenomenon being due to a single-step Ca^{2+}-dependent inactivation process, but is consistent with recovery from H-mode.

Physiological Relevance of Ca^{2+} Spikes

It is also important to note that the heightened sensitivity to small increases in $[Ca^{2+}]$ implicit in the adaptation proposal, is the exact opposite of the low sensitivity that is now thought must occur in cardiac RyRs in-vivo in order to explain local control of Ca^{2+} release.[19] In fact, somewhat ironically, it seems likely that cardiac RyRs in-vivo are normally activated by 'Ca^{2+} spike' stimuli that occur when an adjacent L-type Ca^{2+} channel or neighboring RyR activates. Thus, the key findings of the flash photolysis experiments are perhaps more appropriately viewed as indicating the response of RyRs to physiological stimuli, rather than identifying a new negative control mechanism. As 'adaptation' occurs more rapidly in the presence of physiological $[Mg^{2+}]$,[3] Ca^{2+} spike stimuli appear able to induce brief bursts of activity in RyRs in bilayers that may be comparable with the properties of RyRs in-vivo. This might be important in the generation of the so-called Ca^{2+} sparks observed in skeletal and cardiac muscle,[10] although some additional inactivation process may also be involved (see Section E).

Recent 'Adaptation' Data

Recently, it has also been claimed that 'adaptation' does not occur with flash photolysis of caged Sr^{2+}, indicating that Ca^{2+} is essential for the decline in activity.[20] However, even ignoring the problems with the photolysis procedure, there can be considerable doubt about the concentration changes involved in these experiments, because the associated steady-state experiments using complex buffering indicated that Sr^{2+} activated the RyR with ten times *higher* affinity than Ca^{2+} (K_D 0.1 and 1 μM, respectively). This conflicts with previous work using simpler buffering which found that the affinity for Sr^{2+} was 100 times lower than this.[21] Other recent flash photolysis experiments show that the 'adaptation' phenomenon is lost when RyRs are treated with FK-506 to remove the FK-506-binding protein (FKBP).[22] Again these experiments ignore the problem of the Ca^{2+} spikes, as well as the fact that FK-506 appears to potentiate activation by a direct action on the RyR, independent of FKBP dissociation.[23,24] Thus, the apparent lack of 'adaptation' may well just simply reflect an increase in steady-state Ca^{2+} sensitivity after FK-506 treatment, and furthermore may be quite unrelated to FKBP binding.

Models of Adaptation

Current models describing 'adaptation' do not take into account the real Ca^{2+} stimulus.[25–29] Furthermore, except for the model by Cheng *et al.*[29] they are also fundamentally inconsistent with the reported single channel behavior, in that they are unable to account for the ability of a RyR to 'adapt' to one stimulus and then respond again immediately afterwards to a larger stimulus.[2] The models are only able to show the same *average* behavior by utilizing a population of RyRs, some of which respond to the first stimulus and then 'adapt', with others responding only to the second stimulus (see Ref. 10). The model of Cheng *et al.*[29] is able to reproduce this aspect of the data by assuming that each channel has multiple Ca^{2+}-dependent activation and inactivation sites. However, it does not mimic the observed responses if the true Ca^{2+}-spike/step stimulus is used in the modeling, and it incorrectly predicts that the 'adaptation' rate depends on the 'step' [Ca^{2+}]. In contrast, a model in which the Ca^{2+} spike induces a transition to a long-lived H-mode (see above) can account for all the characteristics of single channel 'adaptation'.

INACTIVATION IN ISOLATED RyRs

The properties of RyR inactivation following both voltage- and $[Ca^{2+}]$-steps are distinct from those of adaptation or Ca^{2+}/Mg^{2+} inhibition and these processes result from different regulation mechanisms, as discussed below.

Inactivation After Voltage Steps

At least three different types of voltage-dependent inactivation have been identified in single channel studies. A slow voltage-dependent inactivation at positive potentials has been reported in RyRs from sheep heart,[30] rabbit[31] and chicken skeletal muscle.[32] The channels inactivate by entering long-lived closed states and recover within seconds of steps to negative holding potentials. At +40 mV inactivation occurred over several seconds and it proceeded faster at more positive voltages. In sheep cardiac and rabbit skeletal RyRs.[14A] Inactivation does not occur (30s sample period) if P_o is less than 0.4 and of those RyRs with higher P_o only half showed inactivation. Inactivation could be removed or made faster with ligands that respectively lowered or raised P_o.[14A] Ca^{2+} is apparently not required for this type of inactivation because it occurs in cardiac RyRs that are activated by ryanodine alone (~1 nM Ca^{2+}) or by caffeine plus ATP (~10 nM Ca^{2+}).[14A]

Another inactivation phenomenon, which was called desensitization,[33] was only seen at large negative potentials (>-80 mV). At -100 mV half the RyRs inactivated to 55% of their initial P_o with a time constant of 3.9s. Similar inactivation rates have also been seen in RyRs from sheep cardiac and pig skeletal muscle after voltage steps from +40mV to potential differences at or more negative than –40mV.[14A] RyRs were frequently seen to transiently activate after voltage steps to both positive and negative potentials. The inactivation processes occurring at positive and negative potentials have very similar properties (although opposite voltage-dependencies). It is not clear whether inactivation occurring at opposite voltage-extremes stem from a single mechanism which senses the magnitude of the transmembrane electric field (e.g. a dielectric based mechanism) or from two independent inactivation mechanisms which have opposite voltage-dependencies.

Another apparently different, much faster type of inactivation at positive potentials was named rectification by Ma[34] and was seen in only 20% of RyRs. It was 10–100 times faster than that described above and had time constants which varied from 100ms at 40mV to 10ms at 100mV. Recovery rates had the opposite voltage dependence and varied from 100ms at -40mV to 10ms at -180mV. During recovery, RyRs opened to their maximum conductance level via stochastic transitions between three subconductance states that appeared to be equally spaced. Rectification could be removed by the presence of perchlorate ions and restored by exogenous FKBP. A similar effect was probably seen by Chen *et al.*[35] who reported that FKBP induced rectification in purified rabbit skeletal RyRs.

Inactivation Following [Ca^{2+}] Steps

Three laboratories[13–15] have used fast solution change methods to produce [Ca^{2+}] steps and shown rapid activation of RyRs (millisecond rise times) which, in some conditions, was followed by a decline in channel activity when steady [Ca^{2+}] was maintained for seconds. Two studies[14,15] observed inactivation at +40 mV but not at –40 mV. Another study[13] found inactivation at –50 mV which may arise from a different mechanism and so is considered separately.

Inactivation following [Ca^{2+}] — steps at positive potentials

Under experimental conditions that could induce inactivation (see below), only about half of the RyRs exhibited inactivation (56% in cardiac RyRs,[14] 70% in skeletal RyRs[15]). Sheep cardiac and rabbit skeletal RyRs inactivated within seconds of [Ca^{2+}] steps by entering into long-lasting closures.[14,15] The inactivation rates varied considerably between individual RyRs (500 ms to 15 s) in both skeletal and cardiac RyRs.[15] The inability to reactivate RyRs by repeatedly lowering and reapplying Ca^{2+} (10 s cycles) showed that recovery from inactivation could not be induced simply by withdrawing Ca^{2+} (see Ref. 14). Furthermore, when fast flow methods are used to produce a second, larger [Ca^{2+}]-step (e.g. from 1 to 100 mM), mimicking the double-[Ca^{2+}] step used in flash photolysis experiments,[2] the RyRs could still not be reactivated.[14A] In contrast, voltage steps to –40 mV could remove such inactivation (Ref. 14 and Laver, upublished data). It was pointed out by Sitsapesan *et al.*[14] (and data of Laver and Lamb[14A] supports this) that inactivation following [Ca^{2+}] steps could be related to the inactivation seen following voltage-steps in other studies (see above).

The occurrence of inactivation in RyRs depended on their initial response to activation. In one study, only 17% of RyRs activated to relatively low P_o (~0.2) by Ca^{2+} alone showed inactivation, whereas 56% of the RyRs that were activated to a high P_o (P_o >0.75 with Ca^{2+} in the presence of another channel activator, such as ATP or EMD41000) subsequently inactivated.[14] In other studies,[14A,15] both skeletal and cardiac RyRs needed high P_o (>0.4) in order to show inactivation, and in these experiments RyRs could be activated to a high P_o by Ca^{2+} alone, and the presence of a second activating ligand was not needed to see inactivation. RyR inactivation could be slowed or abolished by channel inhibitors, such as Mg^{2+}, that lowered P_o.[14A] Even though inactivation appeared to be correlated with P_o, it seems that it is not exclusively P_o-dependent, as only half of the RyRs showing a high P_o were seen to inactivate.[14,15] Furthermore, it was found [14] that inactivation would only occur when RyRs had a relatively long mean open time when activated by Ca^{2+} plus ATP or EMD 41000 compared with that for Ca^{2+} alone. Hence, inactivation might occur only when the RyRs are in a special, high P_o gating mode. Laver and Lamb[14A] found a similar effect, in both cardiac and skeletal RyR, in that it was necessary for RyRs to be in a gating mode that had relatively long channel openings in order for inactivation to occur.

Sitsapesan *et al.*[14] have suggested that inactivation after [Ca^{2+}] steps arises from a voltage-dependent (and time-dependent) mechanism. The occurrence of inactivation after [Ca^{2+}] steps, its voltage-dependence and its kinetics, are the same as for the slow inactivation seen after steps to positive potentials at steady [Ca^{2+}] (see above and Refs. 30–32). The idea that the same mechanism produces inactivation after both voltage- and [Ca^{2+}]-steps is supported by the close correlation in the occurrence of these phenomena in individual RyRs.[14A] Since the inactivation after voltage-steps does not require Ca^{2+} (see above) it seems that inactivation with [Ca^{2+}] steps does not depend on [Ca^{2+}] per se, but rather on other aspects of channel activation. The findings[14A,15] point to a three state inactivation model (Scheme 1).

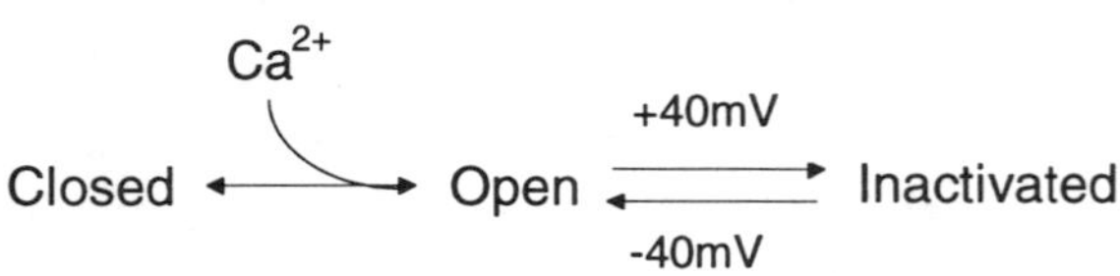

Scheme 1

Inactivation following [Ca^{2+}] — steps at negative potentials

Schiefer *et al.*[13] observed inactivation of dog cardiac RyRs at –50 mV after [Ca^{2+}] steps from 10 nM to high levels. By analogy with inactivation after [Ca^{2+}] steps at positive potentials, the inactivation seen in the experiments of Schiefer *et al.*[13] may arise from the same mechanism that causes inactivation after voltage steps to negative potentials reported by Ma.[33] Inactivation was detected only at [Ca^{2+}] above 3 mM and became more pronounced as the [Ca^{2+}] was increased to 1 mM. Inactivation rate, determined over a 500 ms interval, depended on [Ca^{2+}] and estimates ranged from at 220 ms (at 1 μM Ca^{2+}) to 500 ms (at 10 mM Ca^{2+}). Since the initial peak P_o was relatively independent of [Ca^{2+}] over this range, the inactivation was considered to be Ca^{2+}-dependent. Evidence for a refractory period after inactivation could be seen from the finding that a 30 Hz train of steps between 10 nM and 1 mM Ca^{2+} reduced the ability of subsequent [Ca^{2+}] steps to open the RyRs. It was found that when the [Ca^{2+}] was lowered to sub-activating levels after the RyR had inactivated there were always channel openings, indicating that RyRs recovered via their open state.[13] The timing of these recovery openings allowed the authors to estimate a recovery rate of ~840 ms. The findings were interpreted in terms of a four state model where the channels could inactivate from the open state by two Ca^{2+}-dependent mechanisms. However, the model as it stands predicts virtually complete inactivation with rates 10–100 times faster than the measured rates at 100 and 1000 μM Ca^{2+} respectively. Instead, a reasonable fit to the data can be obtained if the rate constants to the proposed high Ca^{2+}-affinity inactivated state are assumed to be independent of [Ca^{2+}].

Inactivation is Distinct From Ca^{2+}/Mg^{2+} Inhibition

The kinetics and specificity of RyR inactivation is very different to that seen for Ca^{2+}/Mg^{2+} inhibition of RyRs (Section C). Firstly, inactivation at positive potentials does not depend on the presence of Ca^{2+} (or another divalent metal cation) whereas Ca^{2+}/Mg^{2+} inhibition does. Secondly, Ca^{2+}/Mg^{2+} inhibition sets in at sub mM to mM concentrations of divalent ions and shows no specificity between Ca^{2+} and Mg^{2+} (see Section C) whereas mM Mg^{2+} slows inactivation. Thirdly, inactivation takes seconds to develop and is mediated by closures lasting seconds, whereas Ca^{2+}/Mg^{2+} inhibition is associated with a large increase in the frequency of closures lasting 0.1 to 10 ms[30,36] and only takes tens of milliseconds for onset and recovery.[15]

Inactivation by Flow is Different from 'Adaptation' by Flash

The 'adaptation' phenomenon seen in single cardiac and skeletal RyRs[2,3,18] after flash-induced release of caged Ca^{2+} follows a broadly similar time-course to inactivation seen after flow induced [Ca^{2+}] steps.[13–15] However, as mentioned above, 'adaptation' and inactivation are fundamentally different mechanisms. The findings of the solution flow experiments [13–15] indicate that RyRs inactivate rather than 'adapt' to the steady, elevated [Ca^{2+}] following the step. RyR inactivation is followed by a refractory period of the order of one to many seconds in which repeated [Ca^{2+}] steps or steps to even higher [Ca^{2+}] [14A] fail to reactivate RyRs. Inactivation of RyRs does not depend on [Ca^{2+}] per se, whereas adaptation is dependent entirely on the [Ca^{2+}] stimulus. Other differences in the findings of the flash and flow studies also indicate that decline of RyR activity in these two types of studies are different phenomena. Firstly, the range of [Ca^{2+}] over which the two phenomena are observed are quite different. 'Adaptation' occurs predominantly at 1 μM Ca^{2+}or below,[2] where such steady levels of Ca^{2+} are sub-activating. Inactivation, on the other hand, occurs only at activating [Ca^{2+}] and at concentrations up to mM.[13–15] Secondly, the kinetics of the two phenomena appear quite different. Inactivation always occurs by interruption of high P_o activity with long-lasting closures, whereas adaptation frequently appears more like a gradual decrease in RyR activity. Thirdly, Mg^{2+}-inhibition of RyRs either slows or removes inactivation[14A] whereas Mg^{2+} speeds the adaptation phenomenon.[3] Finally, flow induced [Ca^{2+}] steps are unable to activate RyRs at concentrations lower than they would normally be activated in steady state,[13] whereas flash induced [Ca^{2+}] steps *appear* to sensitize RyR to activating Ca^{2+} by a factor of ~10.

Ca^{2+}/Mg^{2+} INHIBITION IN ISOLATED RyRs

Ca^{2+}/Mg^{2+} inhibition is very often confused with Ca^{2+}-specific inactivation (e.g. Section E) (see Ref. 37). Ca^{2+}/Mg^{2+} inhibition in single RyRs and SR vesicles can be observed as a decline in Ca^{2+}-dependent activity at high (mM) [Ca^{2+}], and this is often described simply as "Ca^{2+} inactivation". However, it is clear that other divalent metal cations, in particular Mg^{2+}, cause similar inhibition by acting at the *same* site.[21,36,38–40] This is not simply some average inhibitory behavior occurring in SR vesicles, but has been demonstrated at the single channel level in individual RyRs.[36,38] At physiological ionic strength, the dissociation constant (K_D) of this divalent cation inhibitory site on the mammalian skeletal muscle RyR is ~0.1 mM. Most importantly, as the normal intracellular [Mg^{2+}] is ~1 mM (see Ref. 41), this site can be expected to be almost fully saturated with Mg^{2+} in a skeletal muscle fiber in vivo. Consequently, even if the [Ca^{2+}] locally near the RyR were to rise to millimolar levels during peak Ca^{2+} release, this would only cause an extremely small increase in the occupancy of this non-specific inhibitory site. Thus, this site will be exerting a potent inhibitory effect on RyR activity in a resting skeletal muscle fiber, and *this type of inhibition* would be only marginally increased in absolute terms by any rise in Ca^{2+}, no matter how large. Consequently, it seems most appropriate to refer to this phenomenon as Mg^{2+} inhibition or Ca^{2+}/Mg^{2+} inhibition,[37] because although Ca^{2+} can act on this site in some experimental situations, it is Mg^{2+} that is predominantly affecting this site in-vivo. It is also clear then that this inhibition is quite distinct from the so-called 'Ca^{2+} inactivation' that is reported in experiments in skeletal muscle (see Section E), because such inactivation takes place in the presence of physiological [Mg^{2+}] and consequently seems unlikely to be mediated by the Ca^{2+}/Mg^{2+} inhibition site.[37]

In cardiac muscle, this inhibitory site also shows little specificity between Ca^{2+} and Mg^{2+} (see Ref. 36), but the K_D is much higher than in skeletal muscle, with estimates ranging from about ~0.4 to 1 mM in SR vesicle experiments[42,43] to ~5–15 mM in single channel observations (at 250 mM K^+ or Cs^+).[30,44] It is not clear precisely to what extent the Ca^{2+}/Mg^{2+} inhibitory site will be occupied by Mg^{2+} in cardiac muscle in-vivo, though it is likely to be comparatively little.[44] Nevertheless, irrespective of whether or not the site is substantially occupied in the presence of physiological [Mg^{2+}], it can still be concluded that Ca^{2+} will have comparatively minor effects on this site, because even if it reaches high concentrations locally near the site at the peak of Ca^{2+} release, this will at most cause a mild increase in the level of occupancy, given that there is millimolar Mg^{2+} already present.

Mg^{2+} also has a second important inhibitory action on skeletal and cardiac RyRs, in that it competes with Ca^{2+} for the activation site.[36,39,40,45,46] This second effect is emphasized in the literature so much that the important action of Mg^{2+} on the Mg^{2+}/Ca^{2+} site is often overlooked. Ca^{2+} has ~1000 fold higher affinity for the activation site than Mg^{2+} (see Refs. 36 and 38), so the presence of physiological [Mg^{2+}] (1 mM) will significantly inhibit the ability of Ca^{2+} to activate the RyR when the [Ca^{2+}] is ~1 μM or below, but will have little or no effect at higher [Ca^{2+}]. Thus, if [Ca^{2+}] rises sufficiently high (e.g > 5 μM) near the activation site, it will displace any Mg^{2+} on this site. So even a relatively small, local rise in [Ca^{2+}], be it from Ca^{2+} inflow, neighboring RyRs or a brief opening of the particular RyR itself, will allow Ca^{2+} to exert its maximum stimulatory effect on the RyR via this activation mechanism (e.g. see Fig. 3, Ref. 39). In contrast, even a much larger local rise in [Ca^{2+}] will not remove the inhibitory effect of Mg^{2+} on the Ca^{2+}/Mg^{2+} site, but instead slightly increase the net inhibition.

It should be noted that even in the presence of adenine nucleotide, an important stimulator of channel activity,[46] Mg^{2+} still exerts a powerful inhibitory effect on the mammalian skeletal RyR via the Ca^{2+}/Mg^{2+} site. At physiological ionic strength with 5 mM AMP and maximum Ca^{2+} activation (20 μM), ryanodine binding is reduced more than 10 fold by 1 mM Mg^{2+}.[40] Similarly, Ca^{2+}-induced Ca^{2+} release from SR vesicles is reduced about 6 fold in the presence of 1 mM Mg^{2+}.[39] These values were obtained in the presence of 100 mM Cl^- which is known to potentiate the ability of Ca^{2+} to activate Ca^{2+} release.[40,47,48] Thus, it would appear that in a mammalian skeletal muscle with physiological [Mg^{2+}], no matter how high Ca^{2+} rises, it will be relatively ineffective at initiating Ca^{2+} release because of the potent inhibition exerted by Mg^{2+} via the Ca^{2+}/Mg^{2+} inhibition site. In other words, Ca^{2+}-induced Ca^{2+} release (CICR) appears to be very damped down in mammalian skeletal muscle and consequently unable to contribute substantially to the high rate of Ca^{2+} release occurring during normal excitation-contraction (E-C) coupling,[45] unless the Ca^{2+}/Mg^{2+} inhibition is reduced (see later).

In amphibian and avian fibers, where there are two different isoforms of the RyR, α and β, there is some apparent disagreement about the level of Ca^{2+}/Mg^{2+} inhibition. One report[49] found that only half of the RyRs were appreciably inhibited by high [Ca^{2+}] (0.5 mM) and another report[50] indicated that high [Ca^{2+}] had a smaller inhibitory effect on ryanodine binding to the β isoform than to the α isoform. On the other hand, Murayama & Ogawa[51] found that, at 1 mM, Ca^{2+} and Mg^{2+} both caused substantial inhibition of ryanodine binding to both isoforms in the presence of adenine nucleotide, provided the ionic strength was close to physiological. In single channel measurements of avian RyRs in the presence of ATP, Ca^{2+} activation of both the α and β isoforms was almost completely abolished by 0.5mM Mg^{2+}, but a similar high [Ca^{2+}] was relatively ineffective.[32] It is possible that many of these differences could be attributed to the substantial reduction in Ca^{2+}/Mg^{2+} inhibition at increased ionic strength[38,40,51,52] and to its complete loss after some treatments,[30] as well as to the potentiating effect of Cl^-.[40,47,48] Irrespective of this, it is clear from other experiments that CICR in amphibian fibers is very greatly depressed at physiological [Mg^{2+}], even in the presence of ATP.[45] Addition of 0.55 mM Mg^{2+} completely blocked all CICR in frog muscle triads stimulated with 10 μM Ca^{2+} and 2 mM ATP.[53] Furthermore, using a skinned fiber preparation in which normal E-C coupling mechanism was still functional, Lamb & Stephenson[54] showed that some CICR could be observed in the presence of 1mM Mg^{2+}, but that the rate of release was more than 30 times lower than that elicited by depolarization. As the rise in [Ca^{2+}] near the activation site must have been considerable during even the subdued level of CICR, the inhibitory effect of Mg^{2+} on the activation site should concomitantly have become very small, thus indicating that the inhibition of CICR was brought about by Mg^{2+} acting on the Ca^{2+}/Mg^{2+} inhibitory site.

Thus, it appears that in both mammalian and amphibian muscle fibers, CICR is greatly inhibited at physiological [Mg^{2+}] owing to the presence of Mg^{2+} on the Ca^{2+}/Mg^{2+} site, and raising the [Ca^{2+}] to extremely high levels will not reduce this inhibitory effect. This would seem to argue against a *simple* CICR mechanism inducing a large part of the Ca^{2+} release in normal E-C coupling.[55,56] Furthermore, in order to account for normal E-C coupling, it appears that voltage-sensor activation must either overcome or remove this inhibitory effect of Mg^{2+} on the RyRs. As the voltage-sensors are unable to activate the RyRs in the presence of 10 mM Mg^{2+}, Lamb & Stephenson[37,41,57,58] have proposed that the voltage-sensors normally activate the RyRs by lowering the affinity of the Ca^{2+}/Mg^{2+} site by 10 or 20 fold, causing Mg^{2+} to rapidly vacate the site when there is only 1 mM Mg^{2+} present, thereby allowing the stimulatory action of ATP to activate the channel to some degree. Once this has happened, Ca^{2+} is able to further activate the same channel, as well as other channels where the Mg^{2+} inhibition has been relieved, greatly reinforcing channel activation. It is important to note that this mechanism is distinctly different from simple CICR, in that deactivation of the voltage-sensors will rapidly stop all Ca^{2+} release by allowing Mg^{2+} to bind again to the Ca^{2+}/Mg^{2+} site and once again close the channel even in the presence of a high local $[Ca^{2+}]$.

In cardiac muscle, it appears that there is substantially less inhibition via the Ca^{2+}/Mg^{2+} inhibitory site, so if the $[Ca^{2+}]$ rises high enough near the RyRs to overcome the inhibitory action of Mg^{2+} on the activation site, it should trigger strong CICR, unless some additional inhibitory process subsequently limits it (see Sections D & E). It is possible that phosphorylation of the RyRs in cardiac muscle[59] further reduces the inhibitory effect of Mg^{2+}, perhaps allowing the RyRs to respond maximally to a Ca^{2+} stimulus even in the presence of 1 mM Mg^{2+}. In skeletal muscle, however, it is unlikely that there is a comparable effect, even though phosphorylation has been observed to remove Mg^{2+} inhibition on the RyR,[60] because this would mean that the channels would be substantially open in a resting muscle due to the stimulatory effect that ATP exerts even at very low $[Ca^{2+}]$.[39,57] Furthermore, in skeletal muscle it is known that the voltage-sensor can activate and deactivate Ca^{2+} release on a millisecond time-scale, and it seems unlikely that this is achieved by phosphorylation and dephosphorylation.

INHIBITION OF RyRs BY REDUCTION IN SR LUMINAL $[Ca^{2+}]$

It has been shown in RyR bilayer experiments that raising the $[Ca^{2+}]$ on the SR luminal side, over the range 0.01 to 1 mM, increases the open probability of both cardiac and skeletal (sheep) RyRs, provided that there was a stimulatory cofactor, such as ATP or sulmazole, present on the cytoplasmic side.[61,62] Similarly, experiments with isolated triads from skeletal muscle of rabbits and frogs have shown that the rate constant of Ca^{2+} efflux, and hence presumably the open probability of the RyRs, increases with increased luminal

[Ca^{2+}].[63] These studies concluded that luminal Ca^{2+} directly regulates RyR gating. A recent report [64] has suggested that the stimulatory effect of SR luminal Ca^{2+} is not due to some luminal site of action, but is instead caused by Ca^{2+} passing through the RyR and acting at the cytoplasmic activation site. Irrespective of the exact mechanism, it appears that SR luminal Ca^{2+} has an important stimulatory effect on RyR open probability, or put conversely, that depletion of SR Ca^{2+} in-vivo should have an inhibitory effect on Ca^{2+} release in addition to its effect of reducing the Ca^{2+} gradient. It is yet to been shown that this inhibitory effect also occurs with physiological [Mg^{2+}] on the cytoplasmic side, when Mg^{2+} is competing with Ca^{2+} for the activation site and, in skeletal muscle in particular, is also acting on the Ca^{2+}/Mg^{2+} inhibitory site.

Ca^{2+} INACTIVATION AND Ca^{2+} SPARKS IN CARDIAC AND SKELETAL MUSCLE

In cardiac muscle, it has been proposed that Ca^{2+} not only activates Ca^{2+} release, but also inactivates it,[65] although the latter is not generally agreed.[66,67] In skeletal muscle, depolarization-induced Ca^{2+} release is reported to show an initial peak and then decrease ~ 2 to 5 fold over the following 10 to 20ms, and it has been proposed that this decrease is due to released Ca^{2+} inactivating further release to some extent.[55,68–71] In both cardiac and skeletal muscle it appears unlikely that such 'Ca^{2+}-inactivation' is mediated by the Ca^{2+}/Mg^{2+} inhibitory site (see Section C). Ca^{2+}-specific inactivation in skeletal muscle might be due to some calmodulin-dependent mechanism.[72]

Alternatively, it is possible that the reported inactivation is not in fact dependent on Ca^{2+} but instead depends on channel opening per se.[73] This might explain why liberation of caged Ca^{2+} could not induce such inactivation[74] and also account for why the RyRs appear to inactivate after passing a relatively constant number of Ca^{2+} ions over ~2 ms and then remain inactivated for ~50 ms.[70,71] If the Ca^{2+} efflux transiently generates a substantial SR potential change (i.e. negative inside), it could be proposed that this latter type of inactivation is the same as the voltage- and P_o-dependent inactivation mechanism observed in lipid bilayer experiments[14,14A,34] (see Section B). However, it would also have to be proposed that the onset and recovery of this latter mechanism occur ~100 to 1000 times faster in a muscle fibre than presently observed in lipid bilayers. If instead these are distinctly different mechanisms, one must ask why the RyR inactivation apparently occurring in a functioning muscle, be it dependent on P_o or Ca^{2+}, has not been reported in single channel measurements? Interestingly, the inactivation behavior suggested by Jong *et al.*[70,71] to occur in skeletal muscle fibers seems quite consistent with the properties of Ca^{2+} sparks,[74A] where Ca^{2+} release evidently ceases after several milliseconds despite the expected tendency for it to be self-regenerating, suggesting that the sparks are terminated by the onset of an inactivation process.

'ADAPTATION' IN CARDIAC AND SKELETAL MUSCLE

'Adaptation' has also been described in a number of cellular preparations by Györke and colleagues.[1,75–77] However, in view of the very different time-courses and probable mechanisms involved, these studies have to be divided into two separate groups. The first group describe an 'adaptation' process occurring over 10 to 100 s in skeletal and cardiac muscle SR vesicles[75] and in myotubes.[77] It was shown that in the presence of a low concentration of an agonist such as caffeine, Ca^{2+} release eventually declined, but could be increased again by increasing the agonist concentration. Importantly, the same effect was seen with many different agonists, which indicates that the RyRs were not 'adapting' to stimulus itself , but instead to something associated directly or indirectly with the release process. Moreover, it is probably quite inappropriate to regard the decrease in Ca^{2+} release as 'adaptation', because it has been shown in many preparations that a *constant* concentration of caffeine can induce *cyclic* release of Ca^{2+} (see Refs. 78–80). Clearly in such cases, the preparation has not 'adapted' to the stimulus, because it subsequently responds vigorously despite the continuous presence of the stimulus. Instead, it seems more appropriate to regard the decrease in Ca^{2+} release as being due to the onset of some inhibitory process which subsequently dissipates (either with time alone or because Ca^{2+} release has ceased), allowing the original stimulus to once again induce Ca^{2+} release. This could be, for example, due to Ca^{2+} efflux causing some degree of Ca^{2+}-dependent or opening-dependent inactivation of the RyRs (see Sections B & E) or SR depletion (Section D), the latter being possible because some of the released Ca^{2+} may either have remained bound in the cytoplasm or been taken up into different SR compartments. It is also possible that the action of the SR pump contributes to the effect.[81] This type of 'adaptation' is somewhat analogous to the relative refractory period occurring after an action potential, where a subsequent response can be elicited either by a larger depolarizing stimulus at early times or the original stimulus at later times.

The second group of experiments showing 'adaptation' report a decline in Ca^{2+} release in ~10 ms in mammalian cardiac cells[1] and crustacean skeletal muscle.[76] In these studies, Ca^{2+} release was triggered by voltage-activated Ca^{2+} currents, and fluorescent dyes were used to indicate intracellular $[Ca^{2+}]$. These studies showed that in the presence of a steady inward trigger current, the [Ca^{2+}] initially increased rapidly and then settled at a steady submaximal level, but could subsequently be rapidly increased to a higher level by a larger trigger current. This behavior was attributed to 'adaptation' of the individual RyRs. However, as there is no evidence that individual RyRs can genuinely 'adapt' (see Sections A&B), it seems the effects are probably due to some other cause. One possibility is that the Ca^{2+} released by the first stimulus declined due to RyR inactivation and/or local SR depletion and that the second stimulus caused Ca^{2+} release from a different pool of RyRs. Györke & Györke[77] have argued against this, but their data referred to a very different 'adaptation' process (see above), and only showed reactivation within 1 μm wide regions and consequently did not indicate whether or not the same pool of RyRs was involved. Another possibility is that the two stimuli both induced activation of members of a homogenous pool of RyRs, with the overall response determined by the transient redistribution of the channels between resting, open and inactivated states, which includes the case described by Zahradnikova & Zahradnik[28] where there are different activity modes. As noted in Section A, this latter model does not explain 'adaptation' in a single RyR, but it is quite adequate to explain the activity in a population of RyRs

REPOLARIZATION-INDUCED STOP OF Ca^{2+} RELEASE (RISC) IN SKELETAL MUSCLE

Finally, it has been reported by Suda and colleagues [82–84] that repolarization is able to stop caffeine-induced Ca^{2+} release in cultured mammalian muscle cells. This implies that deactivation of the voltage-sensors exerts an inhibitory effect on the RyRs preventing them being directly activated by caffeine. In these studies, changes in Ca^{2+} release were inferred from the rate of change of the intracellular [Ca^{2+}], without taking into account the effect of Ca^{2+} uptake by the SR pump. It was found that the phenomenon only occurred if the voltage-sensors were inducing part of the total SR Ca^{2+} release at the time of repolarization. It now appears that all the results can be simply explained as being due to the withdrawal of a stimulus rather than to voltage-sensor deactivation actually inhibiting the RyRs. Caffeine directly activates the RyR, and its effectiveness is influenced by both the cytoplasmic [Ca^{2+}] and the degree of Ca^{2+} loading in the SR,[45] with the latter probably being related to the luminal Ca^{2+} effects described in Section D. Suda and colleagues[82–84] showed that when caffeine and a depolarizing stimulus were together inducing Ca^{2+} release, stopping the depolarizing stimulus caused the [Ca^{2+}] to decrease more rapidly than it did when Ca^{2+} release was stimulated by caffeine alone. Depolarization and caffeine together evidently induce faster Ca^{2+} release than either stimulus alone. Such increased Ca^{2+} release could be expected to cause an increased level of inhibition on the release process, be it by Ca^{2+}-dependent or opening-dependent inactivation (Section B) or SR depletion effects (Section D). Furthermore, if the [Ca^{2+}] is either steady or declining, Ca^{2+} uptake must also have increased and be roughly matching the increased Ca^{2+} release. Thus, if the depolarizing stimulus is suddenly withdrawn, Ca^{2+} reuptake will considerably exceed the rate of Ca^{2+} release that is then elicited by caffeine alone, particularly if there has been any increase in total inhibitory action on RyRs. Consequently, it is possible to account for all the results without supposing that the deactivation of the voltage-sensors has some direct inhibitory effect on the RyRs.

CONCLUDING REMARKS

A number of quite different negative control mechanisms have been observed in isolated RyRs, including Ca^{2+}/Mg^{2+} inhibition, voltage- and P_o- dependent inactivation, and control by SR luminal [Ca^{2+}], but there is considerable evidence against the existence of a specific 'adaptation' mechanism in isolated RyRs. The adaptation phenomenon is likely to be due to prolonged activation and subsequent deactivation in response to the Ca^{2+} spike stimulus produced by the flash photolysis technique. When true Ca^{2+} 'steps' are produced by rapid solution exchange the decline in RyR activity is due to voltage- and P_o-dependent inactivation.

At the cellular level, Mg^{2+} inhibition can be understood in terms of the Ca^{2+}/Mg^{2+} inhibition mechanism plus an additional inhibitory action of Mg^{2+} at the Ca^{2+}-activation site. 'Ca^{2+} inactivation' described in skeletal and cardiac muscle cells is often confused with Ca^{2+}/Mg^{2+} inhibition. Such 'Ca^{2+} inactivation' may be a true Ca^{2+}-specific process or might instead depend on channel opening. In either case, it is quite possibly the same phenomenon as the apparent inactivation occurring with Ca^{2+} sparks. Other negative control processes observed in muscle cells, such as 'adaptation' and RISC, are probably not due to novel single channel inhibitory processes but instead are due to a composite of other processes. Finally, it must be said that many aspects of 'inactivation' observed at the cellular level, such as with 'Ca^{2+} sparks' and 'Ca^{2+} inactivation', cannot be accounted for by the properties of isolated RyRs as presently described.

REFERENCES

1. Yasui K, Palade P, Györke S. Negative control mechanism with features of adaptation controls Ca^{2+} release in cardiac myocytes. *Biophys J* 1994; **67**: 457–460.

2. Györke S, Fill M. Ryanodine receptor adaptation: control mechanism of Ca^{2+} release in heart. *Science* 1993; **260**: 807–809.
3. Valdivia HH, Kaplan JH, Ellis-Davies GCR *et al.* Rapid adaptation of cardiac ryanodine receptors: modulation by Mg^{2+} and phosphorylation. *Science* 1995; **267**: 1997–2000.
4. Zucker RS. The calcium concentration clamp: spikes and reversible pulses using the photolabile chelator DM-nitrophen. *Cell Calcium* 1993; **14**: 87–100.
5. Lamb GD, Fryer MF, Stephenson DG. Ca^{2+}-induced Ca^{2+} release in response to flash photolysis. *Science* 1994; **263**: 986–987.
6. Escobar AL, Cifuentes F, Vergara JL. Detection of Ca^{2+}-transient elicited by flash photolysis of DM-nitrophen with a fast calcium indicator. *FEBS Let.* 1995; **364**: 335–338.
7. Lamb GD, Stephenson DG. Activation of ryanodine receptors by flash photolysis of caged Ca^{2+}. *Biophys J* 1995; **68**: 946–948.
8. Ellis-Davies-Davies GCR, Kaplan JH, Barsotti RJ. Laser photolysis of caged calcium: rates of calcium release by nitophenyl-EGTA and DM-nitrophen. *Biophys J* 1996; **70**: 335–338.
9. Velez P, Györke S, Escobar AL *et al.* Adaptation of single cardiac ryanodine receptor channels. *Biophys J* 1997; **72**: 691–697.
10. Lamb GD. Ryanodine receptor 'adaptation': a flash in the pan? *J Muscle Res Cell Mot* 1997; **18**: 611–616.
11. Zahradnikova A, Györke S. Rapid activation and deactivation of the cardiac ryanodine receptor by Ca^{2+} spikes induced by photolysis of caged Ca^{2+}. *Biophys J* 1997; **72**: A375.
12. Györke S, Fill M. 1994 Ca^{2+}-induced Ca^{2+} release in response to flash photolysis (reply). *Science* 1994; **263**: 987–988.
13. Schiefer A, Meissner G, Isenberg G. Ca^{2+} activation and Ca^{2+} inactivation of canine reconstituted cardiac sarcoplasmic reticulum Ca^{2+} release channels. *J Physiol* 1995; **489**: 337–348.
14. Sitsapesan R., Montgomery RAP, Williams AJ. New insights into the gating mechanisms of cardiac ryanodine receptors revealed by rapid changes in ligand concentration. *Circ Res* 1995; **77**: 765–772.
14A. Laver Dr, Lamb GD. Inactivation of Ca^{2+} release channels (ryanodine receptors RyR1 and RyR2) with rapid steps in [Ca^{2+}] and voltage. *Biophys J* 1998; **74**:2352–2364.
15. Laver DR, Curtis BA. Response of ryanodine receptor channels to Ca^{2+} steps produced by rapid solution exchange. *Biophys J* 1996; **71**: 732–741.
16. Zahradnikova A, Zahradnik I. Description of modal gating of the cardiac calcium release channel in planar lipid membranes. *Biophys J* 1995; **69**: 1780–1788.
17. Armisén R, Sierralta J, Vélez P, Naranjo D *et al.* Modal gating in neuronal and skeletal muscle ryanodine-sensitive Ca^{2+} release channels. *Am J Physiol* 1996; **271**: C144–C153.
18. Györke S, Velez P. Suarez-Isla B *et al.* Activation of single cardiac and skeletal ryanodine receptor channels by flash photolysis of caged Ca^{2+}. *Biophys J* 1994; **66**: 1879–1886.
19. Cheng H, Lederer MR, Xiao RP *et al.* Excitation-contraction coupling in heart: new insights from Ca^{2+} sparks. *Cell Calcium* 1996; **20(2)**: 129–140.
20. Velez P, Escobar A, Zhogbi ME *et al.* Divalent cation sensitivity of ryanodine receptor (RyR) activation and adaptation. *Biophys J* 1997; **72**: A375.

21. Soler F, Fernandez-Belda F, Gomez-Fernandez JC. The Ca^{2+} release channel in junctional sarcoplasmic reticulum: gating and blockade by cations. *Int J Biochem* 1992; **24**: 903–909.
22. Xiao RP, Valdivia HH, Bogdanov K *et al.* The immunophilin FK506-binding protein modulates Ca^{2+} release channel closure in rat heart. *J Physiol* 1997; **500 (2)**: 343–354.
23. Lamb GD, Stephenson DG. Effects of FK506 and rapamycin on excitation-contraction coupling in skeletal muscle fibres of the rat. *J. Physiol* 1996; **494**: 569–576.
24. Ahern GP, Junankar PR, Dulhunty AF. Ryanodine receptors from rabbit skeletal muscle are reversibly activated by rapamycin. *Neuroscience Let* 1997; **225**: 81–84.
25. Tang Y, Othmer HG. A model of calcium dynamics in cardiac myocytes based on the kinetics of ryanodine-sensitive calcium channels. *Biophys J* 1994; **67**: 2223–2235.
26. Sachs F, Qin F, Palade P. Models of Ca^{2+} release channel activation. *Science* 1995; **267**: 2010–2011.
27. Keizer J, Levine L. Ryanodine receptor adaptation and Ca^{2+}-induced Ca^{2+} release-dependent Ca^{2+} oscillations. *Biophys J* 1996; **71**: 3477–3487.
28. Zahradnikova A, Zahradnik I. A minimal gating model for the cardiac calcium release channel. *Biophys J* 1996; **71**: 2996–3012.
29. Cheng H, Fill M, Valdivia H *et al.* Models of Ca^{2+} release channel activation. *Science* 1995; **267**: 2009–2010.
30. Laver DR, Roden LD, Ahern GP *et al.* Cytoplasmic Ca^{2+} inhibits the ryanodine receptor from cardiac muscle. *J Memb Biol* 1995; **147**: 7–22.
31. Imredy JP, Chen W, Zhang L *et al.* Voltage-dependent rectification of the recombinant rabbit skeletal muscle ryanodine Expressed in HEK 293 cells. *Biophys J* 1996; **70**: A124.
32. Percival AL, Williams AJ, Kenyon JL *et al.* Chicken skeletal muscle ryanodine receptor isoforms: ion channel properties. *Biophys J* 1994; **67**: 1834–1850.
33. Ma J. Desensitization of the skeletal muscle ryanodine receptor: evidence for heterogeneity of calcium release channels. *Biophys J* 1995; **68**: 893–899.
36. Ma J, Bhat MB, Zhao J. Rectification of skeletal muscle ryanodine receptor mediated by FK506 binding protein. *Biophys J* 1995; **69**: 2398–1404.
35. Chen W, Zhang L, MacLennan DH. Asymmetrical blockade of the Ca^{2+} release channel (ryanodine receptor) by 12-KDa FK506 binding protein. *Proc Natl Acad Sci USA* 1994; **91**: 11953–11957.
36. Laver DR, Baynes TM, Dulhunty AF. Magnesium inhibition of ryanodine - receptor calcium channels: Evidence for two independent mechanisms. *J Mem Biol* 1997; **156**: 213–229.
37. Lamb GD. Ca^{2+} inactivation, Mg^{2+} inhibition and malignant hyperthermia. *J Muscle Res Cell Mot* 1993; **14**: 554–556.
38. Laver DR, Owen VJ, Junankar PR *et al.* Reduced inhibitory effect of Mg^{2+} on ryanodine receptor-Ca^{2+} release channels in malignant hyperthermia. *Biophys J* 1997; **73**: 1917–1923.
39. Meissner G, Darling E, Eveleth J. Kinetics of rapid Ca^{2+} release by sarcoplasmic reticulum. Effect of Ca^{2+}, and Mg^{2+}, and adenine nucleotides. *Biochem* 1986; **25**: 234–244.
40. Meissner G, Rios E, Tripathy A *et al.* Regulation of skeletal muscle Ca^{2+} release channel (ryanodine receptor) by Ca^{2+} and monovalent cations and anions. *J Biol Chem* 1997; **272**: 1628–1638.

41. Lamb GD, Stephenson DG. Importance of Mg^{2+} in excitation-contraction coupling in skeletal muscle. *News Physiol Sci* 1992; **7**: 270–274.
42. Meissner G, Henderson JS. Rapid calcium release from cardiac sarcoplasmic reticulum vesicles is dependent on Ca^{2+} and is modulated by Mg^{2+}, adenine nucleotide, and calmodulin. *J Biol Chem* 1987; **262 (7)**: 3065–3073.
43. Chu A, Fill M, Stefani E, Entman KL. Cytoplasmic Ca^{2+} does not inhibit the cardiac muscle sarcoplasmic reticulum ryanodine receptor Ca^{2+} channel, although Ca^{2+} induced Ca^{2+} inactivation of Ca^{2+} release is observed in native vesicles. *J Mem Biol* 1993; **135**: 49–59.
44. Xu L, Mann G, Meissner G. Regulation of cardiac Ca^{2+} release channel (ryanodine receptor) by Ca^{2+}, H^+, Mg^{2+}, and adenine nucleotides under normal and simulated ischemic conditions. *Circ Res* 1996; **79**: 1100–1109.
45. Endo M. Calcium release from sarcoplasmic reticulum. *Current Topics in Membranes and Transport* 1985; **25**: 181–230.
46. Meissner G. Ryanodine receptor/ Ca^{2+} release channels and their regulation by endogenous effectors. *Ann Rev Physiol* 1994; **56**: 485–508.
47. Hasselbach W, Migala A. Modulation by monovalent anions of calcium and caffeine induced calcium release from heavy sarcoplasmic vesicles. *Z Naturforsch* 1992; **47c**: 440–448.
48. Fruen BR, Kane PK, Mickelson JR *et al.* Chloride-dependent sarcoplasmic reticulum Ca^{2+} release correlates with increased Ca^{2+} activity of ryanodine receptors. *Biophys J* 1996; **71**:2522–2530.
49. Bull R, Marengo JJ. Sarcoplasmic reticulum release channels from frog skeletal muscle display two types of calcium dependence. *FEBS Lett* 1993; **331**: 223–227.
50. O'Brien J, Valdivia HH, Block BA. Physiological differences between the α and β ryanodine receptors of fish skeletal muscle. *Biophys J* 1995; **68**: 271–482.
51. Murayama T, Ogawa Y. Similar Ca^{2+} dependences of [^{3}H] ryanodine binding to α- and β-ryanodine receptors purified from bullfrog skeletal muscle in an isotonic medium. *FEBS* 1996; **380**: 267–271.
52. Shomer NH, Louis CF, Fill M *et al.* Reconstitution of abnormalities in the malignant hyperthermia-susceptible pig ryanodine receptor. *Am J Physiol* 1993; **33**: C125–C135.
53. Donoso P, Hidalgo C. pH-sensitive calcium release in triads from frog skeletal muscle. *J Biol Chem* 1993; **268**: 25432–25438.
54. Lamb GD, Stephenson DG. Control of calcium release and the effect of ryanodine in skinned muscle fibres of the toad. *J Physiol* 1990; **423**: 519–542.
55. Rios E, Pizarro G. Voltage sensors and calcium channels of excitation-contraction coupling. *News in Physiol Sci* 1988; 3: 223-227.
56. Jacquemond V, Csernoch L, Klein MG *et al.* Voltage-gated and calcium-gated calcium release during depolarization of skeletal muscle fibres. *Biophys J* 1991; **60**: 867–873.
57. Lamb GD, Stephenson DG. Effect of Mg^{2+} on the control of Ca^{2+} release in skeletal muscle fibres of the toad. *J Physiol* 1991; **434**: 507–528.
58. Lamb GD, Stephenson DG. Effect of intracellular pH and [Mg^{2+}] on excitation-contraction coupling in skeletal muscle fibres of the rat. *J Physiol* 1994; **489**: 331–339.
59. Hain J, Onoue H, Mayrleitner M *et al.* Phosphorylation modulates the function of the calcium release channel of sarcoplasmic reticulum from cardiac muscle. *J Biol Chem* 1995; **270 (5)**: 2074–2081.

60. Hain J, Nath S, Mayrleitner M *et al.* Phosphorylation modulates the function of the calcium release channel of sarcoplasmic reticulum from skeletal muscle. *Biophys J* 1994; **67**: 1823–1833.
61. Sitsapesan R, Williams AJ. Regulation of the gating of the sheep cardiac sarcoplasmic reticulum Ca^{2+}-release channel by luminal Ca^{2+}. *J Memb Biol* 1994; **137**: 215–226.
62. Sitsapesan R, Williams AJ. The gating of the sheep skeletal sarcoplasmic reticulum Ca^{2+} release channel is regulated by luminal Ca^{2+}. *J Memb Biol* 1995; **146**: !33—144.
63. Donoso P, Prieto H, Hidalgo C. Luminal calcium regulates calcium release in triads isolated from frog and rabbit skeletal muscle. *Biophys J* 1995; **68**: 507–515.
64. Tripathy A, Meissner G. Sarcoplasmic reticulum lumenal Ca^{2+} has access to cytosolic activation and inactivation sites of skeletal muscle Ca^{2+} release channel. *Biophys J* 1996; **70**: 2600–2615.
65. Fabiato A. Time and calcium dependence of activation and inactivation of calcium-induced release of calcium from the sarcoplasmic reticulum of a skinned cardiac purkinje cell. *J Gen Physiol* 1985; **85**: 247–289.
66. Cleemann L, Morad M. Role of Ca^{2+} channel in cardiac excitation-contraction coupling in the rat: evidence from Ca^{2+} transients and contraction. *J Physiol* 1991; **432**: 283–312.
67. Bers DM. *Excitation-contraction coupling and cardiac contractile Force.* Kluwer Academic Publishers, 1991: 258pp.
68. Simon BJ, Klein MG, Schneider MF. Calcium dependence of inactivation of calcium release from the sarcoplasmic reticulum in skeletal muscle fibres. *J Gen Physiol* 1991; **97**: 437–471.
69. Schneider MF. Control of calcium release in functioning skeletal muscle fibers. *Ann Rev Physiol* 1994; **56**: 463–484.
70. Jong DS, Pape PC, Chandler WK. Reduction of calcium inactivation of sarcoplasmic reticulum calcium release by Fura-2 in voltage-clamped cut twitch fibres from frog muscle. *J Gen Physiol* 1993; **102**: 333–370.
71. Jong DS, Pape PC, Baylor SM *et al.* Calcium inactivation of calcium release in frog cut muscle fibres that contain millimolar EGTA or Fura-2. *J Gen Physiol* 1995; **106**: 337–388.
72. Tripathy A, Xu L, Meissner G. Calmodulin activation and inhibition of skeletal muscle Ca^{2+} release channel (ryanodine receptor). *Biophys J* 1995; **69**: 106–119.
73. Pizarro G, Shirokova N, Tsugorka A *et al.* 'Quantal' calcium release operated by membrane voltage in frog skeletal muscle. *J Physiol* 1997;**501 (2)**: 289—303.
74. Hill D, Simon B. Use of "caged calcium" in skeletal muscle to study calcium-dependent inactivation of SR calcium release. *Biophys J* 1991; **59**: A293.
74A. Shirokova N, Rios E. Small evert Ca^{2+} release: a probable precursor of Ca^{2+} sparks in frog skeletal muscle. *J. Physiol* 1997; 502:3–11.
75. Dettbarn C, Györke S, Palade P. Many agonists induced "quantal Ca^{2+} release or adaptive behavior in muscle ryanodine receptors. *Mol Pharm* 1994; **46**: 502–507.
76. Györke S, Palade P. Ca^{2+} -dependent negative control mechanism for Ca^{2+} -induced Ca^{2+} release in crayfish muscle. *J Physiol* 1994; **476 (2)**: 315–322.

77. Györke I, Györke S. Adaptive control of intracellular Ca^{2+} release in C2C12 mouse myotubes. *Pflügers Arch* 1996; **431**: 838—843.
78. Lüttgau HC, Oetliker H. The action of caffeine on the activation of the contractile mechanism in striated muscle fibres. *J Physiol* 1968; **194**: 51–74.
79. Herrmann-Frank A. Caffeine- and Ca^{2+}-induced mechanical oscillations in isolated muscle fibres of the frog. *J Muscle Res Cell Mot* 1989; **10**: 437–445.
80. Yamazawa T, Takeshima H, Sakurai T *et al.* Subtype specificity of the ryanodine receptor for Ca^{2+} signal amplification in excitation-contraction coupling. *EMBO J* 1996; **15 (22)**:6172–6177.
81. Steenbergen JM, Fay FS. The quantal nature of calcium release to caffeine in single smooth muscle cells results from activation of the sarcoplasmic reticulum Ca^{2+}-ATPase. *J Biol Chem* 1996; **271**: 1821–1824.
82. Suda N, Penner R. Membrane repolarization stops caffeine-induced Ca^{2+} release in skeletal muscle cells. *Proc Natl Acad Sci USA* 1994; **91**: 5725–5729.
83. Suda N. Involvement of dihydropyridine receptors in terminating Ca^{2+} release in rat skeletal myotubes. *J Physiol* 1995; **481 (1)**: 105–112.
86. Suda N, Heinemann C. RISC (Repolarization -induced stop of caffeine-contracture) is not due to store depletion in cultured murine skeletal muscle. *Pflügers Arch* 1996; **432**: 948–951.

Chapter 15

Regulation of SR Calcium Release in Intact Ventricular Myocytes

Donald M. Bers

Much has been discovered studying sarcoplasmic reticulum (SR) Ca^{2+} release channels in SR vesicles and lipid bilayers. The focus here is on how SR Ca^{2+} release is regulated in intact mammalian ventricular myocytes, using fluorescent Ca^{2+} indicators, voltage clamp and confocal microscopy. Three major factors appear to contribute to the probability of spontaneous localized SR Ca^{2+} release events (or Ca^{2+} sparks) in resting myocytes: (1) cytosolic [Ca^{2+}], (2) SR Ca^{2+} content and (3) time after previous activity (i.e. recovery from adapted or inactivated state). These same 3 factors function during excitation-contraction (E-C) coupling and can explain rest potentiation of twitches, increased fractional SR Ca^{2+} release at higher SR Ca^{2+} loads and Ca^{2+} overload. Since SR Ca^{2+} release is sensitive to both I_{Ca} and SR Ca^{2+} load, we have controlled (and measured) these parameters. At constant SR Ca^{2+} load and Ca^{2+} current (I_{Ca}) in intact cells we have found that SR Ca^{2+} release is increased by Ca^{2+}-Calmodulin dependent protein kinase (CaMKII) and the immunosupressant drug FK-506 (which may interfere with the interaction between the Ca^{2+} release channel and the FK binding protein) and is reduced by the Ca^{2+} channel agonist Bay K 8644, CaMKII inhibitors and during ventricular hypertrophy and failure. Thus the regulation of the SR Ca^{2+} release channel in the intact cell is an important factor in cellular cardiac function.

INTRODUCTION

We have learned a tremendous amount about the fundamental properties and regulation of the cardiac SR Ca^{2+} release channel or ryanodine receptor (RyR) from biochemical studies in isolated microsomes or SR vesicles from the heart[1] (and for a review see Ref. 2). In addition, the ability to incorporate the cardiac RyR into lipid bilayers and record current carried by single RyR channels has provided unique insights into the properties of these channels, including their conductance, open time, open probability and factors which alter channel gating properties.[2–10] For example, the regulation of RyR channel gating by Ca^{2+}, ATP, Mg^{2+}, phosphorylation and other factors have been carefully studied in this type of well controlled environment. These important types of studies are the focus of several other chapters in this book and have been essential in better understanding of Ca^{2+}-induced Ca^{2+}-release (CICR) as the critical mechanism in cardiac excitation-contraction (E-C) coupling.

While these fundamental studies have been very important, it is equally important to evaluate how the SR Ca^{2+} release channel behaves in intact cells in the normal physiological environment. This is clearly important because the RyR channel appears to be affected by so many different things. Indeed, rather subtle changes in gating properties could have a major impact on the physiological function of the SR Ca^{2+} release channel and the cellular Ca^{2+} transient and contraction which ensue in the heart. In this chapter I will discuss work focusing on factors that alter or regulate SR Ca^{2+} release channel function in intact mammalian ventricular myocytes. This work complements and extends work in more isolated systems and has the inherent disadvantage that it is harder to maintain control of all of the potentially relevant factors in the complex environment of the cell. On the other hand, the importance of this kind of data makes these kind of studies crucial. These studies have generally involved measurements of intracellular $[Ca^{2+}]$ ($[Ca^{2+}]_i$) with fluorescent Ca^{2+} indicators and have often been coupled with whole cell voltage clamp to allow control of membrane potential (E_m) while allowing measurement of transmembrane currents (e.g. Ca^{2+} current, I_{Ca} and Na^+/Ca^{2+} exchange $I_{Na/Ca}$). These studies help build a more comprehensive understanding of cardiac E-C coupling and the regulation of SR Ca^{2+} release.

METHODS

Ventricular myocytes isolated from rat, rabbit and ferret hearts using standard methods were studied in a chamber on the stage of an inverted microscope equipped for epifluorescence measurement of $[Ca^{2+}]_i$ using the indicator indo-1, cell shortening using a video edge detector and ionic currents using whole cell voltage clamp and incorporating a rapid solution switching device. Some experiments were also done using laser scanning confocal fluorescence microscopy (LSCFM). The details of the experimental methods are described more completely in the individual papers from which the figures below and the data discussed were obtained.[11–13] Most of the experiments described were done with a default rate of stimulation (0.5 Hz), at 23°C and a modified normal Tyrode's (NT) solution containing (in mM): 140 NaCl, 6 KCl, 1 $MgCl_2$, 2 $CaCl_2$ (1 in rat experiments), 10 glucose and 5 HEPES at pH 7.40.

RESULTS AND DISCUSSION

Ca^{2+} Sparks: Spontaneous Openings of SR Ca^{2+} Release Channels in Resting Cells

During rest there must be a finite rate of Ca^{2+} leak from the SR. In the steady state this resting Ca^{2+} leak may be compensated by Ca^{2+} re-uptake by the SR Ca-ATPase, such that the SR Ca^{2+} content does not change. In some cells (e.g. rat ventricular myocytes) this appears to be the case, while in other cells the rate of resting Ca^{2+} leak exceeds the rate of reuptake such that a net loss of Ca^{2+} occurs during rest (e.g. in rabbit or guinea-pig ventricular myocytes). This issue will be discussed more explicitly below (in *Restitution of E-C coupling...*).

Cheng *et al.*[14] first described local spontaneous SR Ca^{2+} release events in ventricular myocytes using confocal microscopy and called them Ca^{2+} sparks. These Ca^{2+} sparks are due to SR Ca^{2+} release via RyR channels and represent release through a single channel, or more likely, via a small cluster of RyR release channels working as a single functional unit.[15–16] Ca^{2+} sparks can also be activated by Ca^{2+} entry via L-type Ca^{2+} current (I_{Ca}).[17] Indeed, the global Ca^{2+} transient during the normal twitch is also made up of a very large number of Ca^{2+} sparks occurring throughout the cell almost simultaneously. During the twitch the Ca^{2+} sparks are synchronized by the action potential and L-type Ca^{2+} current which triggers SR Ca^{2+} release via Ca^{2+}-induced Ca^{2+}-release.

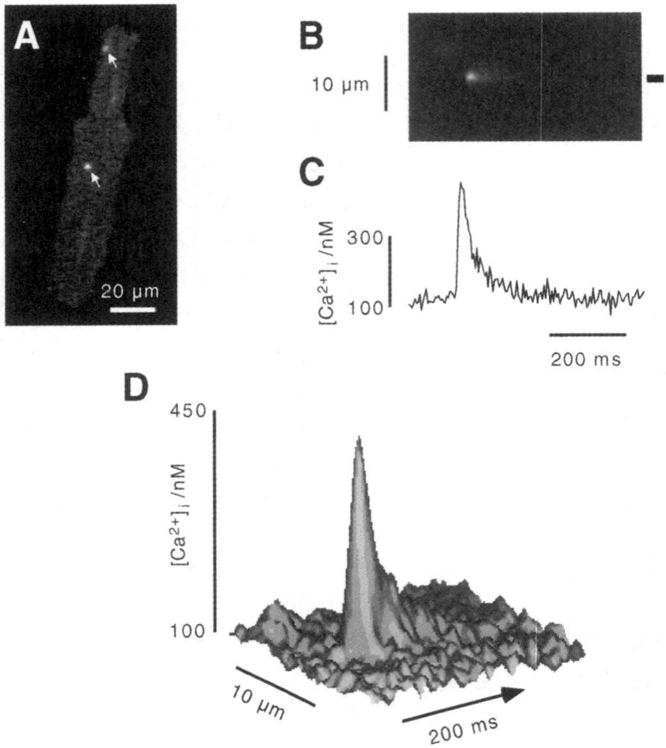

Fig. 15.1 Ca^{2+} sparks in an isolated mouse ventricular myocyte. **A.** Two dimensional laser scanning confocal fluorescence image of a single myocyte loaded with the Ca^{2+}-sensitive indicator fluo-3, exhibiting two Ca^{2+} sparks (arrows). **B.** Line scan image along the long axis of a mouse ventricular myocyte (only part of the scanned cell length is shown). Scans were repeated every 4 ms and stacked from left to right. Distance along the cell is in the vertical direction. **C.** Line graph of $[Ca^{2+}]_i$ measured at the spot indicated by the bar in B for the same line scan image (~1 μm). **D.** Surface plot of $[Ca^{2+}]_i$ during a Ca^{2+} spark, indicating the temporal and spatial spread of Ca^{2+} (original figure prepared by Dr. Jörg Hüser).

Figure 15.1 shows examples of spontaneous Ca^{2+} sparks in a resting mouse ventricular myocyte. Panel A shows a 2-dimensional image of a myocyte where we happened to catch two Ca^{2+} sparks (indicated by the arrows). To increase temporal resolution it has been customary to use the line scan mode of the confocal microscope. In this cell we scanned the whole length of the cell every 2 ms along a line avoiding nuclei. Figure 15.1B shows a line scan image with a single prominent Ca^{2+} spark, where distance along the cell length is shown in the vertical dimension and time along the horizontal dimension. Figure 15.1C shows the $[Ca^{2+}]_i$ in the narrow region of the cell where the Ca^{2+} spark occurs (indicated by the bar in Fig. 15.1B). The surface plot in Fig. 15.1D shows the time and spatial dependence of local $[Ca^{2+}]_i$ during this single Ca^{2+} spark. The decline of local $[Ca^{2+}]_i$ during the Ca^{2+} spark is largely due to Ca^{2+} diffusing away from the site of release, but SR Ca^{2+} reuptake also contributes significantly to $[Ca^{2+}]_i$ decline.[18] Thus the rate of SR Ca^{2+} pumping can affect the spatial and temporal spread of Ca^{2+} sparks, and may thereby influence their ability to activate neighboring regions.

Returning to the issue of resting Ca^{2+} leak from the SR, it seems likely that much, if not virtually all of this resting leak is through occasional openings of SR Ca^{2+} release channels and Ca^{2+} sparks. Bassani and Bers[19] measured the rate of unidirectional Ca^{2+} leak from the SR in intact rat and rabbit ventricular myocytes using global Ca^{2+} measurements (0.3 mmol/l cytosol/s). This flux could be explained by a spark frequency of 2–10 $pL^{-1}\cdot s^{-1}$ with each corresponding to a 2 pA Ca^{2+} flux lasting 3 ms[8] or 4 pA for 10 ms.[14] Furthermore 10 Ca^{2+} sparks pL^{-1} s^{-1} would correspond to 300 Ca^{2+} sparks per cell per second (for a typical 30 pL cell).

The normal Ca^{2+} transient associated with the twitch in ventricular myocytes is also composed of a temporal and spatial summation of many Ca^{2+} sparks which are synchronized by the action potential and Ca^{2+} influx via Ca^{2+} channels.[17,20,21] To produce the peak SR Ca^{2+} release flux (3 mM/s) which has been estimated by Wier *et al.*[22] would require simultaneous activation of about 9,000 Ca^{2+} sparks per cell (as above). If a Ca^{2+} spark flux of 40 fC (=4 pA × 10 ms) were due to a single RyR this would require only 0.5% of the ~1.7 million RyRs in a 30 pL rat ventricular myocyte to open.[23] Even if there were 6 RyR required to produce the Ca^{2+} spark flux above, this would still only require 3–4% of the total number of cellular RyR. Thus twitch activation only requires a small percentage of available release channels to function at any given twitch. Interestingly, estimates of the fraction of L-type Ca^{2+} channels required to produce the measured cellular Ca^{2+} current are also only a few percent (e.g. 3%).[24] Santana *et al.*[25] also inferred that the opening of a single L-type Ca^{2+} channel can trigger a Ca^{2+} spark and that the spark probability depends on the square of the local $[Ca^{2+}]$ in the space between the sarcolemmal Ca^{2+} channel and the RyR. Thus spark frequency and SR Ca^{2+} release clearly depend critically on the local cytosolic $[Ca^{2+}]$ and this is a cornerstone of Ca^{2+}-induced Ca^{2+}-release and local control in E-C coupling in cardiac muscle.

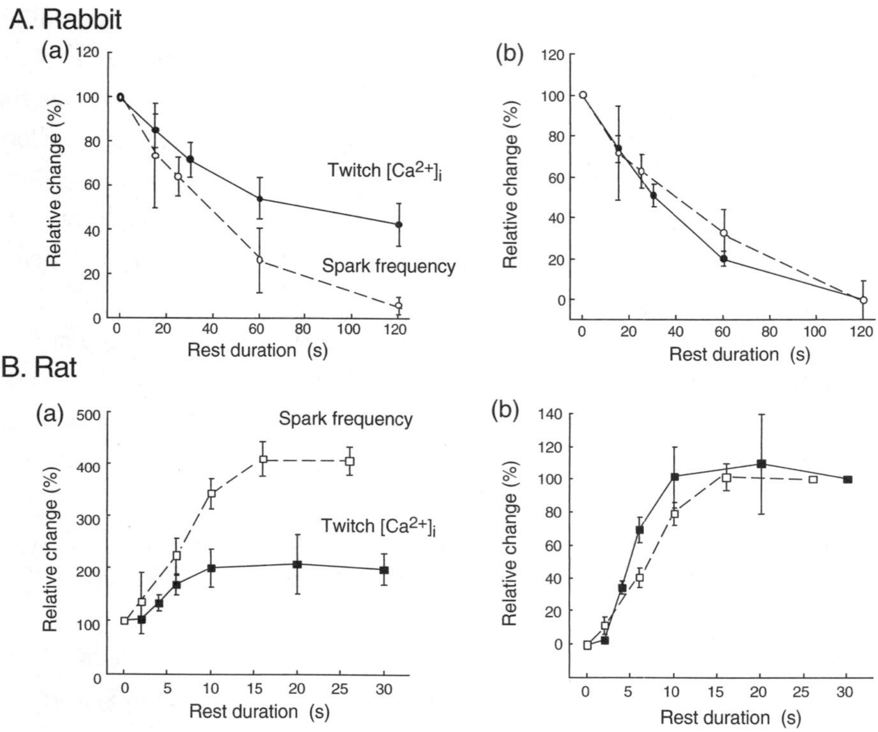

Fig. 15.2 Ca^{2+} spark frequency in rabbit and rat ventricular myocytes during rest, after 1 Hz stimulation in rabbit (**A**) and rat (**B**) ventricular myocytes. The Ca^{2+} spark frequency was normalized to the number of sparks observed during the first 2 sec of rest n rabbit and rat (solid symbols and lnes). These initial values were ~90 and 40 sparks $\bullet pL^{-1} \bullet s^{-1}$ for rabbit and rat, respectively, but by 15 s of rest this difference was reversed. The amplitude of the first post-rest global twitch Ca^{2+} transient was normalized to the 1 Hz steady state twitch Ca^{2+} transient (broken lines). In (b) the data from (a) were further normalized to minimum and maximum values to allow a clearer kinetic comparison between post-rest twich $[Ca^{2+}]_i$-transients and spark frequency (from Satoh *et al.*[13] with permission).

Let us now consider some other factors that can alter Ca^{2+} spark probability. Satoh *et al.*[13] studied the frequency of Ca^{2+} sparks in both rabbit and rat ventricular myocytes and some results are shown in Fig. 15.2. In rabbit ventricular myocytes there is a gradual decrease in Ca^{2+} spark frequency after the last steady state stimulated twitch, whereas in the rat just the opposite is the case. The decrease in Ca^{2+} spark frequency in rabbit is paralleled by a gradual decline in the amplitude of post-rest twitch global Ca^{2+} transients (a process classically referred to as rest decay). This decline was also paralleled by a gradual decline in SR Ca^{2+} content assessed by caffeine-induced Ca^{2+} transients. This resting loss in SR Ca^{2+} can be completely prevented by switching to a Na-free, Ca^{2+}-free solution ($0Na^+$,$0Ca^{2+}$) during the rest period and this changes the decline in Ca^{2+} spark frequency in rabbit to a gradual increase, very much as in the control rat. Thus it appears that the decline in Ca^{2+} spark frequency during rest in rabbit can be attributed to a time-dependent decrease in SR Ca^{2+} content. When the SR Ca^{2+} load is changed in other ways (e.g. changing stimulation frequency or extracellular [Ca^{2+}]) the Ca^{2+} spark frequency changes in concert, even when there is no change in the diastolic $[Ca^{2+}]_i$ or time since the last stimulation. Clearly intra-SR Ca^{2+} can modulate Ca^{2+} spark frequency and this is consistent with results from single channel recordings where luminal Ca^{2+} appears to increase open probability under a number of circumstances.[26,27]

In control rat ventricular myocytes in Fig. 15.2 (and rabbit cells rested in $0Na^+$, $0Ca^{2+}$)[13] the Ca^{2+} spark frequency increases during the first 10–20 sec of rest to a maximal value and this is associated with a rest potentiation of the post-rest twitch Ca^{2+} transient. This occurs without any change in SR Ca^{2+} content during the rest as assessed by caffeine-induced Ca^{2+} transients. It can also be seen without any increase in resting $[Ca^{2+}]_i$. In fact, Satoh *et al.*[13] demonstrated that the very small increase in diastolic $[Ca^{2+}]_i$ in this circumstance in rat cells was the result of the increasing Ca^{2+} spark frequency (rather than causing the increase in sparks). Thus it is clear that there is a time dependent recovery after the last activation for the microscopic SR Ca^{2+} release units to be available to produce Ca^{2+} sparks. This component of restitution is relatively slow (several s). Cheng *et al.*[28] also showed a fast component of restitution with a half-time of ~0.5 s. These two phases of restitution probably coexist and might reflect different processes.

In conclusion, it is clear that the probability of an SR Ca^{2+} release event depends on at least three key factors: (1) local $[Ca^{2+}]_i$ as in Ca^{2+}-induced Ca^{2+}-release, (2) intra-SR Ca^{2+} load, perhaps attributable to a luminal effect of Ca^{2+} and (3) recovery of the SR Ca^{2+} release channel from inactivation or adaptation. However, the interaction of these factors is probably complex. Evidence of this is that the relative quiescent period attributed to recovery of SR Ca^{2+} release channels can be eliminated when SR Ca^{2+} load is further increased to the stage approaching Ca^{2+} overload.[13]

Restitution of E-C Coupling in Intact Ventricular Myocytes at the Whole Cell Level

As shown in Fig. 15.2, the microscopic effect on Ca^{2+} spark probability has exact parallels in terms of the global cellular Ca^{2+} transient. Thus rabbit and guinea-pig ventricular myocytes typically exhibit rest decay of twitches that is paralleled by decline in SR Ca^{2+} content.[29,30] This can be seen in the two normal Tyrode (NT) curves in Fig. 15.3A (where twitch refers to post-rest twitch amplitude and Caff reflects the SR Ca^{2+} content). When the myocytes are incubated in $0Na^+$, $0Ca^{2+}$ solution during the rest, the gradual decline in SR Ca^{2+} content is prevented and rabbit exhibits rest potentiation of twitches very much like that shown for the control rat ventricular myocytes in NT in Fig. 15.3B (which do not lose SR Ca^{2+} during rest in NT). To cause the rat SR to lose Ca^{2+} during rest, the ability of Na^+/Ca^{2+} exchange to extrude Ca^{2+} from the cells must be enhanced. In Fig. 15.3B the rat ventricular myocytes have been depleted of intracellular Na^+ and then exposed to a Ca^{2+}-free, 140 mM Na^+ solution during rest (optimizing Ca^{2+} extrusion by Na^+/Ca^{2+} exchange) and these cells then exhibit rest decay of twitches and SR Ca^{2+} content similar to that seen in the control rabbit.

In ferret ventricular myocytes under control conditions (NT) there is a monotonic decline in SR Ca^{2+} content and a biphasic twitch response (Fig. 15.3C). That is, rest potentiation is seen at short rest intervals, but rest decay is seen at longer rests. It is pretty clear now that most of the early rest potentiation in ferret is due to the recovery of E-C coupling from refractoriness, because there is no increase in SR Ca^{2+} content, action potential or Ca^{2+} current.[29] This is like the control rat and rabbit when SR Ca^{2+} depletion is prevented by $0Na^+$, $0Ca^{2+}$. At longer rests in ferret, the SR Ca^{2+} content is declining and this becomes the factor responsible for the eventual rest decay of twitches. Indeed, the curve fit for the ferret twitch data in Fig. 15.3C is a simple product of two exponentials, where one is the slow decline of SR Ca^{2+} (τ~250 s) and this is multiplied by a faster recovering E-C coupling "fraction of SR Ca^{2+} release" exponential (τ~6 s). Thus, restitution of E-C coupling and resting loss of SR Ca^{2+} have opposite effects, with recovery of E-C coupling dominant at short times and the SR Ca^{2+} content at long times in ferret. In contrast, control rabbit is dominated by rest decay and control rat by rest potentiation. Guinea-pig ventricle behaves much like rabbit in this regard, mouse is like rat and feline, canine and human ventricle seem to be most like ferret.[2]

An issue that may merit clarification is why rabbit and ferret SR lose Ca^{2+} during rest and why $0Na^+$, $0Ca^{2+}$ solution prevents resting loss (or gain) of SR Ca^{2+}. Na^+/Ca^{2+} exchange is the main cellular competitor with the SR Ca-ATPase.[2] Thus as Ca^{2+} leaks from the SR its fate is to be transported either back into the SR by the Ca-ATPase or out of the cell by

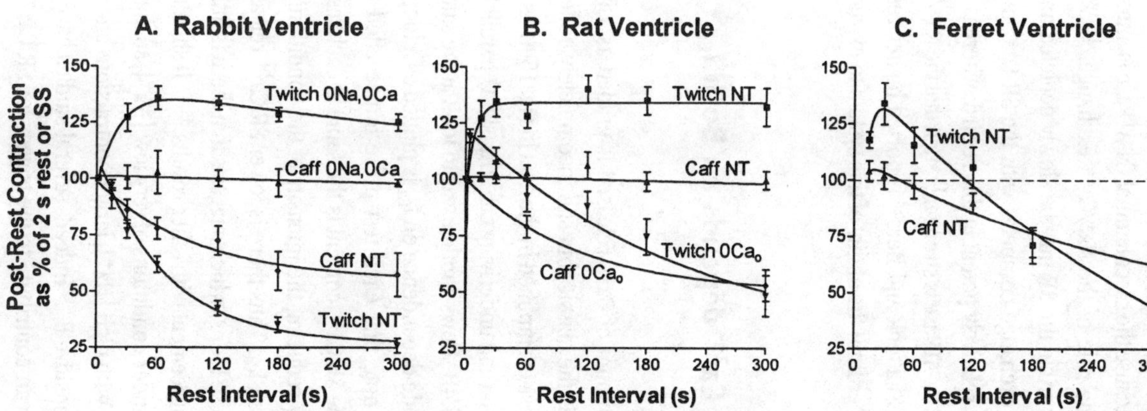

Fig. 12.3 Rest decay and rest potentiation of twitches and SR Ca^{2+} content in ventricular myocytes. Post-rest twitches and caffeine-induced contractures (Caff) were measured either in a modified normal Tyrode's solution (NT), when Na^+/Ca^{2+} exchange was blocked during the rest ($0Na^+,0Ca^{2+}$) or after pre-depletion of $[Na^+]_i$ and rest in Ca^{2+}-free, 140 mM Na solution ($0Ca^{2+}{}_o$). Caff data are indicative of SR Ca^{2+} load. Data were fit with the expression $A\bullet \exp(-t/\tau_{RD})\,(1-\exp(-t/\tau_{RP})) + B$, where τ_{RP} is the time constant for recovery of E-C coupling, τ_{RD} is the time constant of rest decay of SR Ca^{2+} content, B is a residual baseline and A is a proportionality constant. Data is from Bers *et al.*[29] and Bassani and Bers[30] and has been combined and reanalyzed.

Na^+/Ca^{2+} exchange. In 0Na^+, 0Ca^{2+} solution the Na^+/Ca^{2+} exchange is completely blocked, so virtually all of the Ca^{2+} leaked by the SR is resequestered and there is almost no loss in Ca^{2+} content with time (as in Fig. 15.3A). An exception to this is the ferret ventricle, which still loses SR Ca^{2+} in 0Na^+, 0Ca^{2+}, because the sarcolemmal Ca-ATPase is unusually powerful in this species and partly serves the role of the Na^+/Ca^{2+} exchange.[31] When Na^+/Ca^{2+} exchange is functional in rabbit ventricle the SR may be slowly depleted and the rate will depend on how well the Na^+/Ca^{2+} exchange competes with the SR Ca^{2+} pump. In rabbit ventricle the Na^+/Ca^{2+} exchanger is relatively potent in this regard, whereas in rat the Na^+/Ca^{2+} exchanger is weaker and the SR Ca-ATPase is stronger[12] such that the Na^+/Ca^{2+} exchange has much less impact. In addition Shattock and Bers[32] reported higher resting $[Na^+]_i$ in rat ventricle which would even further limit the ability of this system to extrude Ca^{2+} during rest.

Fractional Release of SR Ca^{2+} depends on Both I_{Ca} Trigger and SR Ca^{2+} Load

For a given SR Ca^{2+} content the amount of Ca^{2+} released is graded as a function of the amount of triggering Ca^{2+}.[33] This is the classic, but still incompletely understood graded behavior of Ca^{2+}-induced Ca^{2+}-release exemplified by the data in Fig. 15.4. As the test voltage is varied the amplitude of contraction (and underlying Ca^{2+} transient) parallel the amplitude of the Ca^{2+} current (I_{Ca}). Figure 15.4B shows another way to look at the same type of data. In this case the contraction amplitude dependence on I_{Ca} is plotted as a loop with different test potentials. It is important to note that each test pulse is preceded by a series of conditioning steps to ensure that the SR Ca^{2+} content is the same at each test pulse. When the SR Ca^{2+} load was decreased by reducing the intensity of the conditioning train, there was a smaller contraction for a given I_{Ca}. Thus there is lower SR Ca^{2+} release at the lower load. There is little hysteresis in these loops, which suggests that there is little intrinsic E_m-dependence of SR Ca^{2+} release, in agreement with other results.[34,35] It should also be noted that these experiments were done under conditions of very low $[Na^+]_i$ so that Ca^{2+} entry via Na^+/Ca^{2+} exchange was prevented. With high $[Na^+]_i$ larger contractions and Ca^{2+} transients can be observed for a given I_{Ca} at positive E_m. Indeed, several studies have suggested that Ca^{2+} entry via Na^+/Ca^{2+} exchange can contribute to the trigger for SR Ca^{2+} release under certain conditions.[36–38] It is not clear how much this contributes to physiological triggering of SR Ca^{2+} release.

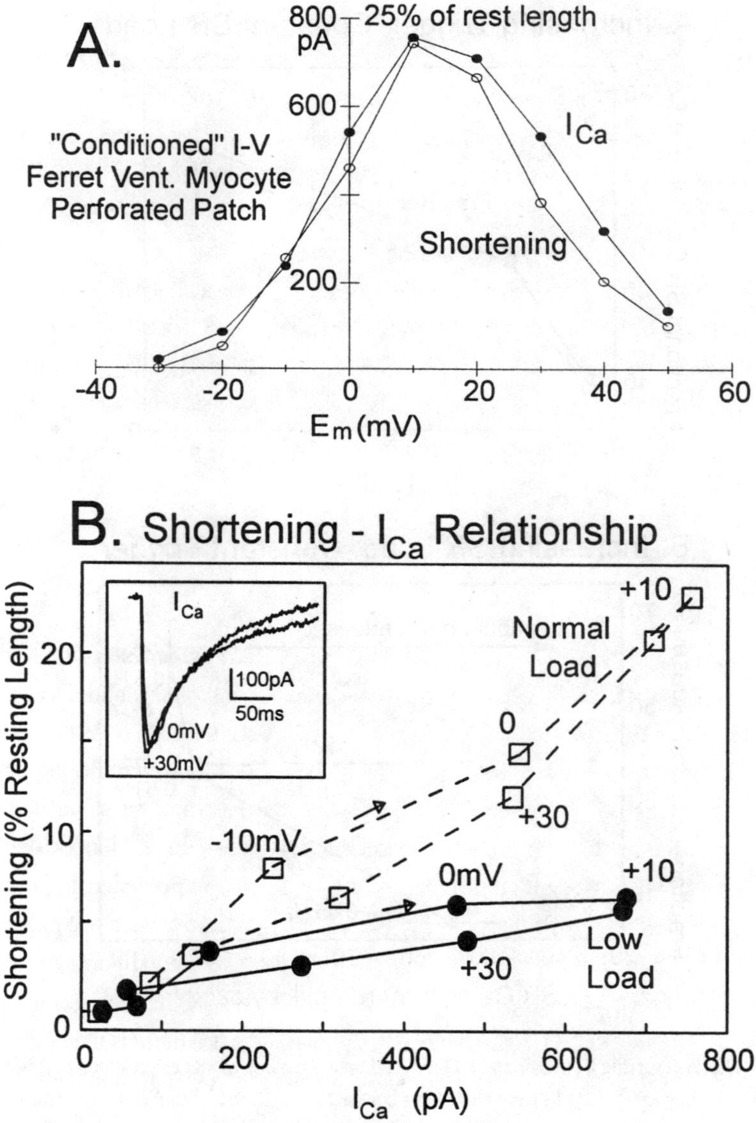

Fig. 15.4 Bell shaped I_{Ca}- and contraction-voltage relationships (**A**) and the dependence of contraction on I_{Ca} (**B**). A set of five conditioning pulses (from −70 mV to 0 mV) preceded the test pulses to different potentials (E_m). Contraction and I_{Ca} were recorded simultaneously. **B.** Contraction amplitude is plotted as a function of I_{Ca} for Normal load as in A and for a Low SR Ca^{2+} load (after a weaker train of conditioning pulses). The inset shows raw currents traces from the Normal curve at −30 and 0 mV (from Bassani *et al.*[40] with permission.)

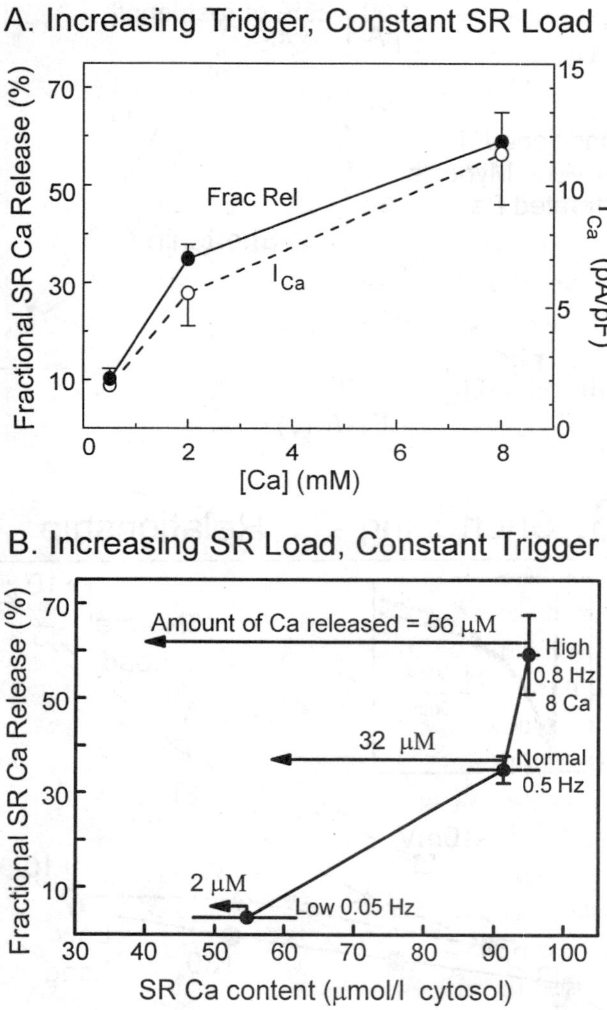

Fig. 15.5 Effects of varying I_{Ca} trigger and SR Ca^{2+} load on the fraction of SR Ca^{2+} released during a twitch in ferret ventricular myocytes. **A.** Cells were loaded to a constant level (by stimulation at 0.5 Hz) and $[Ca^{2+}]_o$ was changed just prior to the test contraction where the fraction of SR Ca^{2+}release is measured. Thus increasing trigger (at constant SR Ca^{2+} load) increases fractional release. Parallel voltage clamp experiments were done to evaluate how I_{Ca} changes over this range of $[Ca^{2+}]_o$. **B.** Using a constant trigger (with 2 mM $[Ca^{2+}]_o$) the SR of cells were Ca^{2+} loaded to different levels by varying frequency (as indicated) and also increasing $[Ca^{2+}]_o$ at the highest load. SR Ca^{2+} content was measured by translating the peak and resting $[Ca^{2+}]_i$ during caffeine-induced contractures to total cytosolic $[Ca^{2+}]$, using measured values for cytosolic Ca^{2+} buffering[78] (from Bassani *et al.*[4] with permission).

These types of studies have been very useful and important, however, they have not provided quantitative information about exactly what fraction of SR Ca^{2+} content is released at a twitch and how this is modified by changes in triggering I_{Ca} and by the SR Ca^{2+} load. Bassani *et al.*[39,40] developed an experimental strategy to examine these issues, using Ca^{2+} transients in intact ferret ventricular myocytes evoked by action potentials and application of caffeine. Figure 15.5 shows some of their results. In Fig. 15.5A the SR Ca^{2+} load was kept constant and the fractional SR Ca^{2+} release was measured at different levels of Ca^{2+} current trigger (in this case by changing $[Ca^{2+}]_o$). Just as when I_{Ca} was changed by altering test potential, increasing trigger I_{Ca} increases the fractional SR Ca^{2+} release. This is expected according to Ca^{2+}-induced Ca^{2+}-release, but this data provides an indication of what fraction of SR Ca^{2+} gets released under normal conditions (~35% in ferret) and over what dynamic range it is modulated over a six-fold range of I_{Ca} trigger (from 10 to 60%).

Figure 15.5B shows the effect of varying SR Ca^{2+} load with a constant I_{Ca} trigger. The Normal SR Ca^{2+} load at 0.5 Hz stimulation was 91 μmol/l cytosol and 35% was released at the normal twitch. When cells were paced at 0.8 Hz and at high $[Ca^{2+}]_o$ (8 mM) the SR Ca^{2+} load reached a maximum of 95 μmol/l cytosol when the cells were on the verge of Ca^{2+} overload (i.e. just below the frequency where spontaneous SR Ca^{2+} release and waves occurred). It was somewhat surprising that the twitch Ca^{2+} transients were so much larger with only a 4% increase in SR Ca^{2+} load. Indeed the amount of SR Ca^{2+} released was almost doubled (from 32 to 56 μmol/l cytosol), but most of this increase was due to an increase in fractional SR Ca^{2+} release (from 35 to 59%). It is possible that the SR Ca^{2+} load is limited by the thermodynamic capacity of the SR Ca^{2+}-ATPase to generate a free trans-SR $[Ca^{2+}]$ gradient.[41] It is also possible that under our normal conditions the intra-SR Ca^{2+} buffers such as calsequestrin become saturated so that little increase in SR Ca^{2+} content can occur with more aggressive stimulation, but the intra-SR free $[Ca^{2+}]$ ($[Ca^{2+}]_{SR}$) may rise. This rise in $[Ca^{2+}]_{SR}$ may in turn exert some effect to increase RyR gating and thereby fractional release. Whether luminal Ca^{2+} exerts this effect directly on the RyR or via some less direct means cannot be assessed in these whole cell measurements, but luminal Ca^{2+} has been reported to increase RyR open probability in single channels.[26,27]

This luminal Ca^{2+} effect may be an important integral control of SR Ca^{2+} release. Furthermore, what is usually referred to as spontaneous SR Ca^{2+} release with Ca^{2+} overload, may really be release triggered by increasing intra-SR Ca^{2+}. It should also be noted that the increase in fractional SR Ca^{2+} release works synergistically with increases in Ca^{2+} load and may be the more important factor at relatively high SR Ca^{2+} loads.

When SR Ca^{2+} load is decreased by reduced frequency of stimulation the fractional SR Ca^{2+} release is dramatically reduced. In Fig. 15.5B reduction of SR load by only 40%, almost abolished SR Ca^{2+} release. That is, only 4% of the SR Ca^{2+} load was released (2 mmol/l cytosol) and this amount is even much less than that which enters the cell via I_{Ca} (5–12 mmol/l cytosol).[42] Thus at low levels of SR Ca^{2+} load, the SR nearly stops participating in E-C coupling even though there is a substantial pool of Ca^{2+} available for release (55 mmol/l cytosol).

The foregoing sections makes clear that at both microscopic and whole cell levels, the SR Ca^{2+} release by the RyR in its normal cellular environment is regulated by $[Ca^{2+}]_i$, $[Ca^{2+}]_{SR}$ and recovery time from a previous activation. In addition, it is well known that the RyR can be modulated by many other endogenous as well as exogenous moieties. In the following sections effects of several specific modulators on SR Ca^{2+} release in the intact ventricular myocyte will be discussed.

CaMKII Increase Fractional SR Ca Release for a given I_{Ca} Trigger and SR Ca^{2+} Load

CaMKII is a ubiquitous Ca^{2+}-dependent second messenger system that has been implicated in the control of processes from gene transcription to contraction to active memory. The cardiac RyR has been identified as a target for specific CaMKII phosphorylation[43] and several reports have examined the effects of CaMKII on the cardiac RyR in bilayers and SR vesicles.[43,44] However, these results have not provided clear understanding of how this modulation may occur in the intact cell. Given the cyclical nature of the cardiac Ca^{2+} transient and its physiological modulation by hormones, it is important to evaluate effects of CaMKII on SR Ca^{2+} release in the intact cell.

Li *et al.*[45] recently evaluated the effects of endogenous CaMKII on E-C coupling in intact ventricular myocytes, using the perforated patch clamp method to prevent intracellular dialysis. Given the known influence of SR Ca^{2+} load and I_{Ca} on E-C coupling, these two parameters must be measured and controlled in the cellular experiments so that the influence of other agents on the process of SR Ca^{2+} release can be appropriately evaluated. Thus, if we can make the SR Ca^{2+} load the same and the I_{Ca} trigger the same, then we can examine how agents (such as the CaMKII inhibitor KN-93) alter the relationship between trigger Ca^{2+} and SR Ca^{2+} release.

Figure 15.6 shows voltage clamp experiments from Li *et al.*[45] showing how the SR Ca^{2+} load after a conditioning train was matched in the absence and presence of the CaMKII

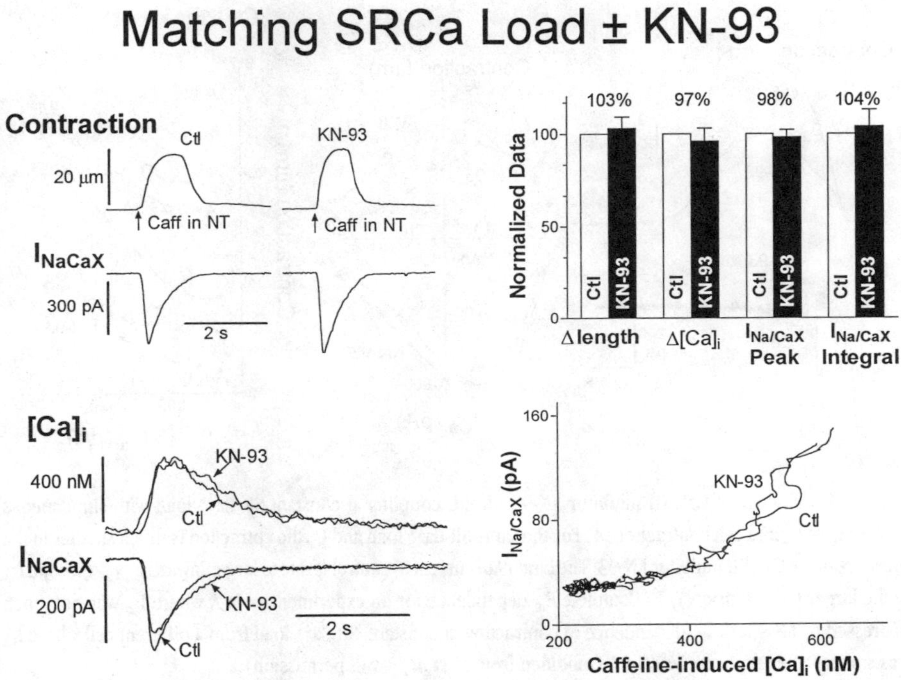

Fig. 15.6 Matching SR Ca^{2+} loads for the study of CaMKII effects on E-C coupling. After a train of conditioning pulses, caffeine was applied to release SR Ca^{2+} (left). The amplitude of the resulting contraction, Ca^{2+} transient and Na^+/Ca^{2+} exchange current ($I_{Na/CaX}$) and the integral of Ca^{2+} extruded by $I_{Na/CaX}$ are all quantitative indices of the SR Ca^{2+} content. The bar graph shows pooled results for these measures in the absence and presence of KN-93. The lower right graph shows that the $I_{Na/CaX}$ dependence on $[Ca^{2+}]_i$ was not affected by KN-93 (Modified from Li *et al.*[45] with permission).

inhibitor KN-93. After a series of 5 conditioning voltage clamp pulses 10 mM caffeine was rapidly applied causing release of the SR Ca^{2+} content, and maintained caffeine exposure prevented reaccumulation of Ca^{2+} by the SR. The peak of the caffeine-induced contracture and Ca^{2+} transient give two estimates of SR Ca^{2+} content (see Fig. 15.6, left). In addition during the caffeine application almost all of the SR Ca^{2+} is extruded by Na^+/Ca^{2+} exchange[11,12,46] which produces an inward current ($I_{Na/CaX}$) which can be integrated to provide an additional measure of the SR Ca^{2+} content.[47,48] Because KN-93 slightly decreases I_{Ca}, a slightly more aggressive series of conditioning pulses was often required to obtain matching SR Ca^{2+} loads with KN-93 vs control. The bar graph in Fig. 15.6 indicates that the

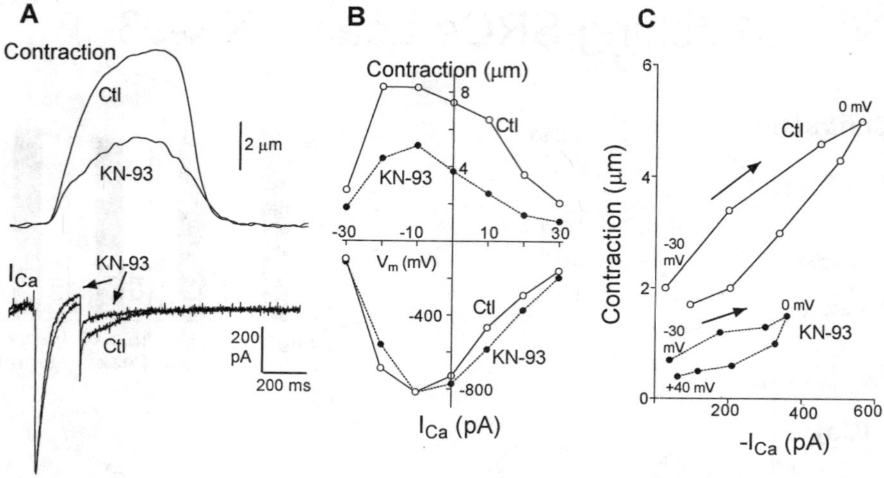

Fig. 15.7 Influence of CaMKII inhibitor KN-93 on E-C coupling at constant SR Ca^{2+} load with simulaneous recording of I_{Ca} and cell contraction. **A.** For the same SR Ca^{2+} load and I_{Ca} the cotraction is much smaller in the presence of the CaMKII inhibitor KN-93. The larger tail current in control is due to larger inward $I_{Na/CaX}$ (secondary to the larger Ca^{2+} transient). **B.** Complete E_m-dependence for an experiment as in A where I_{Ca} was not much depressed by KN-93. **C.** I_{Ca}-dependence of contraction at constant SR Ca^{2+} load from a different cell where I_{Ca} was smaller after exposure to KN-93 (modified from Li *et al.*[45] with permission).

SR Ca^{2+} load achieved with KN-93 was the same as in control (and KN-93 did not alter the $[Ca^{2+}]_i$ dependence of $I_{Na/CaX}$, lower right).

Now at constant SR Ca^{2+} load the relationship between trigger I_{Ca} and SR Ca^{2+} release can be assessed by different test pulses after the respective conditioning trains. In Fig. 15.7A it can be seen that for a given I_{Ca} (and SR Ca^{2+} load), the contraction produced when CaMKII was blocked is much smaller. The mean twitch contractions and Ca^{2+} transients were 39 ± 4% and 49 ± 6% of control in the presence of KN-93 (for matched SR Ca^{2+} load and I_{Ca}). Figure 15.7B and C show the full range of data from two different cells (one where I_{Ca} was hardly altered, B, and one where it was, C). This effect was partially reversible and could not be mimicked by KN-92 an analog that does not inhibit CaMKII. Furthermore, if weaker conditioning trains were used in general, then KN-93 had no (or little) effect. This is consistent with endogenous CaMKII normally being activated by the conditioning pulses. Closely complementary studies by duBell *et al.*[49] showed that dialysis of voltage clamped ventricular myocyte with exogenous phosphatases also inhibited E-C coupling. These results

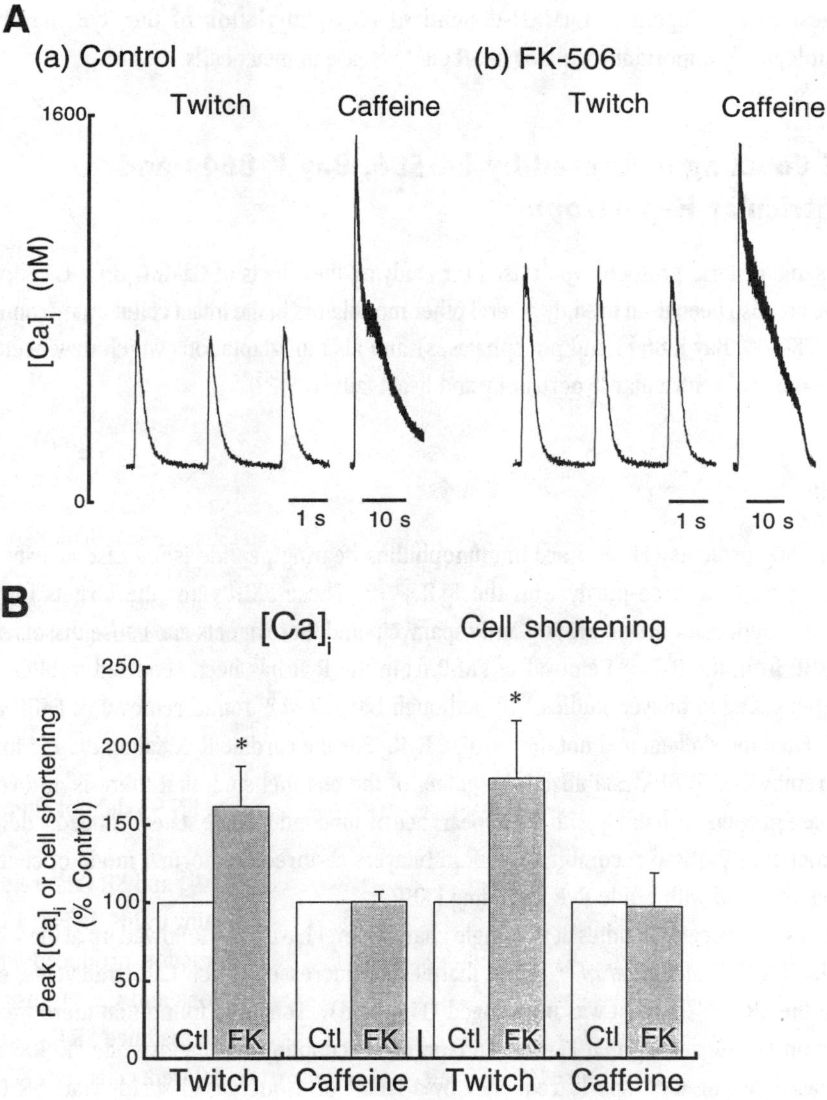

Fig. 15.8 Effect of FK-506 on Ca^{2+} transients and contractions during twitches and caffeine-induced contractures in rat ventricular myocytes. **A.** FK-506 (5 μM) increased twitch Ca^{2+} transients (measured with fluo-3) even when SR Ca^{2+} content was not altered (based on caffeine-induced Ca^{2+} transients). B. This data is from a subset of six cells that did not show the common increase in SR Ca^{2+} content with FK-506 (allowing simpler conclusions). The observation that neither caffeine-induced Ca^{2+} transients or contractures were modified also suggests no change in myofilament Ca^{2+} sensitivity with FK-506 (from McCall *et al.*[52] with permission).

suggest that endogenous CaMKII-dependent phosphorylation of the RyR may be a physiologically important regulator of SR Ca^{2+} release in intact cells.

E-C Coupling is Altered by FK-506, Bay K 8644 and Ventricular Hypertrophy

The same general protocol we devised for studying the effects of CaMKII on E-C coupling above has also been used to study several other modulators in the intact cellular environment (e.g. FK-506, Bay K 8644 and phosphatases) and also to adaptations which may occur in the etiology of ventricular hypertrophy and heart failure.[49–55]

FK-506

FK binding proteins (FKBPs) are immunophillins bearing peptide isomerase activity and which bind to and co-purify with the RyR.[56–58] These FKBPs are the targets for the immunosuppresant drugs, FK-506 and rapamycin and these agents can cause dissociation of FKBP from the RyR.[56] Removal of FKBP from the RyR has been reported to alter RyR channel gating in bilayer studies,[59–62] although Barg *et al.*[63] found removal of FKBP only modulated the skeletal and not the cardiac RyR. For the cardiac RyR Kaftan *et al.*[62] found that removal of FKBP destabilized the gating of the channel such that there is an overall increase in open probability with the appearance of subconductance states. Indeed, addition of exogenous FKBP to recombinant RyR in bilayers restored the normal mode of channel gating observed with crude RyR including FKBP.

These provocative studies at the single channel level have been followed up at the whole cell level.[49,52,64] McCall *et al.*[52] found that FK-506 increased twitch Ca^{2+} transients, even when the SR Ca^{2+} content was unchanged (Fig. 15.8). They also found that there was no effect on I_{Ca}, suggesting that FK-506 enhanced E-C coupling. On the average FK-506 also increased resting Ca^{2+} spark frequency by a factor of 4, increased steady state SR Ca^{2+} content (i.e. the cells in Fig. 15.8 were a selected subset) and also depressed Na^+/Ca^{2+} exchange. Given the higher Ca^{2+} spark rate and greater SR Ca^{2+} release during twitches, one might have expected lower SR Ca^{2+} load, but it is possible that the inhibition of the Na^+/Ca^{2+} exchange allows more Ca^{2+} to remain in the cell and SR. Xiao *et al.*[64] found similar results with higher concentrations of FK-506 (50 μM) with respect to Ca^{2+} transients, I_{Ca} and Ca^{2+} sparks, but they found that the Ca^{2+} sparks were longer in duration, with a population of lower amplitude events. This corresponded to some of the RyR bilayer findings and was consistent with a slowing of RyR adaptation. At lower FK-506 concentration (5–10 μM) McCall *et al.*[52] did not find measurable changes in Ca^{2+} spark amplitude or duration. In addition, duBell *et al.*[54] showed that 50 μM FK-506 inhibits outward K currents in ventricular myocytes, causing prolongation of the action potential duration. This may also contribute to the positive inotropic action of FK-506 reported in the other studies. Thus FKBP may have a stabilizing effect on the RyR in intact cells. This may limit SR Ca^{2+} release during twitches and also limit the spontaneous SR Ca^{2+} release which can be arrhythmogenic.

Bay K 8644

Bay K 8644 is a well known L-type Ca^{2+} channel agonist which prolongs the openings of single Ca^{2+} channel.[65] Hryshko *et al.*[66,67] first reported that Bay K 8644 greatly accelerated rest decay of twitches in canine ventricle. McCall *et al.*[50] followed up these observations in ferret ventricle and found that this accelerated rest decay was due to rapid depletion of SR Ca^{2+} during rest (almost like that seen with ryanodine). They also demonstrated that Bay K 8644 increased ryanodine binding in intact myocytes, but this effect was abolished by cellular disruption. Finally, they demonstrated that the acceleration of resting SR Ca^{2+} loss by Bay K 8644 was still present in the complete absence of extracellular Ca^{2+} and when Na^+/Ca^{2+} exchange was prevented. This proved that Ca^{2+} entry could not be triggering SR Ca^{2+} release. Satoh *et al.*[68] also showed that Bay K 8644 dramatically increased the frequency of Ca^{2+} sparks in ferret ventricular myocyte (even in Ca^{2+}-free solution) and this effect could be inhibited by nifedipine. This increased SR Ca^{2+} spark/leak rate then explained the faster resting SR Ca^{2+} loss. These studies led to the hypothesis depicted in Fig. 15.9 that there may be a relatively direct Ca^{2+}-independent pathway whereby Bay K 8644 bound to the dihydropyridine receptor (DHPR) communicates to the resting cardiac RyR to increase its open probability.[50] This intriguing hypothesis probably merits further study.

McCall *et al.*[51] extended these studies with Bay K 8644 to do the type of E-C coupling experiments described above for CaMKII (in Fig. 15.7). Figure 15.10 shows that, somewhat surprisingly, Bay K 8644 depressed E-C coupling. That is, for a given I_{Ca} and SR Ca^{2+} content, the contraction was smaller after 100 nM Bay K 8644. One possible explanation for this effect is that once Bay K 8644 binds to the DHPR (indirectly increasing diastolic RyR openings) the RyR becomes less sensitive to activating Ca^{2+} coming from the sarcolemmal

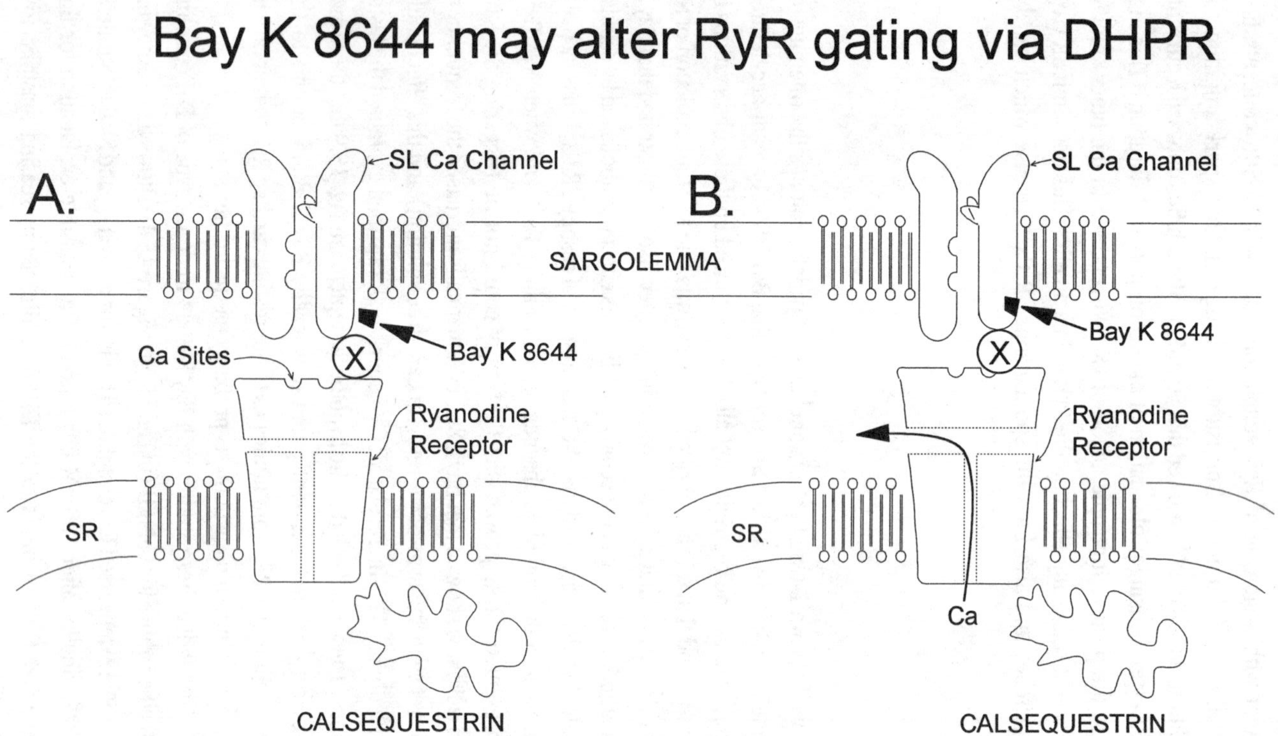

Fig. 15.9 Simplified schematic diagram of how Bay K 8644 might cause increased RyR gating via an effect on the dihydridine receptor. Bay K 8644 may bind to the dihydropyridine receptor on the L-type Ca^{2+} channel and this may alter RyR state by a relatively direct (Ca^{2+} influx independent) coupling, perhaps through an intermediate protein (X).

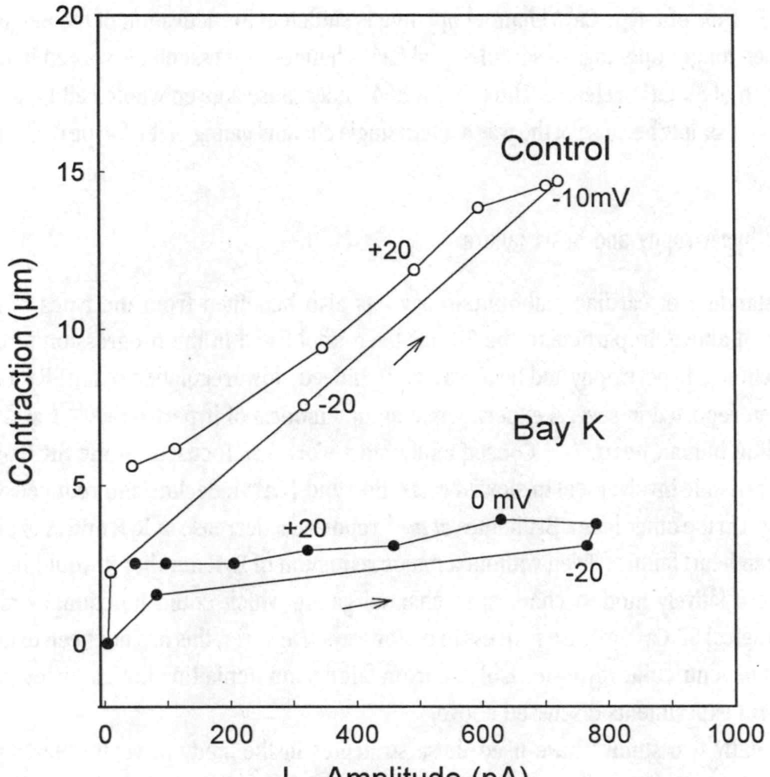

Fig. 15.10 Effects of Bay K 8644 on E-C coupling in ferret ventricular myocytes. Conditioned current-voltage relationships were assessed in the absence and presence of 100 nM Bay K 8644 (as in Figs. 4 and 7). The SR Ca^{2+} content was verified to be the same for both coditions. Lower $[Ca^{2+}]_o$ or manipulation of E_m was used to obtain comparable load and I_{Ca} in the presence of Bay K 8644 compared to control (from McCall and Bers[51] with permission).

Ca^{2+} channel (cartooned as linker X partially blocking a possible Ca^{2+} activating site in Fig. 15.9). The increased spark frequency with Bay K 8644 might also slow restitution of the RyR, as discussed above. While this is an interesting, untested possibility there is also a simple alternative explanation. Since Bay K 8644 increases single channel open times dramatically from 0.5–1 ms to 10–20 ms,[65] we may expect fewer individual sarcolemmal Ca^{2+} channels to have opened for a given peak I_{Ca} (because more will remain open). If the first 0.5–1 ms of L-type Ca^{2+} channel opening is sufficient for activation of the neighboring RyR, then longer openings of sarcolemmal Ca^{2+} channels are essentially wasted in terms of activation of SR Ca^{2+} release. Thus, Bay K 8644 may cause a given whole cell I_{Ca} to be less efficacious simply because of the way it alters single channel gating of the L-type Ca^{2+} channel.

Cardiac hypertrophy and heart failure

Understanding of cardiac pathophysiology has also benefited from the types of studies described above. In particular, the SR has been implicated in the progression of cellular dysfunction in hypertrophy and heart failure.[69] Indeed, downregulation of the SR Ca-ATPase has been reported in several experimental animal models of hypertrophy[70–72] and also in the failing human heart.[73–75] Consequently most work has focused on the SR Ca-ATPase and its possible involvement in slowed relaxation and $[Ca^{2+}]_i$ decline and reduced SR Ca^{2+} loading. On the other hand, Brilliantes *et al.*[76] reported a decrease in RyR mRNA expression in human heart failure. Even without a major reduction of RyR number it would be easy to imagine relatively modest changes in channel gating which could functionally alter the physiological SR Ca^{2+} release process in major ways. However, this has not been extensively studied in ventricular myocytes isolated from failing and nonfailing human using the kind of control experiments discussed above.

Recently two studies have used these strategies in the study of ventricular myocytes isolated from hypertrophic and failing rat ventricles.[53,55] Gómez *et al.*[55] studied both hypertensive salt-sensitive Dahl rats (Dahl-SS) with ventricular hypertrophy and a line of spontaneous hypertensive rats which develop congestive heart failure (SH-HF). The hypertrophic myocytes isolated from Dahl-SS hearts had normal spontaneous Ca^{2+} sparks, SR Ca^{2+} content, I_{Ca} density and RyR number and single channel properties. However, a given I_{Ca} produced a smaller numbers of Ca^{2+} sparks as well as reduced global Ca^{2+} transients compared to controls. They inferred that this reflected a depression of E-C coupling in the hypertrophied myocytes (as with KN-93 in Fig. 15.7) and myocytes from the SH-HF showed the same effect. In the Dahl-SS rats the depression of E-C coupling could be overcome by activation with β-adrenergic agonists, but this was not the case with the SH-HF rats. They speculated that there might be a spatial defect (larger gap between RyR and DHPR) in hypertrophy and failure, such that a given I_{Ca} influx could not activate the appropriate RyRs as well as in control. There is no structural evidence for this speculative hypothesis, but functionally any effect that reduces the apparent sensitivity of SR Ca release to a given I_{Ca} trigger would fit the data (and the geometric idea is intriguing). Boosting I_{Ca} with isoproterenol could then allow better activation of SR Ca^{2+} release and restoration of E-C coupling.

McCall *et al.*[53] and Delbridge *et al.*[77] studied a more moderate model of hypertrophy in the rat (due to chronic pressure overload induced by coarctation of the abdominal aorta for 16 weeks). Interestingly, they also found normal I_{Ca} density, numbers and affinity of RyR and SR Ca^{2+} content in these hypertrophied left ventricular myocytes. In the normal $[Ca^{2+}]_o$ there was also no depression of E-C coupling (in experiments like those in Fig. 15.7). However, when $[Ca^{2+}]_o$ was reduced to 0.5 mM, there was a substantial depression of E-C coupling in hypertrophic myocytes vs those from sham operated animals (although SR Ca^{2+} loads were still comparable between the groups). This is entirely consistent with the results of Gómez *et al.*[53] with more severe hypertrophy and heart failure. With less severe hypertrophy the defect may be more subtle and the normal I_{Ca} trigger may be just strong enough to mask the developing pathophysiology. Indeed, this could mark the state, prior to overt heart failure.

Thus, it is becoming increasingly clear that defects in the regulation of SR Ca^{2+} release must be considered in pathophysiological alterations in cardiac function. Even in the case of reduced SR Ca-ATPase expression and SR Ca^{2+} accumulation, this bears consideration. This is especially the case because any reduction in optimal steady state SR Ca^{2+} loading (secondary to reduced SR Ca-ATPase activity) will not only reduce the amount of Ca^{2+} available for release, but also the fraction of SR Ca^{2+} released (e.g. as shown in Fig. 15.5B).

CONCLUSIONS

The foregoing discussions should emphasize that it is important to examine the behavior of SR Ca^{2+} release and the RyR in the normal cellular environment. Complementary electrophysiological and Ca^{2+} measurements in intact single cells with studies of more isolated and inherently controllable systems (such as SR vesicles and RyR in bilayers) will continue to be valuable in building a more integrated and comprehensive understanding of how SR Ca^{2+} release is regulated in physiological and pathophysiological contexts. The studies described here have shown that the RyR is sensitive to $[Ca^{2+}]_i$, $[Ca^{2+}]_{SR}$ and time after the last activation, where increasing any of these increases SR Ca^{2+} release. Endogenous CaMKII and FK-506 also appear to increase the efficacy of E-C coupling while Bay K 8644 and cardiac hypertrophy or failure may lead to a decrease in the efficacy of a given I_{Ca} trigger to release SR Ca^{2+} (even with the same SR Ca^{2+} load).

ACKNOWLEDGMENTS

The talented contributions of Drs. Rosana Bassani, José Bassani, Lothar Blatter, Kenneth S. Ginsburg, Li Li, Eileen McCall, Hiroshi Satoh and Weilong Yuan to the work described here is most gratefully acknowledged. I also thank Dr. Jörg Hüser for preparing Fig. 1 from his experiments and Ms. Christina Hovance and Mr. Steven Scaglione for careful work in isolating cardiac myocytes. This work was supported by grants from the USPHS (HL-30077, HL-44583 and HL-52478).

REFERENCES

1. Meissner G and Henderson JS. Rapid calcium release from cardiac sarcoplasmic reticulum vesicles is dependent on Ca^{2+} and is modulated by Mg^{2+}, adenine nucleotide, and calmodulin. *J Biol Chem* 1987; **262**:3065–3073.
2. Bers DM. Excitation-Contraction Coupling and Cardiac Contractile Force. Kluwer Academic Press, Dordrecht, Netherlands. 1991:258 pp.
3. Meissner G. Ryanodine activation and inhibition of the Ca^{2+} release channel of sarcoplasmic reticulum. *J Biol Chem* 1986; **261**:6300–6306.
4. Rousseau E and Meissner G. Single cardiac sarcoplasmic reticulum Ca^{2+}-release channel: activation by caffeine. *Am J Physiol* 1989; **256**:H328–H333.
5. Rousseau E, Smith JS, Henderson JS *et al.* Single channel and $^{45}Ca^{2+}$ flux measurements of the cardiac sarcoplasmic reticulum calcium channel. *Biophys J* 1986; **50**:1009–1014.
6. Rousseau E, Smith JS, and Meissner G. Ryanodine modifies conductance and gating behavior of single Ca^{2+} release channel. *Am J Physiol* 1987; **253**:C364–C368.
7. Györke S, and Fill M. Ryanodine receptor adaptation: control mechanism of Ca^{2+}-induced Ca^{2+} release in heart. *Science* 1993; **260**:807–809.
8. Tinker A, Lindsay ARG, and Williams AJ. Cation conductance in the calcium release channel of the cardiac sarcoplasmic reticulum under physiological and pathophysiological conditions. *Cardiovasc Res* 1993; **27**: 1820–1825.
9. Valdivia HH, Kaplan JH, Ellis-Davies GCR *et al.* Rapid adaptation of cardiac ryanodine receptors: modulation by Mg^{2+} and phosphorylation. *Science* 1995; **267**:1997–1999.
10. Schiefer A, Meissner G, and Isenberg G. Ca^{2+} activation and Ca^{2+} inactivation of cardiac sarcoplasmic reticulum Ca^{2+}-release channels. *J Physiol* 1995; **489**:337–348.
11. Bassani RA, Bassani JWM, and Bers DM. Mitochondrial and sarcolemmal Ca transport can reduce $[Ca]_i$ during caffeine contractures in rabbit cardiac myocytes. *J Physiol* 1992; **453**: 591–608.
12. Bassani JWM, Bassani RA, and Bers DM. Relaxation in rabbit and rat cardiac cells: species-dependent differences in cellular mechanisms. *J Physiol* 1994; **476**:279–293.
13. Satoh H, Blatter LA, and Bers DM. Effects of $[Ca]_i$, Ca^{2+} load and rest on Ca^{2+} spark frequency in ventricular myocytes. *Am J Physiol* 1997; **272**:H657–668.

14. Cheng H, Lederer WJ, and Cannell MB. Calcium sparks: elementary events underlying excitation-contraction coupling in heart muscle. *Science* 1993; **262**:740–744.
15. Tsugorka A, Rios E, and Blatter LA. Imaging elementary events of calcium release in skeletal muscle cells. *Science* 1995; **269**:172–1726.
16. Parker I, Zang WJ, Wier WG. Ca^{2+} sparks involving multiple Ca^{2+} release sites along Z-lines in rat heart cells. *J Physiol* 1996;**497**:31–38.
17. Cannell MB, Cheng H, and Lederer WJ. The control of calcium release in heart muscle. *Science* 1995; **268**:1045–1049.
18. Gómez AM, Cheng H, Lederer WJ *et al.* Ca^{2+} diffusion and SR transport both contribute to the $[Ca^{2+}]_i$ decline during Ca^{2+} sparks in rat ventricular myocytes. *J Physiol* 1996;**496**: 575–581.
19. Bassani RA and Bers DM. Rate of diastolic Ca^{2+} release from the sarcoplasmic reticulum of intact rabbit and rat ventricular myocytes. *Biophys J* 1995; **68**:2015–2022.
20. López-López JR, Shacklock PS, Balke CW *et al.* Local calcium transients triggered by single L-type calcium channel currents in cardiac cells. *Science* 1995; **268**:1042–1045.
21. Cannell MB, Cheng H, and Lederer WJ. Spatial non-uniformities in $[Ca^{2+}]$i during excitation-contraction coupling in cardiac myocytes. *Biophys J* 1994; **67**:1942–1956.
22. Wier WG, Egan TM, López-López JR *et al.* Local control of excitation-contraction coupling in rat heart cells. *J Physiol* 1994; **474**:463–471.
23. Bers DM and Stiffel VM. The ratio of ryanodine:dihydropyridine receptors in cardiac and skeletal muscle and implications for E-C coupling. *Am J Physiol* 1993; **264**:C1587—C1593.
24. Lew WYW, Hryshko LV, and Bers DM. Dihydropyridine receptors are primarily functional L-type Ca^{2+} channels in rabbit ventricular myocytes. *Circ Res* 1991; **69**:1139–1145.
25. Santana LF, Cheng H, Gómez AM *et al.* Relation between the sarcolemmal Ca^{2+} current and Ca^{2+} sparks and local control theories for cardiac excitation-contraction coupling. *Circ Res* 1996; **78**:166–171.
26. Sitsapesan R and Williams AJ. Regulation of the gating of the sheep cardiac sarcoplasmic reticulum Ca^{2+}-release channel by luminal Ca^{2+}. *J Memb Biol* 1994; **137**:215–226.
27. Lukyanenko, Györke I, and Györke S. Regulation of calcium release by calcium inside the sarcoplasmic reticulum in ventricular myocytes. *Pflügers Arch* 1996; **432**:1047–1054.
28. Cheng H, Lederer MR, Lederer WJ *et al.* Calcium sparks and $[Ca^{2+}]_i$ waves in cardiac myocytes. *Am J Physiol* 1996;**270**:C148–C159.
29. Bers DM, Bassani RA, Bassani JWM *et al.* Paradoxical twitch potentiation after rest in cardiac muscle: increased fractional release of SR calcium. *J Molec Cell Cardiol* 1993; **25**:1047–1057.
30. Bassani RA, Bers DM. Na-Ca Exchange is Required for Rest-Decay but not for Rest-Potentiation of Twitches in Rabbit and Rat Ventricular Myocytes. *J Molec Cell Cardiol* 1994;**26**:1335–1347.
31. Bassani RA, Bassani JWM, Bers DM. Relaxation in ferret ventricular myocytes: Unusual interplay among calcium transport systems. *J Physiol* 1994;**476**:295–308.
32. Shattock MJ and Bers DM. Rat vs. rabbit ventricle: Ca^{2+} flux and intracellular Na^+ assessed by ion-selective microelectrodes. *Am J Physiol* 1989; **256**:C813–C822.
33. Fabiato A. Time and calcium dependence of activation and inactivation of calcium-induced release of calcium from the sarcoplasmic reticulum of a skinned canine cardiac Purkinje cell. *J Gen Physiol* 1985b; **85**:247–290.

34. Niggli, E. and W.J. Lederer. Voltage-independent calcium release in heart muscle. *Science* 250: 565–568, 1990.
35. Näbauer M, Callewart G, Cleemann L *et al.* Regulation of calcium release is gated by calcium current, not gating charge, in cardiac myocytes. *Science* 1989; **244**:800–803.
36. Bers DM, Christensen DM, and Nguyen TX. Can Ca^{2+} entry via Na-Ca exchange directly activate cardiac muscle contraction? *J Molec Cell Cardiol* 1988; **20**:405–414.
37. Leblanc N and Hume JR. Sodium current-induced release of calcium from cardiac sarcoplasmic reticulum. *Science* 1990; **248**: 372–376.
38. Levi AJ, Spitzer KW, Kohmoto O *et al.* Depolarization-induced Ca^{2+} entry via Na-Ca exchange triggers SR release in guinea pig cardiac myocytes. *Am J Physiol* 1994; **266**:H1422–H1433.
39. Bassani JWM, Bassani RA, and Bers DM. Twitch-dependent SR Ca^{2+} accumulation and release in rabbit ventricular myocytes. *Am J Physiol* 1993; **265**:C53–C540.
40. Bassani JWM, Yuan W, and Bers DM. Fractional SR Ca^{2+} release is altered by trigger Ca and SR Ca content in cardiac myocytes. *Am J Physiol* 1995; **268**:C1313–C1319.
41. Shannon TR and Bers DM. Assessment of intra-SR free $[Ca^{2+}]$ and buffering in rat heart. *Biophys J* 1997; **73**:1524–1531.
42. Yuan W, Ginsburg KS, and Bers DM. Comparison of sarcolemmal Ca^{2+} channel current in rabbit and rat ventricular myocytes. J Physiol 1996; 493:733-746.
43. Witcher DR, Kovacs RJ, Schulman H *et al.* Unique phosphorylation site on the cardiac ryanodine receptor regulates calcium channel activity. *J Biol Chem* 1991; **266**:1114–11152.
44. Hain J, Onoue H, Mayrleitner M *et al.* Phosphorylation modulates the function of the calcium release channel of sarcoplasmic reticulum from cardiac muscle. *J Biol Chem* 1995; **270**: 2074–2081.
45. Li L, Satoh H, Ginsburg KS *et al.* The effects of CaMKII on cardiac excitation-contraction coupling in ferret ventricular myocytes. *J Physiol* 1997; **501**:17–32.
46. Negretti N, O'Neill SC, and Eisner DA. The relative contributions of different intracellular and sarcolemmal systems to relaxation in rat ventricular myocytes. *Cardiovasc Res* 1993; **27**: 1826–1830.
47. Varro A, Negretti N, Hester SB *et al.* An estimate of the calcium content of the sarcoplasmic reticulum in rat ventricular myocytes. *Pflügers Arch* 1993; **423**:15–160.
48. Delbridge LMD, Bassani JWM, and Bers DM. Steady-state twitch Ca^{2+} fluxes and cytosolic Ca^{2+} buffering in rabbit ventricular myocytes. *Am J Physiol* 1996; **39**:C192–C199.
49. DuBell WH, Lederer WJ, and Rogers TB. Dynamic modulation of cardiac excitation-contraction coupling by protein phosphatases in rat ventricular myocytes. *J Physiol* 1996; **493**:793–800.
50. McCall E, Hryshko LV, Stiffel VM *et al.* Functional linkage between the cardiac dihydropyridine and ryanodine receptor: acceleration of rest decay by Bay K 8644. *J Molec Cell Cardiol* 1996; **28**:79–93.
51. McCall E and Bers DM. Bay K 8644 depresses excitation-contraction coupling in cardiac muscle. *Am J Physiol* 1996; **270**:C878–C884.
52. McCall E, Li L, Satoh H *et al.* Effects of FK-506 on contraction and Ca^{2+} transients in cardiac myocytes. *Circ Res* 1996; **79**:1108–1119.

53. McCall E and Bers DM. The effect of hypertrophy on excitation-contraction coupling in a rat model. *J Mol Cell Cardiol* 1996; **28**:A145.
54. DuBell WH, Wright PA, Lederer WJ *et al.* Effects of the immunosupressant FK506 on excitation-contraction coupling and outward K^+ currents in rat ventricular myocytes. *J Physiol* 1997; **501**:509–516.
55. Gómez AM, Valdivia HH, Cheng H *et al.* Defective excitation-contraction coupling in experimental hypertrophy and heart failure. *Science* 1997; **276**:800–805.
56. Jayaraman T, Brillantes AM, Timerman AP *et al.* FK506 binding protein associated with the calcium release channel (ryanodine receptor). *J Biol Chem* 1992; **267**:9474–9477.
57. Timerman AP, Ogunbunmi E, Freund E *et al.* The calcium release channel of sarcoplasmic reticulum is modulated by FK-506 binding protein. *J Biol Chem* 1993; **268**:2299–22999.
58. Timerman AP, Jayaraman T, Wiederrecht G *et al.* The ryanodine receptor from canine heart sarcoplasmic reticulum is associated with a novel FK-506 binding protein. *Biochem Biophys Res Commun* 1994; **198**:701—706.
59. Brillantes AB, Ondrias K, Scott A *et al.* Stabilization of calcium release channel (ryanodine receptor) function by FK-506 binding protein. *Cell* 1994; 77:513–523.
60. Ahern GP, Junankar PR, and Dulhunty AF. Single channel activity of the ryanodine receptor calcium release channel is modulated by FK-506. *FEBS Letts* 1994; **352**:369–374.
61. Chen SRW, Chang L, and MacLennan DH. Assymetrical blockade of the Ca^{2+} release channel (ryanodine receptor) by 12-KDa FK-506 binding protein. *Proc Natl Acad Sci* 1994; **91**: 11953–11957.
62. Kaftan E, Marks AR, and Ehrlich BE. Effects of rapamycin on ryanodine receptor/calcium release. *Circ Res* 1996;**78**:990–997.
63. Barg S, Copello JA, and Fleischer S. Different interactions of cardiac and skeletal muscle ryanodine receptors with FK-506 binding protein isoforms. *Am J Physiol* 1997; **272**:C172–C1733.
64. Xiao RP, Valdivia HH, Bogdanov K *et al.* The immunophilin FK506-binding protein modulates Ca^{2+} release channel closure in rat heart. *J Physiol* 1997; **500**:343–354.
65. Hess P, Lansman JB, and Tsien RW. Different modes of Ca^{2+} channel gating behavior favored by dihydropyridine Ca^{2+} agonists and antagonists. *Nature* 1984; **311**: 538–544.
66. Hryshko LV, Bouchard R, Chau T *et al.* Inhibition of rest potentiation in canine ventricular muscle by BAY K 8644: comparison with caffeine. *Am J Physiol* 1989; **257**: H399–H406.
67. Hryshko LV, Kobayashi T, and Bose D. Possible inhibition of canine ventricular sarcoplasmic reticulum by BAY K 8644. *Am J Physiol* 1989: **257**:H407–H414.
68. Satoh H, Bers DM, and Blatter LM. Modulation of spatial and temporal characteristics of calcium sparks: effects of BayK 8644, caffeine and ryanodine. *Biophys J* 1996; **70**:274.
69. Arai M, Matsui H, and Periasamy M. Sarcoplasmic reticulum gene expression in cardiac hypertrophy and heart failure. *Circ Res* 1994; **74**:555–564.
70. de la Bastie D, Levitsky D, Rappaport L *et al.* Function of the sarcoplasmic reticulum and expression of its Ca^{2+}-ATPase gene in pressure overload-induced cardiac hypertrophy in the rat. *Circ Res* 1990; **66**:554–564.

71. Feldman AM, Weinberg EO, Ray PE *et al.* Selective changes in cardiac gene expression during compensated hypertrophy and the transition to cardiac decompensation in rats with chronic aortic banding. *Circ Res* 1993; **73**:184–192.
72. Qi M, Shannon TR, Euler DE *et al.* Downregulation of sarcoplasmic reticulum Ca^{2+}-ATPase during the progression of pressure overload left ventricular hypertrophy. *Am J Physiol* 1997; **272**:H2416–H2424.
73. Mercadier JJ, Lompre AM, Duc P *et al.* Altered sarcoplasmic reticulum Ca^{2+}-ATPase gene expression in the human ventricle during end-stage heart failure. *J Clin Invest* 1990; **85**: 305–309.
74. Takahashi T, Allen PD, and Izumo S. Expression of A-, B-, and C-Type Natriuretic Peptide Genes in Failing and Developing Human Ventricles: correlation with expression of the Ca^{2+}-ATPase gene. *Circ Res* 1992; **71**:9–17.
75. Hasenfuss G, Reinecke H, Studer R *et al.* Relation between myocardial function and expression of sarcoplasmic reticulum Ca^{2+} ATPase in failing and nonfailing human myocardium. *Circ Res* 1994; **75**:434–442.
76. Brillantes AM, Allen P, Takahashi T *et al.* Differences in cardiac calcium release channel (ryanodine receptor) expression in myocardium from patients with end-stage heart failure caused by ischemic versus dilated cardiomyopathy. *Circ Res* 1992; **71**:18–26.
77. Delbridge LMD, Satoh H, Yuan W *et al.* Cardiac myocyte volume, Ca^{2+} fluxes and sarcoplasmic reticulum loading in pressure overload hypertrophy. *Am J Physiol* 1997; **272**:H2425–H2435.
78. Hove-Madsen L, Bers DM. Passive Ca^{2+} buffering and SR Ca^{2+} uptake in permeabilized rabbit ventricular myocytes. *Am. J. Physiol.* 1993;**264**: C677–C686.

Index

C

I

J

K

L

M

N

O

P

Q

R